Lecture Notes in Computer Science 13463

More information about this series at https://link.springer.com/bookseries/558

Josep Domingo-Ferrer · Maryline Laurent (Eds.)

Privacy in Statistical Databases

International Conference, PSD 2022
Paris, France, September 21–23, 2022
Proceedings

Springer

Editors
Josep Domingo-Ferrer 🆔
Universitat Rovira i Virgili
Tarragona, Catalonia, Spain

Maryline Laurent 🆔
Télécom SudParis
Palaiseau, France

ISSN 0302-9743   ISSN 1611-3349 (electronic)
Lecture Notes in Computer Science
ISBN 978-3-031-13944-4   ISBN 978-3-031-13945-1 (eBook)
https://doi.org/10.1007/978-3-031-13945-1

This Springer imprint is published by the registered company Springer Nature Switzerland AG
The registered company address is: Gewerbestrasse 11, 6330 Cham, Switzerland

Preface

Privacy in statistical databases is a discipline whose purpose is to provide solutions to the tension between the social, political, economic, and corporate demands of accurate information, and the legal and ethical obligation to protect the privacy of the various parties involved. In particular, the need to enforce privacy regulations, epitomized by the EU General Data Protection Regulation (GDPR), in our world of big data has made this tension all the more pressing. Stakeholders include the subjects, sometimes called respondents (the individuals and enterprises to which the data refer), the data controllers (those organizations collecting, curating, and, to some extent, sharing or releasing the data), and the users (the ones querying the database or the search engine, who would like their queries to stay confidential). Beyond law and ethics, there are also practical reasons for data controllers to invest in subject privacy: if individual subjects feel their privacy is guaranteed, they are likely to provide more accurate responses. Data controller privacy is primarily motivated by practical considerations: if an enterprise collects data at its own expense and responsibility, it may wish to minimize leakage of those data to other enterprises (even to those with whom joint data exploitation is planned). Finally, user privacy results in increased user satisfaction, even if it may curtail the ability of the data controller to profile users.

There are at least two traditions in statistical database privacy, both of which started in the 1970s: the first one stems from official statistics, where the discipline is also known as statistical disclosure control (SDC) or statistical disclosure limitation (SDL), and the second one originates from computer science and database technology. In official statistics, the basic concern is subject privacy. In computer science, the initial motivation was also subject privacy but, from 2000 onwards, growing attention has been devoted to controller privacy (privacy-preserving data mining) and user privacy (private information retrieval). In the last few years, the interest and the achievements of computer scientists in the topic have substantially increased, as reflected in the contents of this volume. At the same time, the generalization of big data is challenging privacy technologies in many ways: this volume also contains recent research aimed at tackling some of these challenges.

Privacy in Statistical Databases 2022 (PSD 2022) was held in Paris, France, under the sponsorship of the UNESCO Chair in Data Privacy, which has provided a stable umbrella for the PSD biennial conference series since 2008. In fact, PSD started in 2004 and PSD 2022 is the tenth conference in the series. Previous PSDs were held in various locations around the Mediterranean, and had their proceedings published by Springer in the LNCS series: PSD 2020, Tarragona, LNCS 12276; PSD 2018, Valencia, LNCS 11126; PSD 2016, Dubrovnik, LNCS 9867; PSD 2014, Eivissa, LNCS 8744; PSD 2012, Palermo, LNCS 7556; PSD 2010, Corfu, LNCS 6344; PSD 2008, Istanbul, LNCS 5262; PSD 2006, Rome, LNCS 4302; and PSD 2004, Barcelona, LNCS 3050. The PSD series took over from the high-quality technical conferences on SDC which started 24 years ago with Statistical Data Protection-SDP (Lisbon, 1998, OPOCE proceedings) and the AMRADS project SDC Workshop (Luxemburg, 2001, LNCS 2316).

The PSD 2022 Program Committee accepted for publication in this volume 25 papers out of 45 submissions. Furthermore, 10 of the above submissions were selected for short oral presentation at the conference. Papers came from authors in 18 different countries and four different continents. Each submitted paper received at least two reviews. The revised versions of the 25 accepted papers in this volume are a fine blend of contributions from official statistics and computer science. Covered topics include privacy models, statistical tables, disclosure risk assessment and record linkage, privacy-preserving protocols, unstructured and mobility data, synthetic data, machine learning and privacy, and case studies.

We are indebted to many people. First, to the Organization Committee for making the conference possible and especially to Jesús Manjón, who helped prepare these proceedings, and to Samia Bouzefrane, who led the local arrangements. In evaluating the papers we were assisted by the Program Committee and by Ziqi Zhang as an external reviewer. We also wish to thank all the authors of submitted papers and we apologize for possible omissions.

This volume is dedicated to the memory of William (Bill) E. Winkler, who was always supportive and served as a PC member of many PSD conferences.

July 2022   Josep Domingo-Ferrer
                                                                          Maryline Laurent

Organization

Program Chair

Josep Domingo-Ferrer — UNESCO Chair in Data Privacy and Universitat Rovira i Virgili, Catalonia, Spain

General Chair

Maryline Laurent — Télécom SudParis, France

Program Committee

Jane Bambauer	University of Arizona, USA
Bettina Berendt	Technical University of Berlin, Germany
Aleksandra Bujnowska	Eurostat, Luxembourg
Jordi Castro	Polytechnic University of Catalonia, Spain
Anne-Sophie Charest	Université Laval, Canada
Peter Christen	Australian National University, Australia
Chris Clifton	Purdue University, USA
Graham Cormode	University of Warwick, UK
Peter-Paul de Wolf	Statistics Netherlands, The Netherlands
Roberto Di Pietro	Hamad Bin Khalifa University, Qatar
Josep Domingo-Ferrer	Universitat Rovira i Virgili, Catalonia, Spain
Jörg Drechsler	IAB, Germany
Khaled El Emam	University of Ottawa, Canada
Mark Elliot	University of Manchester, UK
Sèbastien Gambs	Université du Québec à Montréal, Canada
Sarah Giessing	Destatis, Germany
Hiroaki Kikuchi	Meiji University, Japan
Maryline Laurent	Télécom SudParis, France
Bradley Malin	Vanderbilt University, USA
Anna Monreale	Università di Pisa, Italy
Krishnamurty Muralidhar	University of Oklahoma, USA
David Naccache	École Normale Supérieure, France
Benjamin Nguyen	INSA Centre Val de Loire, France
Anna Oganyan	National Center for Health Statistics, USA
Melek Önen	Eurécom, France
Javier Parra-Arnau	Polytechnic University of Catalonia, Spain

Constantinos Patsakis	University of Piraeus, Greece
Jerome Reiter	Duke University, USA
Yosef Rinott	Hebrew University, Israel
Felix Ritchie	University of the West of England, UK
Steven Ruggles	University of Minnesota, USA
Nicolas Ruiz	OECD and Universitat Rovira i Virgili, Catalonia, Spain
David Sánchez	Universitat Rovira i Virgili, Catalonia, Spain
Eric Schulte-Nordholt	Statistics Netherlands, The Netherlands
Natalie Shlomo	University of Manchester, UK
Aleksandra Slavković	Pennsylvania State University, UK
Jordi Soria-Comas	Catalan Data Protection Authority, Catalonia, Spain
Tamir Tassa	The Open University of Israel, Israel
Vicenç Torra	Umeå University, Sweden
Rolando Trujillo-Rasúa	Deakin University, Australia
Lars Vilhuber	Cornell University, USA

Organization Committee

Samia Bouzefrane	CNAM, France
Joaquín García-Alfaro	Télécom SudParis, France
Jesús Manjón	Universitat Rovira i Virgili, Catalonia, Spain

Contents

Machine Learning and Privacy

Case Studies

Privacy Models

An Optimization-Based Decomposition Heuristic for the Microaggregation Problem

Jordi Castro[1($\boxtimes$)], Claudio Gentile[2], and Enric Spagnolo-Arrizabalaga[1]

[1] Department of Statistics and Operations Research,
Universitat Politècnica de Catalunya, Jordi Girona 1–3,
08034 Barcelona, Catalonia
jordi.castro@upc.edu, gentile@iasi.cnr.it
[2] Istituto di Analisi dei Sistemi ed Informatica "A. Ruberti",
Consiglio Nazionale delle Ricerche, Rome, Italy

Abstract. Given a set of points, the microaggregation problem aims to find a clustering with a minimum *sum of squared errors (SSE)*, where the cardinality of each cluster is greater than or equal to k. Points in the cluster are replaced by the cluster centroid, thus satisfying k-anonymity. Microaggregation is considered one of the most effective techniques for numerical microdata protection. Traditionally, non-optimal solutions to the microaggregation problem are obtained by heuristic approaches. Recently, the authors of this paper presented a mixed integer linear optimization (MILO) approach based on column generation for computing tight solutions and lower bounds to the microaggregation problem. However, MILO can be computationally expensive for large datasets. In this work we present a new heuristic that combines three blocks: (1) a decomposition of the dataset into subsets, (2) the MILO column generation algorithm applied to each dataset in order to obtain a valid microaggregation, and (3) a local search improvement algorithm to get the final clustering. Preliminary computational results show that this approach was able to provide (and even improve upon) some of the best solutions (i.e., of smallest SSE) reported in the literature for the Tarragona and Census datasets, and $k \in \{3, 5, 10\}$.

Keywords: Statistical disclosure control · Microdata ·
Microaggregation problem · Mixed integer linear optimization ·
Column generation · Local search · Heuristics

1 Introduction

A microdata file of p individuals (people, companies, etc.) and d variables (or attributes) is, in practice, a matrix $A \in \mathbb{R}^{p \times d}$ whose element a_{ij} provides the value of attribute j for individual i, and whose row a_i gives the d attributes for

Supported by grant MCIU/AEI/FEDER RTI2018-097580-B-I00.

J. Domingo-Ferrer and M. Laurent (Eds.): PSD 2022, LNCS 13463, pp. 3–14, 2022.
https://doi.org/10.1007/978-3-031-13945-1_1

individual i. Formally, a microdata file is a mapping $M : S \subseteq P \to V_1 \times \ldots \times V_t$, where P is a population, S is a sample of the population and V_i is the domain of the attribute $i \in \{1, \ldots, d\}$.

Microdata files must be protected before being released; otherwise, confidential individual information would be jeopardized. Microaggregation [5,6] is a statistical disclosure control technique, mainly for numeric variables, which is related with k-*anonymity* [20].

The goal of microaggregation is to modify the values of the variables such that the released microdata satisfies k-anonymity. Therefore, it first partitions the individuals (or points in $\mathbb{R}^d$) into subsets of size at least k, called *clusters*, and it then replaces each point in the cluster with the centroid of the cluster in order to minimize the loss of information, called *spread*. In practical cases, the value of k is relatively small (e.g., $3 \leq k \leq 10$, see [6]). A widely used measure for evaluating the spread is the *sum of squared errors* (SSE) [6]:

$$SSE = \sum_{i=1}^{q} \sum_{j=1}^{n_i} (a_{i_j} - \overline{a}_i)^T (a_{i_j} - \overline{a}_i), \tag{1}$$

where q denotes the number of clusters, n_i the size of cluster $\mathcal{C}_i = \{a_{i_j}, j = 1, \ldots, n_i\}$, and $\overline{a}_i = \frac{1}{n_i} \sum_{j=1}^{n_i} a_{i_j}$ its centroid, for $i = 1, \ldots, q$. An equivalent measure that is also widely used in the literature is the *information loss* (IL), which is defined as

$$IL = \frac{SSE}{SST} \cdot 100, \tag{2}$$

where SST is the total sum of squared errors for all the points, that is:

$$SST = \sum_{i=1}^{p} (a_i - \bar{a})^\top (a_i - \bar{a}) \quad \text{where } \bar{a} = \frac{\sum_{i=1}^{p} a_i}{p}. \tag{3}$$

IL always takes values within the range $[0, 100]$; the smaller the IL, the better the clustering. From now on, we will denote as *feasible clustering* a partition into clusters of size at least k.

Finding the partition that minimizes IL (or SSE) and satisfies the cardinality requirement $n_i \geq k$ for $i = 1, \ldots, q$ is a difficult combinatorial optimization problem when $d > 1$ (multivariate data), which is known to be NP-hard [15]. For univariate data (that is, $d = 1$)—which in practice are the exception— microaggregation can be solved in polynomial time using the algorithm of [11], which is based on the shortest path problem.

Microaggregation differs from other related clustering problems, such as k-medians or k-means [10], specifically in that it imposes a lower bound k on the cardinality of each cluster, but no fixed number of clusters. On the other hand, k in k-medians and k-means fixes the number of clusters, while imposing no constraint on the cardinality of each cluster.

There exist various papers on heuristic algorithms for feasible solutions to multivariate microaggregation with reasonable IL. Heuristics like *minimum distance to average* (MDAV) [6,8] and *variable minimum distance to average*

(VMDAV) [19] sequentially build groups of fixed (MDAV) or variable (VMDAV) size based on considering the distances of the points to their centroid. Other approaches first order the multivariate points and apply the polynomial time algorithm of [11] to the ordered set of points, such as in [7], which used several fast ordering algorithms based on paths in the graph that is associated with the set of points, whereas [14] used (slower) Hamiltonian paths (which involve the solution of a traveling salesman problem). The heuristic of [16] also sequentially builds a set of clusters attempting to locally minimize IL. Other approaches, such as those of [2,13], are based on refining the solutions previously provided by another heuristic.

To our knowledge, the only two papers in the literature to apply optimization techniques to microaggregation and formulate it as a combinatorial optimization problem are [1] and [4]. Both of them apply a column generation algorithm inspired by the work in [9]. Those optimization approaches solve the linear relaxation of the integer microaggregation problem, thus computing a (usually tight) lower bound for the problem. They also provide a (usually very good) upper bound solution with an IL that is smaller than the ones reported by other heuristics. Note that having a lower bound of the optimal solution is instrumental in order to know how good are the (upper bound) solutions computed by heuristics, even to perform fair comparisons between them. For instance, the heuristic introduced in [17] reported IL values below the certified lower bound— thus, not possible—, which clearly indicates that the values of the datasets used in that paper were different than those in the rest of the literature (likely due to some sort of normalization of attributes).

The downside of those optimization based techniques is that, when the dataset is large, the column generation may involve a large number of iterations, thus making it computationally very expensive. The main difference between the approaches in [1] and [4] is that the pricing problem of the former involves a nonlinear integer problem while the latter requires a simpler linear integer problem. In practice this means that the pricing subproblem in [1] can be tackled only by means of complete enumeration and only for small values of k, while [4] theoretically offers more flexibility and can deal with larger values of k.

Since optimization-based methods can be inefficient for large datasets but can provide high quality solutions in reasonable time for microdata with a small number of points, this work suggests a new approach consisting of first partitioning the set of points, and then applying an optimization approach to each smaller subset. The initial partitioning of the dataset is done according to a feasible clustering previously computed with the MDAV/VMDAV heuristics. Additionally, a local search improvement heuristic is also applied twice: first, to the solution provided by the MDAV/VMDAV heuristics prior to the partitioning; and second, to the final solution provided by the optimization approach.

This short paper is organized as follows. Section 2 outlines the optimization-based decomposition heuristic. Sections 3 and 4 outline the two main building blocks of the heuristic: the local search improvement algorithm and the mixed integer linear optimization method based on column generation in [4]. Section 5 shows the preliminary results from this approach with the standard Tarragona and Census datasets used in the literature.

2   The Decomposition Heuristic

The decomposition heuristic comprises the following steps:

> **Input:** Microdata matrix $A^0 \in \mathbb{R}^{p \times d}$, minimum cluster cardinality k, upper bound of the number of subsets s in which the microdata will be decomposed.
>
> 1. Standardize attributes/columns of A^0, obtaining matrix $A \in \mathbb{R}^{p \times d}$. Compute the squared Euclidean distance matrix $D \in \mathbb{R}^{p \times p}$, where $D_{ij} = (a_i - a_j)^{\top}(a_i - a_j)$, which is to be used in the remaining steps.
> 2. Apply MDAV and VMDAV microaggregation heuristics using D. Let $\mathcal{C} = \{\mathcal{C}_1, \ldots, \mathcal{C}_q\}$ be the best of the two feasible clusterings provided by MDAV and VMDAV (that is, the one with smallest IL). Here, q represents the number of clusters, and $\mathcal{C}_i$ the set of points in cluster i.
> 3. Apply the local search improvement algorithm (described in Sect. 3) to $\mathcal{C}$, obtaining the updated clustering $\mathcal{C}' = \{\mathcal{C}'_1, \ldots, \mathcal{C}'_q\}$. The updated clustering $\mathcal{C}'$ has the same number of clusters q as $\mathcal{C}$, but the points in subsets $\mathcal{C}'_i$ and $\mathcal{C}_i$ can be different.
> 4. Partition the microdata and distance matrices A and D in $s' \leq s$ subsets $\mathcal{S}_i, i = 1, \ldots, s'$, according to the clustering $\mathcal{C}'$. For this purpose we compute $\bar{p} = round(p/s)$, the minimum number of points in each subset of the partition, and build each subset $\mathcal{S}_i$ by sequentially adding points of clusters $\mathcal{C}'_j, j = 1, \ldots, q$ until the cardinality $\bar{p}$ is reached.
> 5. Apply the mixed integer linear optimization method based on column generation in [4] to each microaggregation subproblem defined by $A_{\mathcal{S}_i}$ and $D_{\mathcal{S}_i}$, $i = 1, \ldots, s'$. Obtain feasible clustering $\mathcal{O}_i$ for points in $\mathcal{S}_i$, $i = 1, \ldots, s'$.
> 6. Perform the union of clusterings $\mathcal{O} = \mathcal{O}_1 \cup \cdots \cup \mathcal{O}_{s'}$. $\mathcal{O}$ is a feasible clustering for the original microdata A.
> 7. Finally, once again apply the local search improvement algorithm from Sect. 3 to $\mathcal{O}$ in order to obtain the final microaggregation $\mathcal{O}'$.
>
> **Return:** Clustering $\mathcal{O}'$.

Step 3 of the algorithm can be skipped, thus obtaining the partition in Step 4 with the clustering $\mathcal{C}$ from Step 2. However, we have observed that better results are generally obtained if the local search improvement heuristic is applied in both Steps 3 and 7, not only in Step 7. Indeed, if efficiency is a concern, it is possible to stop the whole procedure after Step 3, which thus returns cluster $\mathcal{C}'$ as a solution and, in general, significantly outperforms the solution obtained in Step 2. In this way, it is possible to avoid Step 5, which is usually computationally expensive.

Note also that the clustering $\mathcal{C}'$ obtained in Step 3 is used only in Step 4 to decompose the microdata into subsets, but not as a starting solution for

Algorithm *Two-swapping heuristic(C: initial feasible clustering))*
   bestSSE= currentSSE= SSE(C)
  Repeat
    for $i = 1$ **to** p
      for $j = i + 1$ **to** p
        if swapping i and j improves *bestSSE* **then**
           Update best swapping: $i' = i$, $j' = j$, update *bestSSE*
        end__if
      end__for
    end__for
    improvement= (bestSSE < currentSSE)
    if *improvement* **then**
      Update C by swapping points i' and j'
      currentSSE = bestSSE
    end__if
  while improvement
  Return: C updated with swaps
End__algorithm

Fig. 1. Two-swapping local search improvement heuristic

the optimization procedure in Step 5. Therefore, the clustering computed in Steps 5–6 by the optimization procedure might have a larger SSE than C'. On the other hand, by not starting the optimization procedure in Step 5 with the solution C' we have some chances to obtain a different and possibly better local solution. In the current implementation, C' is not used as a starting solution for the optimization algorithm.

The larger the value of s, the faster the algorithm will be, since the minimum number of points $\bar{p} = round(p/s)$ in each subset $S_i, i = 1, \ldots, s'$ will be smaller, and therefore the optimization algorithm of [4] will be more efficient. However, the final IL (SSE) of the final clustering O' also increases with s. Therefore parameter s can be used as a trade-off between efficiency and solution quality.

In the next two sections, we outline the local search improvement heuristic used in Steps 3 and 7, as well as the mixed integer linear optimization method in Step 5.

3   The Local Search Improvement Heuristic

Given a feasible clustering for the microaggregation problem, a local search algorithm tries to improve it by finding alternative solutions in a local neighborhood of the current solution. The local search considered in this work is a two-swapping procedure; in addition to its simplicity, it has proven to be very effective in practice. Briefly, the two-swapping heuristic performs a series of iterations, and at each iteration it finds the pair of points (i, j) located in different clusters that would most reduce the overall SSE if they were swapped. This operation is repeated until no improvement in SSE is detected. The cost per iteration of the heuristic is $O(p^2/2)$. Similar approaches have been used in other clustering techniques, such as in the partitioning around medoids algorithm for k-medoids [12]. The two-swapping algorithm implemented is shown in Fig. 1.

4   The Mixed Integer Linear Optimization Algorithm Based on Column Generation

In this section we quickly outline the optimization method presented in [4]. Additional details can be found in that reference.

The formulation of microaggregation as an optimization problem in [4] is based on the following property of the SSE_h of cluster $\mathcal{C}_h = \{a_{h_i}, i = 1, \ldots, n_h\}$ (see [4, Prop. 3] for a proof):

$$
\begin{aligned}
SSE_h &= \sum_{i=1}^{n_h} (a_{h_i} - \overline{a_h})^\top (a_{h_i} - \overline{a_h}) \\
&= \frac{1}{2n_h} \sum_{i=1}^{n_h} \sum_{j=1}^{n_h} (a_{h_i} - a_{h_j})^\top (a_{h_i} - a_{h_j}) = \frac{1}{2n_h} \sum_{i=1}^{n_h} \sum_{j=1}^{n_h} D_{h_i h_j}.
\end{aligned}
\tag{4}
$$

That is, for computing SSE_h, we do not need the centroid of the cluster, but only the distances between the points in the cluster.

From (4), defining binary variables x_{ij}, $i, j = 1, \ldots, p$ (which are 1 if points i and j are in the same cluster, 0 otherwise), then the microaggregation problem can be formulated as:

$$
\min \quad SSE \triangleq \frac{1}{2} \sum_{i=1}^{p} \frac{\sum_{j=1, j \neq i}^{p} D_{ij} x_{ij}}{\sum_{j=1, j \neq i}^{p} x_{ij} + 1}
\tag{5a}
$$

$$
\text{s. to} \quad x_{ir} + x_{jr} - x_{ij} \leq 1 \quad i, j, r = 1, \ldots, p, \quad i \neq j, r \neq j, i \neq r
\tag{5b}
$$

$$
\sum_{j=1, j \neq i}^{p} x_{ij} \geq k - 1 \quad i = 1, \ldots, p
\tag{5c}
$$

$$
x_{ij} = x_{ji}, x_{ij} \in \{0, 1\}, i, j = 1, \ldots, p.
\tag{5d}
$$

Constraints (5b) are triangular inequalities, that is, if points i and r, and r and j are in the same cluster, then points i and j are also in the same cluster. Constraints (5c) guarantee that the cardinality of the cluster is at least k. The denominator in the objective function (5a) is the cardinality of the cluster that contains point i. Unfortunately, (5) is a difficult nonlinear and nonconvex integer optimization problem (see [4] for details).

A more practical alternative is to consider a formulation inspired by the clique partitioning problem with minimum clique size of [9]. Defining as $\mathcal{C}^* = \{\mathcal{C} \subseteq$

$\{1,\ldots,p\} : k \leq |\mathcal{C}| \leq 2k - 1\}$ the set of feasible clusters, the microaggregation problem can be formulated as:

$$\min \sum_{\mathcal{C} \in \mathcal{C}^*} w_{\mathcal{C}} x_{\mathcal{C}}$$

$$\text{s. to} \sum_{\mathcal{C} \in \mathcal{C}^* : i \in \mathcal{C}} x_{\mathcal{C}} = 1 \; i \in \{1,\ldots,p\} \tag{6}$$

$$x_{\mathcal{C}} \in \{0,1\} \quad \mathcal{C} \in \mathcal{C}^*,$$

where $x_{\mathcal{C}} = 1$ means that feasible cluster $\mathcal{C}$ appears in the microaggregation solution provided, and the constraints guarantee that all the points are covered by some feasible cluster.

From (4), the cost $w_{\mathcal{C}}$ of cluster $\mathcal{C}$ in the objective function of (6) is

$$w_C = \frac{1}{2|C|} \sum_{i \in \mathcal{C}} \sum_{j \in \mathcal{C}} D_{ij}. \tag{7}$$

The number of feasible clusters in $\mathcal{C}^*$—that is, the number of variables in the optimization problem (6)—is $\sum_{j=k}^{2k-1} \binom{p}{j}$, which can be huge. For instance, for $p = 1000$ and $k = 3$ we have $|\mathcal{C}^*| = 8.29 \cdot 10^{12}$. However, the linear relaxation of (6) can be solved using a column generation technique, where the master problem is defined as (6) but it considers only a subset $\bar{\mathcal{C}} \subseteq \mathcal{C}^*$ of the variables/clusters. The set $\bar{\mathcal{C}}$ is updated at each iteration of the column generation algorithm with new clusters, which are computed by a pricing subproblem. The pricing subproblem either detects that the current set $\bar{\mathcal{C}}$ contains the optimal set of columns/clusters or, otherwise, it generates new candidate clusters with negative reduced costs. For small datasets and values of k, the pricing subproblem can be solved by complete enumeration; otherwise, an integer optimization model must be solved. The master problem requires the solution of a linear optimization problem. Both the linear and integer optimization problems were solved with CPLEX in this work. The solution of the linear relaxation of (6) provides a lower bound to the microaggregation problem (usually a tight lower bound). In addition, solving the master problem as an integer problem allows us to obtain a feasible solution to the microaggregation problem (usually of high quality). A thorough description of this procedure, and of the properties of the pricing subproblem, can be found in [4] and [18].

5 Computational Results

The algorithm in Sect. 2 and the local search heuristic in Sect. 3 have been implemented in C++. We used the code of [4] (also implemented in C++) for the solution of the mixed integer linear optimization approach based on column generation, as described in Sect. 4. A time limit of 3600 s was set for the solution of each subproblem with the column generation algorithm in Step 5 of the heuristic in Sect. 2. We tested the decomposition algorithm with the datasets Tarragona and Census, which are standard in the literature [3]. The results for Tarragona and Census are shown, respectively, in Tables 1–2 and 3–4. These tables show, for each instance and value $k \in \{3, 5, 10\}$, the IL and CPU time for the main steps of the decomposition heuristic in Sect. 2. We also tried the different values $s \in \{40, 20, 10, 5, 2\}$ for partitioning the dataset in Step 5. For Step 2, the tables also show which of the MDAV or VMDAV algorithms reported the best solution. The difference between Tables 1 and 2 (also between Tables 3 and 4) is that the former show results with the Step 3, while in the latter this step was skipped. We remind the reader that the decomposition algorithm could be stopped after Step 3 with a feasible and generally good solution. However, the best IL values for each k, which are marked in boldface in the tables, are obtained after Step 7, although this means going through the usually more expensive Step 5.

Table 1. Results for the Tarragona dataset considering Step 3 of the algorithm. The best IL for each k is marked in boldface.

Instance	k	Step 2			Step 3		Step 5			Step 7	
		Alg	IL	CPU	IL	CPU	s	IL	CPU	IL	CPU
Tarragona	3	VMDAV	15.85	0.6	15.00	1.9	40	14.96	0.2	14.85	0.1
							20	14.83	0.4	14.81	0.1
							10	14.68	0.9	14.65	0.3
							5	14.57	2.8	14.53	0.4
							2	14.51	4.2	**14.50**	0.2
Tarragona	5	MDAV	22.46	0.5	20.74	5.2	40	20.74	1.0	20.73	0.5
							20	20.74	7.9	20.73	0.3
							10	20.47	103.8	20.40	1.5
							5	20.32	1119.6	**20.25**	1.6
							2	21.06	7207.8	20.46	4.0
Tarragona	10	MDAV	33.19	0.3	30.77	25.6	40	30.77	12119.7	30.77	0.1
							20	33.03	61600.3	30.87	9.0
							10	31.80	88899.1	**30.56**	13.4
							5	33.32	9.8	30.79	17.4
							2	33.20	22.8	30.80	17.2

From Tables 1, 2, 3 and 4 we conclude that:

– In general, the smaller the k and larger the s, the faster the decomposition heuristic. In a few cases, however, smaller values of s also meant smaller CPU times; for instance, this is observed for Census, $k = 10$, and values $s = 20$ and $s = 10$. The explanation is that the maximum time limit was reached in some pricing subproblems in those runs, and therefore the CPU time increased with s.
– In general, smaller ILs are obtained for smaller s values, as expected.

Table 2. Results for the Tarragona dataset without Step 3 of the algorithm. The best *IL* for each *k* is marked in boldface.

Instance	k	Step 2			Step 5			Step 7	
		Alg	*IL*	CPU	s	*IL*	CPU	*IL*	CPU
Tarragona	3	VMDAV	15.85	0.6	40	15.15	0.2	14.95	1.7
					20	14.86	0.3	14.73	1.3
					10	14.75	1.0	14.66	1.2
					5	14.58	1.9	14.54	0.7
					2	14.51	4.6	**14.50**	0.4
Tarragona	5	MDAV	22.46	0.5	40	21.81	0.9	21.18	4.7
					20	21.14	7.9	20.59	4.7
					10	20.62	86.2	20.36	3.8
					5	20.37	894.5	**20.29**	2.1
					2	21.01	7208.1	20.77	6.0
Tarragona	10	MDAV	33.19	0.3	40	32.05	12597.2	30.82	19.9
					20	33.18	60243.5	30.81	20.8
					10	32.23	230644.6	**30.55**	20.9
					5	33.20	13.0	30.84	20.9
					2	33.21	21.2	30.83	20.6

Table 3. Results for the Census dataset considering Step 3 of the algorithm. The best *IL* for each *k* is marked in boldface.

Instance	k	Step 2			Step 3		Step 5			Step 7	
		Alg	*IL*	CPU	*IL*	CPU	s	*IL*	CPU	*IL*	CPU
Census	3	VMDAV	5.66	1.2	5.25	3.3	40	5.21	0.1	5.20	0.2
							20	5.20	0.2	5.19	0.1
							10	5.21	0.6	5.18	0.3
							5	5.12	3.2	5.07	0.8
							2	4.85	5.2	**4.79**	0.4
Census	5	VMDAV	8.98	1.1	8.12	8.7	40	8.12	2.4	8.12	0.2
							20	8.12	22.5	8.11	0.2
							10	8.14	247.0	8.09	0.8
							5	7.96	5334.6	**7.84**	2.0
							2	9.36	7209.3	8.19	8.5
Census	10	VMDAV	14.04	0.8	12.36	38.4	40	12.36	8452.7	12.36	0.4
							20	12.96	77221.1	**12.32**	7.0
							10	12.84	37037.7	12.40	14.5
							5	13.49	7310.4	12.63	28.8
							2	14.45	26.3	12.46	37.1

Table 4. Results for the Census dataset without Step 3 of the algorithm. The best IL for each k is marked in boldface.

Instance	k	Step 2			Step 5			Step 7	
		Alg	IL	CPU	s	IL	CPU	IL	CPU
Census	3	VMDAV	5.66	1.2	40	5.56	0.1	5.19	3.2
					20	5.51	0.3	5.19	2.8
					10	5.44	0.9	5.19	2.1
					5	5.26	2.3	5.10	1.4
					2	4.87	4.9	**4.81**	0.7
Census	5	VMDAV	8.98	1.1	40	8.94	2.2	8.14	8.3
					20	8.84	22.3	8.10	8.2
					10	8.63	218.6	8.10	5.9
					5	8.22	4050.7	**7.90**	3.9
					2	9.13	7210.2	8.19	8.1
Census	10	VMDAV	14.04	0.8	40	13.88	7647.9	12.47	31.8
					20	14.10	77388.6	12.45	35.1
					10	14.05	50960.1	12.68	33.1
					5	14.02	8802.5	12.55	36.8
					2	14.59	24.6	**12.42**	41.9

- When $k = 10$, Step 5 is faster for $s = 2$ or $s = 5$, which was initially unexpected. The reason is that, when k is large and s is small, the column generation optimization algorithm generates new columns heuristically, reaching the maximum allowed space of 3000 columns; thus the solution of the difficult mixed integer linear pricing subproblems is never performed. However, in those cases, poorer values of IL were obtained.
- In general, the best IL values are obtained when using Step 3, with the exception of Tarragona and $k = 10$, whose best IL was given in Table 4 without Step 3. This can be due to the randomness associated with partitioning the dataset into s subsets.
- The solution times in Step 5 are generally longer when k is large and s is small.

Finally, Table 5 summarizes the best IL results obtained with the new approach, comparing them with—to our knowledge—the best values reported in the literature by previous heuristics (citing the source), and the optimization method of [1]. The approaches implemented by those other heuristics were commented in Sect. 1 of this paper. It can be seen that the new approach provided a better solution than previous heuristics in all the cases. In addition, for $k \in \{3, 5\}$, the new approach also provided IL values close to the ones provided by the optimization method of [1], usually while requiring fewer computational resources. For instance, with our approach, the solutions for Tarragona and $k = 3$ and $k = 5$ required, respectively, 7 and 1127 s; whereas the optimization method of

Table 5. Comparison of best IL values obtained with the new approach vs. those found in the literature (citing the source).

Instance	k	New heuristic IL	Previous heuristics IL	Optimization method of [1] IL
Tarragona	3	14.50	14.77 [14]	14.46
	5	20.25	20.93 [2]	20.16
	10	**30.55**	31.95 [2]	—
Census	3	4.79	5.06 [14]	4.67
	5	7.84	8.37 [13]	7.36
	10	**12.32**	12.65 [13]	—

[1] (running on a different—likely older—computer) needed, respectively, 160 and 4779 s. For Census and $k = 3$ and $k = 5$, the solution times with our approach were, respectively, 10 and 5346 s, whereas that of [1] required, respectively, 3868 and 6788 s. The optimization method of [1] is unable to solve problems with $k = 10$, and in this case our new approach reported—as far as we know—the best IL values ever computed for these instances.

6   Conclusions

We have presented here the preliminary results from a new heuristic approach for the microaggregation problem. This method combines three ingredients: a partition of the dataset; solving each subset of the partition with an optimization-based approach; and a local search improvement heuristic. The results have shown that our new approach provides solutions that are as good as (and in some cases even better than) those reported in the literature by other heuristics for the microaggregation problem, although it may generally require longer executions times. Future research will investigate improving Step 5 of the heuristic algorithm and will further consider more sophisticated large-neighborhood search improvement heuristics.

References

1. Aloise, D., Hansen, P., Rocha, C., Santi, É.: Column generation bounds for numerical microaggregation. J. Global Optim. **60**(2), 165–182 (2014). https://doi.org/10.1007/s10898-014-0149-3
2. Aloise, D., Araújo, A.: A derivative-free algorithm for refining numerical microaggregation solutions. Int. Trans. Oper. Res. **22**, 693–712 (2015)
3. Brand, R., Domingo-Ferrer, J., Mateo-Sanz, J. M.: Reference data sets to test and compare SDC methods for protection of numerical microdata. European Project IST-2000-25069 CASC (2002). http://neon.vb.cbs.nl/casc, https://research.cbs.nl/casc/CASCtestsets.html

4. Castro, J., Gentile, C., Spagnolo-Arrizabalaga, E.: An algorithm for the microaggregation problem using column generation. Comput. Oper. Res. **144**, 105817 (2022). https://doi.org/10.1016/j.cor.2022.105817
5. Defays, D., Anwar, N.: Micro-aggregation: a generic method. In: Proceedings of Second International Symposium Statistical Confidentiality, pp. 69–78 (1995)
6. Domingo-Ferrer, J., Mateo-Sanz, J.M.: Practical data-oriented microaggregation for statistical disclosure control. IEEE Trans. Knowl. Data Eng. **14**, 189–201 (2002)
7. Domingo-Ferrer, J., Martínez-Ballesté, A., Mateo-Sanz, J.M., Sebé, F.: Efficient multivariate data-oriented microaggregation. VLDB J. **15**, 355–369 (2006)
8. Domingo-Ferrer, J., Torra, V.: Ordinal, continuous and heterogeneous k-anonymity through microaggregation. Data Mining Knowl. Disc. **11**, 195–212 (2005)
9. Ji, X., Mitchell, J.E.: Branch-and-price-and-cut on the clique partitioning problem with minimum clique size requirement. Discr. Optim. **4**, 87–102 (2007)
10. Ghosh, J., Liu, A.: K-means. In: The Top Ten Algorithms in Data Mining, pp. 21–35. Taylor & Francis, Boca Raton (2009)
11. Hansen, S., Mukherjee, S.: A polynomial algorithm for optimal univariate microaggregation. IEEE Trans. Knowl. Data Eng. **15**, 1043–1044 (2003)
12. Kaufman, L., Rousseeuw, P.J.: Partitioning around medoids (Program PAM). In: Wiley Series in Probability and Statistics, pp. 68–125. John Wiley & Sons, Hoboken (1990)
13. Khomnotai, L., Lin, J.-L., Peng, Z.-Q., Samanta, A.: Iterative group decomposition for refining microaggregation solutions. Symmetry **10**, 262 (2018). https://doi.org/10.3390/sym10070262
14. Maya-López, A., Casino, F., Solanas, A.: Improving multivariate microaggregation through Hamiltonian paths and optimal univariate microaggregation. Symmetry. **13**, 916 (2021). https://doi.org/10.3390/sym13060916
15. Oganian, A., Domingo-Ferrer, J.: On the complexity of optimal microaggregation for statistical disclosure control. Statist. J. U. N. Econ. Com. Eur. **18**, 345–354 (2001)
16. Panagiotakis, C., Tziritas, G.: Successive group selection for microaggregation. IEEE Trans. Knowl. Data Eng. **25**, 1191–1195 (2013)
17. Soria-Comas, J., Domingo-Ferrer, J., Mulero, R.: Efficient near-optimal variable-size microaggregation. In: Torra, V., Narukawa, Y., Pasi, G., Viviani, M. (eds.) MDAI 2019. LNCS (LNAI), vol. 11676, pp. 333–345. Springer, Cham (2019). https://doi.org/10.1007/978-3-030-26773-5_29
18. Spagnolo-Arrizabalaga, E.: On the use of Integer Programming to pursue Optimal Microaggregation. B.Sc. thesis, University Politècnica de Catalunya, School of Mathematics and Statistics, Barcelona (2016)
19. Solanas, A., Martínez-Ballesté, A.: V-MDAV: a multivariate microaggregation with variable group size. In: Proceedings of COMPSTAT Symposium IASC, pp. 917–925 (2006)
20. Sweeney, L.: k-anonymity: a model for protecting privacy. Int. J. Uncertain Fuzziness Knowl. Based Syst. **10**, 557–570 (2002)

Privacy Analysis with a Distributed Transition System and a Data-Wise Metric

Siva Anantharaman[1]([✉]), Sabine Frittella[2], and Benjamin Nguyen[2]

[1] Université d'Orléans, Orléans, France
siva@univ-orleans.fr
[2] INSA Centre Val de Loire, Blois, France
{sabine.frittella,benjamin.nguyen}@insa-cvl.fr

Abstract. We introduce a logical framework DLTTS (*Distributed Labeled Tagged Transition System*), built using concepts from Probabilistic Automata, Probabilistic Concurrent Systems, and Probabilistic labelled transition systems. We show that DLTTS can be used to formally model how a given piece of *private* information P (e.g. a tuple) stored in a given database D protected by generalization and/or noise addition mechanisms, can get captured progressively by an agent repeatedly querying D, by using additional non-private data, as well as knowledge deducible with a more general notion of adjacency based on metrics defined 'value-wise'; such metrics also play a role in differentially private protection mechanisms.

Keywords: Database · Privacy · Transition system · Probability · Distribution

1 Introduction

Data anonymization has been investigated for decades, and many privacy models have been proposed (k-anonymity, *differential privacy*, ...) whose goals are to protect sensitive information. In this paper, our goal is to propose a logical framework to formally model how the information stored in a database can get captured progressively by any agent repeatedly querying the database. This model can also be used to quantify reidentification attacks on a database.

We assume given a data base D, with its attributes set $\mathcal{A}$, divided in three disjoint groups: *identifiers* $\mathcal{A}^{(i)}$, *quasi-identifiers* $\mathcal{A}^{(qi)}$, and *sensitive attributes* $\mathcal{A}^{(s)}$. The tuples of D will be denoted as t, and the subgroup attributes denoted as t^i, t^{qi}, and t^s. A *privacy policy* on D with respect to an agent (user or adversary) A is a set of tuples $P_A(D)$ (noted P) such that $\forall t \in P$, attributes t^s on any t 'not in the knowledge of A' remain inaccessible ('even after deduction') to A.

The logical framework we propose in this work, to model the evolution of the 'knowledge' that an adversary A can gain by repeatedly querying the given database D – with a view to get access to sensitive information protected by a

© Springer Nature Switzerland AG 2022
J. Domingo-Ferrer and M. Laurent (Eds.): PSD 2022, LNCS 13463, pp. 15–30, 2022.
https://doi.org/10.1007/978-3-031-13945-1_2

privacy policy P –, will be called *Distributed Labeled-Tagged Transition System* (DLTTS). The underlying logic for DLTTS is first-order, with countably many variables and finitely many constants (including certain usual dummy symbols like '$\star, \$, \#$'). The basic signature Σ for the framework is assumed to have no non-constant function symbols. By 'knowledge' of A we mean the data that A retrieves as answers to his/her successive queries, as well as other data that can be deduced/derived, under relational operations on these answers; and in addition, some others derivable from these, using relational combinations with data (possibly involving some of the users of D) from finitely many *external DBs given in advance*, denoted as $B_1, \ldots, B_m$, to which the adversary A is assumed to have free access. These relational and querying operations are all assumed done with a well-delimited fragment of the language SQL; *it is assumed that this fragment of SQL is part of the basic signature Σ*. For the sake of simplicity, we shall see any data tuple (that is not part of the given privacy policy P) directly as a first-order variable-free formula over Σ, its arguments typed implicitly with the appropriate headers of the base D; data tuples t in the policy $P_A(D)$ will generally be written as $\neg t$.

The DLTTS framework will be shown to be particularly suitable for capturing the ideas on acquiring knowledge and on policy violation, in an abstract setup. Section 2 introduces the framework; to start with, only the data as well as the answers to the queries do not involve any notion of randomness. In Sects. 3 and 4, the framework will be extended so as to handle differentially private databases as well.

In the second part of the work (Sect. 5 onwards), we propose a method for *comparing the evolution of knowledge* of an adversary at two different moments of the querying process; the same method also applies for comparing the knowledge evolution of two different adversaries A_1, A_2, both querying repeatedly (and independently) the given database.

Running Example. The central Hospital of a Faculty stores recent consultations by faculty staff. In this record, 'Name' is an identifier, 'Ailment' is sensitive, the others are QI; 'Ailment' is categorical with 3 branches: Heart-Disease, Cancer, and Viral-Infection; this latter in turn is categorical too, with 2 branches: Flu and CoVid. Such taxonomical relations are assumed public. We assume **all** Faculty staff are on the consultation list (Table 1).

Table 1. Hospital's 'Secret' record, and Anonymized 'Public' record

Name	Age	Gender	Dept.	Ailment
Joan	24	F	Chemistry	Heart-Disease
Michel	46	M	Chemistry	Cancer
Aline	23	F	Physics	Flu
Harry	53	M	Maths	Flu
John	46	M	Physics	CoVid

Table 2. Hospital's 'Secret' record, and Anonymized 'Public' record

Nr	Age	Gender	Dept.	Ailment
ℓ_1	$[20-30[$	F	Chemistry	Heart-Disease
ℓ_2	$[40-50[$	M	Chemistry	Cancer
ℓ_3	$[20-30[$	F	Physics	Viral-Infection
ℓ_4	$[50-60[$	M	Maths	Viral-Infection
ℓ_5	$[40-50[$	M	Physics	Viral-Infection

The Hospital intends to keep 'secret' CoVid information on faculty members; its privacy policy is thus: $P = \{\neg(John, 46, M, \#, CoVid)\}$ with $\#$ meaning "any value". A "public" table is published using *generalization*: 'Age' generalized as intervals, and 'Ailment' by an upward push in the taxonomy. (*This database is neither k-anonymous nor ϵ-DP.*)

An adversary A, who met John at a faculty banquet, suspects him to be infected with CoVid; she thus decides to consult the published record of the hospital. Knowing that the 'John' she met is a 'man' and that the table must contain John's health bulletin, A has as choice ℓ_2, ℓ_4 and ℓ_5. A looking for a 'CoVid-infected' man, this choice is reduced to the last two tuples of the table – a priori indistinguishable because of generalization to 'Viral-Infection'. A had the impression that John 'was not too old', so feels that the last tuple is twice more likely; she thus 'decides that John must be from the Physics Dept.', and consults the CoVid-cases statement kept publicly visible at that Dept.; which reads:

Recent CoVid-cases in the Dept: Female 0;   Male 1.

This confirms A's suspicion concerning John.

In this case, the DLTTS framework functions as follows: At starting state s a transition with three branches would a priori be possible, corresponding to the three 'M' lines (ℓ_2, ℓ_4, ℓ_5), which represent the knowledge acquirable respectively along these branches. A looking for a possible CoVid case, rules out the ℓ_2 branch (gives this branch probability 0). For the remaining two branches (ℓ_4 and ℓ_5), A chooses the ℓ_5 branch, considering it twice more likely to succeed than the other (A thought 'John was not too old'). That gives the probability distribution $0, 1/3, 2/3$ respectively on the three possible branches for the transition. If s_0, s_1, s_2 are the respective successor states for the transition considered, the privacy policy of the Hospital (concerning John's CoVid information) is violated at state s_2 (with probability $2/3$); it is not violated at s_1 (probability $1/3$); no information is deduced at state s_0.

The probability distributions on the transitions along the runs would generally depend on some random mechanism, which could also reflect the choices the adversary might make. □

Remark 1: This example shows that the violation of privacy policies needs, in general, some additional 'outside knowledge', when applied to databases anonymized using generalization. □

We may assume wlog that the given external bases $B_1, \ldots, B_m$ – to which A could resort, with relational operations for deducing additional information – are also of the same signature Σ as D; so all the knowledge A can deduce/derive from her repeated queries can be expressed as a first-order variable-free formula over signature Σ.

2 Distributed Labeled-Tagged Transition Systems

The DLTTS framework presented in this section synthesizes ideas coming from various domains, such as the Probabilistic Automata of Segala [13], Probabilistic

Concurrent Systems, Probabilistic labelled transition systems [3,4]. Although the underlying signature for the DLTTS can be rich in general, for the purposes of our current work we shall work with a limited first-order signature denoted Σ (see Introduction). Let $\mathcal{E}$ be the set of all variable-free formulas over Σ, and Ext a given subset of $\mathcal{E}$. We assume given a decidable procedure $\mathcal{C}$ whose role is to 'saturate' any finite set G of variable-free formulas into a finite set $\overline{G}$, by adding a finite (possibly empty) set of variable-free formulas, using *relational operations* on G and Ext. This procedure $\mathcal{C}$ will be internal at every node on a DLTTS; *in addition, there will also be a 'blackbox' mechanism* $\mathcal{O}$, acting as an oracle, telling if the given privacy policy on a given database is violated at the current node. More details will be given in Sect. 5 on the additional role the oracle will play in a privacy analysis procedure (for any querying sequence on a given DB), based on a novel data-based metric defined in that section.

Definition 1. *A Distributed Labeled-Tagged Transition System (DLTTS), over a given signature Σ, is formed of:*

- *a finite (or denumerable) set S of states, an 'initial' state $s_0 \in S$, and a special state $\otimes \in S$ named 'fail':*
- *a finite set Act of action symbols (disjoint from Σ), with a special action $\delta \in Act$ called 'violation';*
- *a (probabilistic) transition relation $\mathcal{T} \subset S \times Act \times Distr(S)$, where $Distr(S)$ is the set of all probability distributions over S, with finite support.*
- *a tag $\tau(s)$ attached to every state $s \in S \smallsetminus \{\otimes\}$, formed of finitely many first-order variable-free formulas over Σ; the tag $\tau(s_0)$ at the initial state is the singleton set $\{\top\}$ (representing the knowledge of the querying agent).*
- *at every state s a special action symbol $\iota = \iota_s \in Act$, said to be* internal *at s, completes/saturates $\tau(s)$ into a set $\overline{\tau}(s)$ with the procedure $\mathcal{C}$, by using relational operations between the formulas in $\tau(s)$ and Ext.*

A (probabilistic) transition $\mathsf{t} \in \mathcal{T}$ will generally be written as a triple $(s, a, \mathsf{t}(s))$; and t will be said to be 'from' (or 'at') the state s, the states of $\mathsf{t}(s)$ will be the 'successors' of s under t. The formulas in the tag $\overline{\tau}(s)$ attached to any state s will all be assigned the same probability as the state s in $Distr(S)$. If the set $\overline{\tau}(s)$ of formulas turns out to be inconsistent, then the oracle mechanism $\mathcal{O}$ will (intervene and) impose $(s, \delta, \otimes)$ as the only transition from s, standing for 'violation' and 'fail', by definition.

DLTTS and Repeated Queries on a Database: The states of the DLTTS stand for the various 'moments' of the querying sequence, while the tags attached to the states will stand for the knowledge A has acquired on the data of D 'thus far'. This knowledge consists partly in the answers to the queries she made so far, then completed with additional knowledge using the internal 'saturation' procedure $\mathcal{C}$ of the framework. In our context, this procedure would consist in relational algebraic operations between the answers retrieved by A for his/her repeated queries on D, all seen as tuples (variable-free formulas), and suitable tuples from the given external databases $B_1, \ldots, B_m$. If the saturated knowledge of A at a current state s on the DLTTS (i.e., the tag $\overline{\tau}(s)$ attached to the current state s) is not inconsistent, then the transition from s to its successor states represents the probability distribution of the likely answers A would expect to get for her next query.

No assumption is made on whether the repeated queries by A on D are treated *interactively, or non-interactively*, by the DBMS. It appears that the logical framework would function exactly alike, in both cases.

Remark 2: (a) If t is a transition from a state s on the DLTTS that models a querying sequence on D by an adversary A, and s' is a successor of s under t, then, the 'fresh' knowledge $\tau(s')$ of A at s', is the addition to A's saturated knowledge $\overline{\tau}(s)$ at s, the part of the response for A's query, represented by the branch from s to s' on t.

(b) It is assumed that 'no infinite set can be generated from a finite set' under the functionalities of SQL (included in Σ) needed in the relational operations for gaining additional knowledge. This corresponds to the *bounded inputs outputs* assumption as in, e.g., [1,2]. □

Proposition 1. *Suppose given a database D, a finite sequence of repeated queries on D by an adversary A, and a first-order relational formula $P = P_A(D)$ over the signature Σ of D, expressing the privacy policy of D with respect to A. Let $\mathcal{W}$ be the DLTTS modeling the querying sequence of A on D, and the evolution of the knowledge of A on the data of D along the branches in $\mathcal{W}$, as described above.*

(i) *The given privacy policy $P_A(D)$ on D is violated if and only if the failure state $\otimes$ on the DLTTS $\mathcal{W}$ is reachable from the initial state of $\mathcal{W}$.*

(ii) *The satisfiability of the set of formulas $\overline{\tau}(s) \cup \{\neg P\}$ is decidable, at any state s on the DLTTS, under the assumptions of Remark 2(b).*

3 ϵ-Indistinguishability, ϵ-Local-Differential Privacy

In order to manage more powerful anonymization schemes (such as local differential privacy), one of our objectives now is to extend the result of Proposition 1 to the case when the violation to be considered can be *up to some given* $\epsilon \geq 0$, the meaning of which we explain next. We stick to the same notation as above. The set $\mathcal{E}$ of all variable-free formulas over Σ is thus a disjoint union of subsets of the form $\mathcal{E} = \cup\{\mathcal{E}_i^{\mathcal{K}} \mid 0 < i \leq n, \mathcal{K} \in \Sigma\}$, the index i in $\mathcal{E}_i^{\mathcal{K}}$ standing for the common length of the formulas in the subset, and $\mathcal{K}$ for the common root symbol of its formulas; each set $\mathcal{E}_i^{\mathcal{K}}$ will be seen as a database of i-tuples.

We first look at the situation where the queries intend to capture certain (sensitive) values on a given tuple t in the database D. Two different tuples in $\mathcal{E}$ might correspond to two likely answers to such a query, but with possibly different probabilities in the distribution assigned for the transitions, by the probabilistic mechanism $\mathcal{M}$.

Given two such instances, and a real $\epsilon \geq 0$, one can define a notion of their ϵ-local-indistinguishability, wrt the tuple t and the mechanism $\mathcal{M}$ answering the queries. This can be done in a slightly extended setup, where the answering mechanism may, *as an option*, also add 'noise' to certain numerical data values, for several reasons one among which is data safety. We shall then assume that the internal knowledge saturation procedure $\mathcal{C}$ of the DLTTS at each of its states incorporates the following noise adding mechanisms: the Laplace, geometric, and exponential mechanisms. With the stipulation that this *optional* noise addition to numerical values can be done in a *bounded* fashion, so as to be from a finite prescribed domain around the values (e.g., as in [9]); it will then be assumed that the tuples formed of such noisy data are also in $\mathcal{E}$.

Definition 2. *(i) Suppose that, while answering a given query on the base D, at two instances v, v', the probabilistic answering mechanism $\mathcal{M}$ outputs the same tuple $\alpha \in \mathcal{E}$. Given $\epsilon \geq 0$, these two instances are said to be ϵ-local-indistinguishable wrt α, if and only if:*

$$Prob[\mathcal{M}(v) = \alpha] \leq e^\epsilon Prob[\mathcal{M}(v') = \alpha] and$$
$$Prob[\mathcal{M}(v') = \alpha] \leq e^\epsilon Prob[\mathcal{M}(v) = \alpha].$$

(ii) The probabilistic answering mechanism $\mathcal{M}$ is said to satisfy ϵ-local differential privacy (ϵ-LDP) for $\epsilon \geq 0$, if and only if: For any two instances v, v' of $\mathcal{M}$ leading to the same output, and any set $\mathcal{S} \subset Range(\mathcal{M})$, we have: $Prob[\mathcal{M}(v) \in \mathcal{S}] \leq e^\epsilon Prob[\mathcal{M}(v') \in \mathcal{S}]$.

We shall be using the following notion of ϵ-indistinguishability, and of ϵ-distinguishability, of two different outputs of the mechanism $\mathcal{M}$: These definitions – as well as that of ϵ-DP given below – are essentially reformulations of the same (or similar) notions defined in [7,8].

Definition 3. *Given $\epsilon \geq 0$, two outputs α, α' of the probabilistic mechanism $\mathcal{M}$ answering the queries of A, are said to be ϵ-indistinguishable, if and only if: For every pair v, v' of inputs for $\mathcal{M}$, such that $Prob[\mathcal{M}(v) = \alpha] = p$ and $Prob[\mathcal{M}(v') = \alpha'] = p'$, we have: $p \leq e^\epsilon p'$ and $p' \leq e^\epsilon p$.*
Otherwise, the outputs α, α' are said to be ϵ-distinguishable.

Remark 3: Given an $\epsilon \geq 0$, one may assume as an option, that at every state on the DLTTS *the retrieval of answers to the current query (from the mechanism $\mathcal{M}$) is done up to ϵ-indistinguishability*; this will then be implicitly part of what was called the saturation procedure $\mathcal{C}$ at that state. The procedure thus enhanced for saturating the tags at the states, will then be denoted as $\epsilon\mathcal{C}$, when necessary (it will still be decidable, under the finiteness assumptions of Remark 2 (b)). Inconsistency of the set of formulas, in the '$\epsilon\mathcal{C}$-saturated' tag at any state, will be checked up to ϵ-indistinguishability, and referred to as ϵ-inconsistency, or ϵ-failure. The notion of privacy policy will not need to be modified; that of its violation will be referred to as ϵ-violation. Under these extensions of ϵ-failure and ϵ-violation, it is clear that Proposition 1 will remain valid.   □

We conclude with two examples of ϵ-local-indistinguishability.

(i) The two sub-tuples ([50–60[, M, Maths) and ([40–50[, M, Physics), from the last two tuples on the Hospital's published record in Example 1 (Table 2), both point to Viral–Infection as output; they can thus be seen as $log(2)$-local-indististinguishable, for the adversary A.

(ii) The 'Randomized Response' mechanism RR [14] can be modelled as follows. Input is (X, F_1, F_2) where X is a Boolean, and F_1, F_2 are flips of a coin (H or T). RR outputs X if $F_1 = H$, $True$ if $F_1 = T$ and $F_2 = H$, and $False$ if $F_1 = T$ and $F_2 = T$. This mechanism is $log(3)$-LDP: the instances $(True, H, H)$, $(True, H, T)$, $(True, T, H)$ and $(True, T, T)$ are $log(3)$-indistinguishable for output $True$; $(False, H, H)$, $(False, H, T)$, $(False, T, H)$ and $(False, T, T)$ are $log(3)$-indistinguishable for output $False$.

4 ϵ-Differential Privacy

The notion of ϵ-*indistinguishability of two given databases D, D'*, is more general than that of ϵ-local-indistinguishability. ϵ-indistinguishability is usually defined for pairs of databases D, D' that are *adjacent* in a certain sense. There is no uniquely defined notion of adjacency on pairs of databases; in fact, several are known, and used in the literature. Actually, a notion of adjacency can be defined in a generic parametrizable manner (as done, e.g., in [5]), as follows. Assume given a map $\mathbf{f}$ from the set $\mathcal{D}$ of all databases of m-tuples (for $m > 0$), into some given metric space (X, d_X). The binary relation on pairs of databases in $\mathcal{D}$, defined by $\mathbf{f}_{adj}(D, D') = d_X(\mathbf{f}(D), \mathbf{f}(D'))$ can be seen as a measure of *adjacency* between D, D', and $\mathbf{f}_{adj}$ is said to define an 'adjacency relation'.

Definition 4.   *Let $\mathbf{f}_{adj}$ be a given adjacency relation on a set $\mathcal{D}$ of databases, and $\mathcal{M}$ a probabilistic mechanism answering queries on the bases in $\mathcal{D}$.*

– *Two databases $D, D' \in \mathcal{D}$ are said to be $\mathbf{f}_{adj}$-indistinguishable under $\mathcal{M}$, if and only if, for any possible output $\mathcal{S} \subset Range(\mathcal{M})$, we have:*

$$Prob[\mathcal{M}(D) \in \mathcal{S}] \leq e^{\mathbf{f}_{adj}(D, D')} Prob[\mathcal{M}(D') \in \mathcal{S}].$$

– *The mechanism $\mathcal{M}$ is said to satisfy $\mathbf{f}_{adj}$-differential privacy ($\mathbf{f}_{adj}$-DP), if and only if the above condition is satisfied for every pair of databases D, D' in $\mathcal{D}$, and any possible output $\mathcal{S} \subset Range(\mathcal{M})$.*

Comments: (i) Given $\epsilon \geq 0$, the 'usual' notions of ϵ-*indistinguishability and ϵ-DP* correspond to the choice of adjacency $\mathbf{f}_{adj} = \epsilon d_h$, where d_h is the Hamming metric on databases [5].

(ii) In Sect. 6, we propose a more general notion of adjacency, based on a different metric defined 'value-wise', to serve other purposes as well.

(iii) On disjoint databases, one can work with different adjacency relations, using different maps to the same (or different) metric space(s),

(iv) The mechanism RR described above is actually $log(3)$-DP, not only $log(3)$-LDP. To check DP, we have to check all possible pairs of numbers of the form $(Prob[\mathcal{M}(x) = y], Prob[\mathcal{M}(x') = y])$, $(Prob[\mathcal{M}(x) = y'], Prob[\mathcal{M}(x') = y])$, $(Prob[\mathcal{M}(x) = y], Prob[\mathcal{M}(x') = y'])$, etc., where the x, x'.... are the input instances for RR, and $y, y', \dots$ the outputs. The mechanism RR has 2^3 possible input instances for (X, F_1, F_2) and two outputs (*True, False*); thus 16 pairs of numbers, the distinct ones being $(1/4, 1/4), (1/4, 3/4), (3/4, 1/4), (3/4, 3/4)$; if (a, b) is any such pair, obviously $a \leq e^{log(3)} b$. Thus RR is indeed $log(3)$-DP. $\square$

5   Comparing Two Nodes on One or More Runs

In the sections above, we looked at the issue of 'quantifying' the indistinguishability of two data tuples or databases, under repeated queries of an adversary A. In this section, our concern will be a bit 'orthogonal': the issue will be that of quantifying how different the probabilistic mechanism's answers can be, at different moments of A's querying sequence.

The quantification looked for will be based on a suitable notion of 'distance' between two *sets of type-compatible tuples*. The 'distance' $d(t, t')$, from any given tuple t in this set to another type-compatible tuple t', will be defined as the value of the direct-sum metric of the distances between each attribute of the pair of tuples (t, t'). If S, S' are any two given finite sets of type-compatible tuples, of data that get assigned to the various attributes (along the queries), we define the distance from the set S to the set S' as the number $\rho(S, S') = min\{\overline{d}(t, t') \mid t \in S, t' \in S'\}$

Some preliminaries are needed before we can define the 'distance' function between the data values under every given header of D. We begin by dividing the headers of the base D into four classes.

- 'Nominal': identities, names, attributes receiving literal data *not in any taxonomy* (e.g., diseases), finite sets of such data;
- 'Numerval' : attributes receiving numerical values, or bounded intervals of (finitely many) numerical values;
- 'Numerical': attributes receiving single numerical values (numbers).
- 'Taxoral': attributes receiving literal data in a taxonomy relation.

For defining the 'distance' between any two values v, v' assigned to an attribute under a given 'Nominal' header of D, for the sake of uniformity we agree to consider every value as a *finite set* of atomic values. (In particular, a singleton value 'x' will be seen as the set $\{x\}$.) Given two such values v, v', note first that the so-called *Jaccard Index* between them is the number $jacc(v, v') = |(v \cap v')/(v \cup v')|$, which is a 'measure of their similarity'; but this index is not a metric: the *triangle inequality* is not satisfied; however, the Jaccard metric $d_{Nom}(v, v') = 1 - jacc(v, v') = |(v \Delta v')/(v \cup v')|$ does satisfy that property, and will suit our purposes. Thus, $d_{Nom}(v, v')$ is a 'measure of dissimilarity' between v and v'.

Let $\mathcal{T}_{Nom}$ be the set of all data assigned to the attributes under the 'Nominal' headers of D, along the sequence of A's queries. Then the above defined binary function d_{Nom} extends to a metric on the set of all type-compatible data-tuples from $\mathcal{T}_{Nom}$, defined as the 'direct-sum' taken over the 'Nominal' headers of D.

If $\mathcal{T}_{Num}$ is the set of all data assigned to the attributes under the 'Numerval' headers along the sequence of queries by A, we also define a 'distance' metric d_{Num} on the set of all type-compatible data-tuples from $\mathcal{T}_{Num}$, in a similar manner. We first define d_{Num} on any couple of values u, v assigned to attributes under a given 'Numerval' header of D, then extend it to the set of all type-compatible data-tuples from $\mathcal{T}_{Num}$ (as the direct-sum taken over the 'Numerval' headers of D). This is done as under the 'Nominal' headers: suffices to visualize any finite interval value as a particular way of presenting a set of numerical values (integers, usually). (In particular, a single value 'a' under a 'Numerval' header will be seen as the interval value $[a]$.) Thus defined the (Jaccard) metric distance $d_{Nom}([a, b], [c, d])$ is a measure of 'dissimilarity' between $[a, b]$ and $[c, d]$.

Between numerical data x, x' under the 'Numerical' headers, the distance we shall work with is the euclidean metric $|x - x'|$, *normalized as:* $d_{eucl}(x, x') = |x - x'|/D$, where $D > 0$ is a fixed finite number, bigger than the maximal euclidean distance between the numerical data on the databases and on the answers to A's queries.

On the data under the 'Taxoral' headers, we choose as distance function a novel metric d_{wp}, between the nodes of any Taxonomy tree, that we define in Lemma 1, Appendix C.

The 'datawise distance functions' defined above are *all with values in the real interval* $[0, 1]$. This is also one reason for our choice of the distance metric on Taxonomy trees. This fact is of importance, for comparing the metric ρ we defined above with the Hamming metric, cf. Sect. 6.

An Additional Role for Oracle $\mathcal{O}$: In Appendix B below, we present a procedure for comparing the knowledge of an adversary A at different nodes of the DLTTS that models the 'distributed sequence' of A's queries on a given database D. The comparison can be with respect to any given 'target' dataset T (e.g., a privacy policy P on D). In operational terms, the oracle mechanism $\mathcal{O}$ keeps the target dataset 'in store'; and as said earlier, a first role for the oracle $\mathcal{O}$ is to keep a watch on the deduction of the target dataset by the adversary A at some node. The additional second role that we assign now to the oracle $\mathcal{O}$, is to publish information on the distance of A's saturated knowledge $\overline{\tau}(s)$, at any given node s, to the target dataset T. This distance is calculated wrt the metric ρ, defined above as the minimal distance $\overline{d}(t, t')$ between $t \in \overline{\tau}(s), t' \in T$, where $\overline{d}$ is the direct sum of the 'column-wise distances' between the data.

A procedure for comparing two nodes on a DLTTS is formally presented in Appendix B. The following example illustrates the role played by the notions developed above, in that procedure.

Example 1 bis. We go back to the Hospital-CoVid example seen earlier, more particularly its Table 2. 'Gender' and 'Dept.'. are the 'Nominal' headers in this record, 'Age' is 'Numerval' and 'Ailment' is 'Taxoral'. We are interested in the second, fourth and fifth tuples on the record, respectively referred to as l_2, l_4, l_5. The 'target set' of (type-compatible) tuple in this example is taken as the (negation of the) privacy policy specified, namely the tuple $T = (John, 46, M, \#, CoVid)$.

We compute now the distance $\bar{d}$ between the target T, and the tuples l_2, l_4, l_5. This involves only the subtuple $L = (46, M, \#, CoVid)$ of T:

- $\bar{d}(l_2, L) = d_{Num}(l_2, L) + d_{Nom}(l_2, L) + d_{wp}(L_2, L)$
  $= (1 - 1/10) + 0 + (1 - 2/5) = 9/10 + 3/5 = 15/10$
- $\bar{d}(l_4, L) = d_{Num}(l_2, L) + d_{Nom}(l_4, L) + d_{wp}(L_4, L)$
  $= (1 - 0) + 0 + (1 - 4/5) = 1 + 1/5 = 6/5$
- $\bar{d}(l_5, L) = d_{Num}(l_5, L) + d_{Nom}(l_5, L) + d_{wp}(L_5, L)$
  $= (1 - 1/10) + 0 + (1 - 4/5) = 9/10 + 1/5 = 11/10$

The tuple l_2 is the farthest from the target, while l_5 is the closest. This 'explains' that the adversary can choose the branch on the transition that leads to a state where l_5 is added to his/her knowledge. This is more formally detailed in the procedure presented in Appendix B.   □

6   New Metric for Indistinguishability and DP

Given a randomized/probabilistic mechanism $\mathcal{M}$ answering the queries on databases, and an $\epsilon \geq 0$, recall that the ϵ-indistinguishability of any two given databases under $\mathcal{M}$, and the notion of ϵ-DP for $\mathcal{M}$, were both defined in Definition 4 (Sect. 4), based first on a hypothetical map $\mathbf{f}$ from the set of all the databases concerned, into some given metric space (X, d_X), and an 'adjacency relation' on databases defined as $\mathbf{f}_{adj}(D, D') = d_X(\mathbf{f}D, \mathbf{f}D')$, which was subsequently instantiated to $\mathbf{f}_{adj} = \epsilon d_h$, where d_h is the Hamming metric between databases. It must be observed here, that *the Hamming metric is defined only between databases with the same number of columns*, usually with all data of the same type.

In this section, our objective is to propose a more general notion of adjacency, based on the metric ρ, defined in Sect. 5 between type-compatible tuples on databases with data of multiple types. In other words, our $\mathcal{D}$ here will be the set of all databases *with data of several possible types* as said in the Introduction, and *not necessarily all with the same number of columns*. We define then a new binary relation $\mathbf{f}^{\rho}_{adj}(D, D')$ between databases D, D' in the set $\mathcal{D}$ by setting $\mathbf{f}^{\rho}_{adj}(D, D') = \rho(D, D')$, visualizing D, D' as sets of type-compatible data tuples.

Given ϵ, we can then define the notion of ϵ_{ρ}-indistinguishability of two databases D, D' under a (probabilistic) answering mechanism $\mathcal{M}$, as well as the notion of ϵ_{ρ}-DP for $\mathcal{M}$, exactly as in Definition 4, by replacing $\mathbf{f}_{adj}$ first with the relation $\mathbf{f}^{\rho}_{adj}$, and subsequently with $\epsilon\rho$. The notions thus defined are *more general* than those presented earlier in Sect. 4 with the choice $\mathbf{f}_{adj} = \epsilon d_h$. An example will illustrate this point.

Example 4. We go back to the 'Hospital's public record', with the same notation. We assume in this example, that the mechanism $\mathcal{M}$ answering a query for 'ailment information involving men' on that record, returns the tuples l_2, l_4, l_5 with the probability distribution $0, 2/5, 3/5$, respectively. Let us look for the minimum value of $\epsilon \geq 0$, for which these three tuples will be ϵ_ρ-indistinguishable under the mechanism $\mathcal{M}$.

The output l_2, with probability 0, will be ϵ_ρ-distinguishable for any $\epsilon \geq 0$. Only the two other outputs l_4, l_5 need to be considered. We first compute the ρ-distances between these two tuples: $\overline{d}(l_4, l_5) = (1 - \frac{1}{20}) + 0 + 1 + 0 = 39/20$. The tuples l_4 and l_5 will be ϵ_ρ-indistinguishable under $\mathcal{M}$ if and only if: $(2/5) \leq e^{(39/20)\epsilon} * (3/5)$ $\quad and \quad$ $(3/5) \leq e^{(39/20)\epsilon} * (2/5)$; i.e., $\epsilon \geq (20/39) * ln(3/2)$. In other words, for any $\epsilon \geq (20/39) * ln(3/2)$, the two tuples l_4 and l_5 will be ϵ_ρ-indistinguishable; and for values of ϵ with $0 \leq \epsilon < (20/39) * ln(3/2)$, these tuples will be ϵ_ρ-distinguishable.

For the ϵ-indistinguishability of these tuples wrt the Hamming metric d_h, we proceed similarly: the distance $d_h(l_4, l_5)$ is by definition the number of 'records' where these tuples differ, so $d_h(l_4, l_5) = 2$. So the condition on $\epsilon \geq 0$ for their ϵ-indistinguishability wrt d_h is: $(3/5) \leq e^{2\epsilon} * (2/5)$, i.e., $\epsilon \geq (1/2) * ln(3/2)$.

In other words, if these two tuples are ϵ_ρ-indistinguishable wrt ρ under $\mathcal{M}$ for some ϵ, they will be ϵ-indistinguishable wrt d_h for the same ϵ. But the converse is not true, since $(1/2) * ln(3/2) < (20/39) * ln(3/2)$. Said otherwise: $\mathcal{M}$ ϵ-*distinguishes more finely with* ρ, *than with* d_h. $\qquad\square$

Remark 4: The statement "$\mathcal{M}$ ϵ-distinguishes more finely with ρ, than with d_h", is *always true, in all situations*, here is why. Two records that differ 'at some given position' on two bases D, D' are always at distance 1 for the Hamming metric d_h, by definition. And, as we pointed out earlier, all the 'record-wise' metrics we have defined above also have their values in $[0, 1]$. So, whatever the type of data at corresponding positions on any two bases D, D', the ρ-distance between the records will never exceed their Hamming distance. That suffices to prove the above statement. The following Proposition formulates this in more precise terms:

Proposition 2. *Let $\mathcal{D}_m$ be the set of all databases with the same number m of columns, and $\mathcal{M}$ a probabilistic mechanism answering queries on the bases in $\mathcal{D}$. Let ρ be the metric (defined above) and d_h the Hamming metric, between the bases in $\mathcal{D}$, and suppose given an $\epsilon \geq 0$.*

- *If two databases $D, D' \in \mathcal{D}_m$ are ϵ_ρ-indistinguishable under $\mathcal{M}$ wrt ρ, then they are also ϵ-indistinguishable under $\mathcal{M}$ wrt d_h.*
- *If the mechanism $\mathcal{M}$ is ϵ_ρ-DP on the bases in $\mathcal{D}_m$ (wrt ρ), then it is also ϵ-DP (wrt d_h) on these bases.*

The idea of 'normalizing' the Hamming metric between numerical databases (with same number of columns) was already suggested in [5]. If only numerical databases are considered, the metric ρ that we have defined above is the same as the 'normalized Hamming metric' of [5]. Our metric ρ is to be seen as a generalization of that notion, to directly handle bases with more general types of data: anonymized, taxonomies, ...

7   Related Work and Conclusion

A starting point for the work presented is the observation that databases could be distributed over several 'worlds' in general, so querying such bases leads to answers which would also be distributed; to such distributed answers one could conceivably assign probability distributions of relevance to the query. The probabilistic automata of Segala [12, 13] are among the first logical structures proposed to model such a vision, in particular with outputs. Distributed Transition Systems (DTS) appeared a little later, with as objective the behavioral analysis of the distributed transitions, based on traces or on simulation/bisimulation, using quasi- or pseudo- or hemi- metrics as in [3, 4, 6]. Our lookout in this work was for a syntax-based *metric in the mathematical sense,* that can directly handle data of 'mixed' types – which can be numbers or literals, but can also be 'anonymized' as intervals or sets; they can also be taxonomically related to each other in a tree structure. (The metric d_{wp} we have defined in Appendix C on the nodes of a taxonomy tree is novel.) Data-wise metrics as defined in our work can express more precisely, in a mathemaical sense, the 'estimation errors' of an adversary wrt the given privacy policies on the database, at any point of his/her querying process. (In [10], such estimations are expressed in terms of suitable 'probability measures'.) Implementation and experimentation are part of future work, where we also hope to define a 'divergence measure' between any two given nodes on a DLTTS in terms of the knowledge distributions at the nodes.

Acknowledgement. Sabine Frittella received financial support for this work, from ANR JCJC 2019 and project PRELAP (ANR-19-CE48-0006).

Appendix A: Proof of Proposition 1

Assertion (i) is restatement. Observe now, that at any state s on $\mathcal{W}$, the tags $\tau(s), \overline{\tau}(s)$ are both *finite sets of first-order variable-free formulas* over Σ, without non-constant function symbols. For, to start with, the knowledge of A consists of the responses received for his/her queries, in the form of a finite set of data tuples from the given databases, and some subtuples. By our assumptions of Remark 2 (b), no infinite set can be generated by saturating this initial knowledge with procedure $\mathcal{C}$. Assertion (ii) follows then from the known result that the inconsistency of any finite set of variable-free Datalog formulas is decidable, e.g., by the analytic tableaux procedure. (Only the absence of variables is essential.)

□

Appendix B: A Non-deterministic Comparison Procedure

- Given: DLTTS associated with a querying sequence, by adversary A on given database D; and *a Target set* of tuples T.
- Given: Two states s, s' on the DLTTS, with respective *saturated* tags l, l', and probabilties p, p'. Target T assumed not in l or l': neither $\rho(l, T)$ nor $\rho(l', T)$ is 0. Also given:

- $config_1$: successor states $s_1, \ldots, s_n$ for a transition t from s, with probability distribution $p_1, \ldots, p_n$; and respective tags $l_1, \ldots, l_n$, with the contribution from t (cf. Remark 2(a)).
- $config_2$: successor states $s'_1, \ldots, s'_m$ for a transition t' from s', with probability distribution $p'_1, \ldots, p'_m$; and respective tags $l'_1, \ldots, l'_m$, with the contribution from t' (cf. Remark 2(a)).

- Objective: *Choose states to compare under s, s' (with probability measures not lower than p, p') in $config_1$, or in $config_2$, or from either.*

(i) Compute $d_i = \rho(l_i, T), i \in 1 \cdots n$, and $d'_j = \rho(l'_j, T), j \in 1 \cdots m$.

$$d_{min}(t, T) = min\{d_i \mid i \in 1 \cdots n\}, \quad d'_{min}(t', T) = min\{d'_j \mid j \in 1 \cdots m\}$$

(ii) Check IF *the following conditions are satified* by $config_1$:

$$d_{min}(t, T) \leq d'_{min}(t', T)$$
$$\exists \text{ an } i, 1 \leq i \leq n, \text{ such that } d_i = d_{min}(t, T), p_i \not\leq p,$$
$$\text{and } p_i \geq p'_j \text{ for any } j, 1 \leq j \leq m, \text{ where } d'_j = d'_{min}(t', T)$$

(iii) IF YES, continue under s with $config_1$, else RETURN.

Appendix C: Taxonomies

Taxonomies are frequent in machine learning. Data mining and clustering techniques employ reasonings based on measures of symmetry, or on metrics, depending on the objective. The Wu-Palmer symmetry measure on tree-structured taxonomies is one among those in use; it is defined as follows [15]: Let T be a given taxonomy tree. For any node x on T, define its depth c_x as the number of nodes from the root to x (both included), along the path from the root to x. For any pair x, y of nodes on T, let c_{xy} be the depth of the common ancestor of x, y that is *farthest* from the root. The Wu-Palmer symmetry measure between the nodes x, y on T is then defined as $WP(x, y) = \frac{2c_{xy}}{c_x + c_y}$. This measure, although considered satisfactory for many purposes, is known to have some disadvantages such as not being conform to semantics in several situations.

What we are interested in, for the purposes of our current paper, is a *metric* between the nodes of a taxonomy tree, which in addition will suit our semantic considerations. This is the objective of the Lemma below, a result that seems to be unknown.

Lemma 1. *On any taxonomy tree T, the binary function between its nodes defined by $d_{wp}(x, y) = 1 - \frac{2c_{xy}}{c_x + c_y}$ (notation as above) is a metric.*

Proof. We drop the suffix wp for this proof, and just write d. Clearly $d(x, y) = d(y, x)$; and $d(x, y) = 0$ if and only if $x = y$. We only have to prove the Triangle Inequality; i.e. show that $d(x, z) \leq d(x, y) + d(y, z)$ holds for any three nodes

x, y, z on $\mathcal{T}$. A 'configuration' can be typically represented in its 'most general form' by the diagram below. The boldface characters X, Y, Z, a, h in the diagram all stand for the *number of arcs* on the corresponding paths. So that, for the depths of x, y, z, and of their farthest common ancestors on the tree, we get:

$$c_x = X + h + 1, \quad c_y = Y + h + a + 1, \quad c_z = Z + h + a + 1,$$
$$c_{xy} = h + 1, \quad c_{yz} = h + a + 1, \quad c_{xz} = h + 1$$

The '+1' in these equalities is because the X, Y, Z, a, h are the *number of arcs* on the paths, while the depths are the number of nodes. The X, Y, Z, a, h must all be integers ≥ 0. For the Triangle Inequality on the three nodes x, y, z on $\mathcal{T}$, it suffices to prove the following two relations:

$$d(x, z) \leq d(x, y) + d(y, z) \quad \text{and} \quad d(y, z) \leq d(y, x) + d(x, z).$$

by showing that the following two algebraic inequalities hold:

(1) $1 - \dfrac{2 * (h + 1)}{(X + Y + 2 * h + a + 2)} + 1 - \dfrac{2 * (h + a + 1)}{(Y + Z + 2 * h + 2 * a + 2)} \geq 1 - \dfrac{2 * (h + 1)}{(X + Z + 2 * h + a + 2)}$

(2) $1 - \dfrac{2 * (h + 1)}{(X + Y + 2 * h + a + 2)} + 1 - \dfrac{2 * (h + 1)}{(X + Z + 2 * h + 2 * a + 2)} \geq 1 - \dfrac{2 * (h + a + 1)}{(Y + Z + 2 * h + 2 * a + 2)}$

The third relation $d(x, y) \leq d(x, z) + d(z, y)$ is proved by just exchanging the roles of Y and Z in the proof of inequality (1).

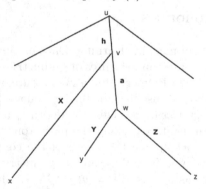

Inequality (1): We eliminate the denominators (all strictly positive), and write it out as an inequality between two polynomials $eq1, eq2$ on X, Y, Z, h, a, which must be satisfied for all their non-negative integer values:

$eq1 : (X + Y + 2 * h + a + 2) * (Y + Z + 2 * h + 2 * a + 2) * (X + Z + 2 * h + a + 2)$

$eq2 : (h + 1) * (Y + Z + 2 * h + 2 * a + 2) * (X + Z + 2 * h + a + 2)$
$\qquad + (h + a + 1) * (X + Y + 2 * h + a + 2) * (X + Z + 2 * h + a + 2)$
$\qquad - (h + 1) * (X + Y + 2 * h + a + 2) * (Y + Z + 2 * h + 2 * a + 2)$

$eq : eq1 - 2 * eq2.$ We need to check: $eq \geq 0$?

The equation eq once expanded (e.g., under *Maxima*) appears as:

$$eq: YZ^2 + XZ^2 + aZ^2 + Y^2Z + 2XYZ + 4hYZ + 2aYZ + 4YZ + X^2Z +$$
$$4hXZ + 2aXZ + 4XZ + a^2Z + XY^2 + 4hY^2 + aY^2 + 4Y^2 + X^2Y +$$
$$4hXY + 2aXY + 4XY + 8h^2Y + 8ahY + 16hY + a^2Y + 8aY + 8Y$$

The coefficients are all positive, and inequality (1) is proved.

Inequality (2): We again proceed as above: we first define the following polynomial expressions:

$eq3 : (X + Y + 2*h + a + 2)*(X + Z + 2*h + a + 2)*(Y + Z + 2*h + 2*a + 2);$

$eq4 : (h + 1)*(Y + Z + 2*h + 2*a + 2)*(2*X + Y + Z + 4*h + 2*a + 4);$

$eq5 : (h + a + 1)*(X + Y + 2*h + a + 2)*(X + Z + 2*h + a + 2);$

If we set $eqn : eq3 + 2*eq5 - 2*eq4$,   we get

$$eqn : -2(h + 1)*(Z + Y + 2h + 2a + 2)*(Z + Y + 2X + 4h + 2a + 4) +$$
$$(Y + X + 2h + a + 2)*(Z + X + 2h + a + 2)(Z + Y + 2h + 2a + 2) +$$
$$2(h + a + 1)*(Y + X + 2h + a + 2)*(Z + X + 2h + a + 2)$$

To prove inequality (2), we need to show that eqn remains non-negative for all non-negative values of X, Y, Z, h, a. For that, we expand eqn (with *Maxima*), to get:

$$eqn : YZ^2 + XZ^2 + aZ^2 + Y^2Z + 2XYZ + 4hYZ + 6aYZ + 4YZ + X^2Z +$$
$$4hXZ + 6aXZ + 4XZ + 8ahZ + 5a^2Z + 8aZ + XY^2 + aY^2 + X^2Y + 4hXY +$$
$$6aXY + 4XY + 8ahY + 5a^2Y + 8aY + 4hX^2 + 4aX^2 + 4X^2 + 8h^2X +$$
$$16ahX + 16hX + 8a^2X + 16aX + 8X + 8ah^2 + 12a^2h + 16ah + 4a^3 + 12a^2 + 8a$$

The coefficients are all positive, so we are done. $\square$.

References

1. Barthe, G., Köpf, B., Olmedo, F., Béguelin, S.Z.: Probabilistic relational reasoning for differential privacy. In: Proceedings of POPL. ACM (2012)
2. Barthe, G., Chadha, R., Jagannath, V., Sistla, A.P., Viswanathan, M.: Deciding differential privacy for programs with finite inputs and outputs. In: LICS 2020: 35th Annual ACM/IEEE Symposium on Logic in Computer Science, Saarbrücken, Germany, 8–11 July 2020 (2020)
3. Castiglioni, V., Chatzikokolakis, K., Palamidessi, C.: A logical characterization of differential privacy via behavioral metrics. In: Bae, K., Ölveczky, P.C. (eds.) FACS 2018. LNCS, vol. 11222, pp. 75–96. Springer, Cham (2018). https://doi.org/10.1007/978-3-030-02146-7_4
4. Castiglioni, V., Loreti, M., Tini, S.: The metric linear-time branching-time spectrum on nondeterministic probabilistic processes. Theor. Comput. Sci. **813**, 20–69 (2020)

5. Chatzikokolakis, K., Andrés, M.E., Bordenabe, N.E., Palamidessi, C.: Broadening the scope of differential privacy using metrics. In: De Cristofaro, E., Wright, M. (eds.) PETS 2013. LNCS, vol. 7981, pp. 82–102. Springer, Heidelberg (2013). https://doi.org/10.1007/978-3-642-39077-7_5
6. de Alfaro, L., Faella, M., Stoelinga, M.: Linear and branching system metrics. IEEE Trans. Softw. Eng. **35**(2), 258–273 (2009)
7. Dwork, C.: Differential privacy. In: Bugliesi, M., Preneel, B., Sassone, V., Wegener, I. (eds.) ICALP 2006. LNCS, vol. 4052, pp. 1–12. Springer, Heidelberg (2006). https://doi.org/10.1007/11787006_1
8. Dwork, C., Roth, A.: The algorithmic foundations of differential privacy. Found. Trends Theor. Comput. Sci. **9**(3–4), 211–407 (2014)
9. Holohan, N., Antonatos, S., Braghin, S., Aonghusa, P.M.: The bounded laplace mechanism in differential privacy. In: Journal of Privacy and Confidentiality (Proc. TPDP 2018), vol. 10, no. 1 (2020)
10. Rebollo-Monedero, D., Parra-Arnau, J., Diaz, C., Forné, J.: On the measurement of privacy as an attacker's estimation error. Int. J. Inf. Secur. **12**(2), 129–149 (2013)
11. Segala, R.: Modeling and verification of randomized distributed real-time systems. Ph.D. thesis, MIT (1995)
12. Segala, R.: A compositional trace-based semantics for probabilistic automata. In: Lee, I., Smolka, S.A. (eds.) CONCUR 1995. LNCS, vol. 962, pp. 234–248. Springer, Heidelberg (1995). https://doi.org/10.1007/3-540-60218-6_17
13. Segala, R., Lynch, N.A.: Probabilistic simulations for probabilistic processes. Nord. J. Comput. **2**(2), 250–273 (1995)
14. Warner, S.L.: Randomized response: a survey technique for eliminating evasive answer bias. J. Am. Stat. Assoc. **60**(309), 63–69 (1965)
15. Wu, Z., Palmer, M.: Verb semantics and lexical selection. In: Proceedings of 32nd Annual Meeting of the Associations for Computational Linguistics, pp. 133–138 (1994)

Multivariate Mean Comparison Under Differential Privacy

Martin Dunsche$^{(\boxtimes)}$, Tim Kutta, and Holger Dette

Ruhr-University, Bochum, Germany
{martin.dunsche,tim.kutta,holger.dette}@ruhr-uni-bochum.de

Abstract. The comparison of multivariate population means is a central task of statistical inference . While statistical theory provides a variety of analysis tools, they usually do not protect individuals' privacy. This knowledge can create incentives for participants in a study to conceal their true data (especially for outliers), which might result in a distorted analysis. In this paper, we address this problem by developing a hypothesis test for multivariate mean comparisons that guarantees differential privacy to users. The test statistic is based on the popular Hotelling's t^2-statistic, which has a natural interpretation in terms of the Mahalanobis distance. In order to control the type-1-error, we present a bootstrap algorithm under differential privacy that provably yields a reliable test decision. In an empirical study, we demonstrate the applicability of this approach.

Keywords: Differential privacy · Private testing · Private bootstrap

1 Introduction

Over the last decades, the availability of large databases has transformed statistical practice. While data mining flourishes, users are concerned about increasing transparency vis-à-vis third parties. To address this problem, new analysis tools have been devised that balance precise inference with solid privacy guarantees.

In this context, statistical tests that operate under *differential privacy* (DP) are of interest: Statistical tests are the standard tool to validate hypotheses regarding data samples and to this day form the spine of most empirical sciences. Performing tests under DP means determining general trends in the data, while masking individual contribution. This makes it hard for adversaries to retrieve unpublished, personal information from the published analysis.

Related Works: In recent years, hypothesis testing under DP has gained increasing attention. In a seminal work [20] introduces a privatization method, for a broad class of test statistics, that guarantees DP without impairing asymptotic performance. Other theoretical aspects such as optimal tests under DP are considered in [3]. Besides such theoretical investigations, a number of privatized tests have been devised to replace classical inference, where sensitive data is at

© Springer Nature Switzerland AG 2022
J. Domingo-Ferrer and M. Laurent (Eds.): PSD 2022, LNCS 13463, pp. 31–45, 2022.
https://doi.org/10.1007/978-3-031-13945-1_3

stake. For example [11] and [17] consider privatizations of classical goodness of fit tests for categorical data, tailored to applications in genetic research, where privacy of study participants is paramount. In a closely related work, [22] use privatized likelihood-ratio statistics to validate various assumptions for tabular data. Besides, [19] propose a method for privatizations in small sample regimes.

A cornerstone of statistical analysis is the study of population means and accordingly this subject has attracted particular attention. For example, [6] develop a private t-test to compare population means under local differential privacy, while [16] consider the multivariate case in the global setting. [12] and [7] construct private confidence intervals for the mean (which is equivalent to the one-sample t-test) under global DP and [21] suggests a differentially private ANOVA. Moreover, [4] present privatizations for a number of non-parametric tests (such as Wilcoxon signed-rank tests) and [10] devise general confidence intervals for exponential families.

A key problem of statistical inference under DP consists in the fact that privatization inflates the variance of the test statistics. If this is not taken into account properly, it can destabilize subsequent analysis and lead to the "discovery" of spurious effects. To address these problems, recent works (such as [11] and [10]) have employed resampling procedures that explicitly incorporate the effects of privatization and are therefore more reliable than tests based on standard, asymptotic theory.

Our Contributions: In this work, we present a test for multivariate mean comparisons under pure-DP, based on the popular Hotelling's t^2-statistic. We retrieve the effect that asymptotic test decisions work under DP, as long as privatizations are weak, whereas for strong privatizations, they yield distorted results (see Sect. 4 for details). As a remedy, we consider a parametric bootstrap that cuts false rejections and is provably consistent for increasing sample size. This method can be extended to other testing problems, is easy to implement (even for non-expert users) and can be efficiently automatized as part of larger data disseminating structures. We demonstrate the efficacy of our approach, even for higher dimensions and strong privatizations, in a simulation study. The proofs of all mathematical results are deferred to the Appendix. The work most closely related to our paper is [16], who consider Hotelling's t^2−statistic for approximate DP and propose a test based on a (heuristic) resampling strategy. In contrast to this paper, we focus on pure-DP, employ a different privatization mechanism and a parametric bootstrap test, for which we provide a rigorous proof of its validity (see Sect. 3.2).

2   Mathematical Background

In this section, we provide the mathematical context for private mean comparisons, beginning with a general introduction into two sample tests. Subsequently, we discuss Hotelling's t^2-test, which is a standard tool to assess mean deviations. Finally, we define the notion of differential privacy and consider key properties, such as stability under post-processing. Readers familiar with any of these topics can skip the respective section.

2.1   Statistical Tests for Two Samples

In this work, we are interested in testing statistical hypotheses regarding the distribution of two data samples (of random vectors) $X_1, ..., X_{n_1}$ and $Y_1, ..., Y_{n_2}$.

Statistical tests are decision rules that select one out of two rivaling hypotheses H_0 and H_1, where H_0 is referred to as the "null hypothesis" (default belief) and H_1 as the "alternative". To make this decision, a statistical test creates a summary statistic $S := S(X_1, ..., X_{n_1}, Y_1, ..., Y_{n_2})$ from the data and based on S determines whether to keep H_0, or to switch to H_1. Typically, the decision to reject H_0 in favor of H_1 is made, if S surpasses a certain threshold q, above which, the value of S seems at odds with H_0. In this situation, the threshold q may or may not depend on the data samples.

Given the randomness in statistical data, there is always a risk of making the wrong decision. Hypothesis-alternative-pairs (H_0, H_1) are usually formulated such that mistakenly keeping H_0 inflicts only minor costs on the user, while wrongly switching to H_1 produces major ones. In this spirit, tests are constructed to keep the risk of false rejection below a predetermined level α, i.e. $\mathbb{P}_{H_0}(S > q) \leq \alpha$, which is referred to as the *nominal level* (or *type-1-error*). Commonly, the nominal level is chosen as $\alpha \in \{0.1, 0.05, 0.01\}$. Notice that α can be regarded as an input parameter of the threshold $q = q(\alpha)$. Even though sometimes an exact nominal level can be guaranteed, in practice most tests only satisfy an asymptotic nominal level, i.e. $\limsup_{n_1, n_2 \to \infty} \mathbb{P}_{H_0}(S > q(\alpha)) \leq \alpha$. Besides controlling the type-1-error, a reasonable test has to be *consistent*, i.e. it has to reject H_0 if H_1 holds and sufficient data is available. In terms of the summary statistic S, this means that S increases for larger data samples and transgresses $q(\alpha)$ with growing probability $\lim_{n_1, n_2 \to \infty} \mathbb{P}_{H_1}(S > q(\alpha)) = 1$.

2.2   Hotelling's t^2-Test

We now consider a specific test for the comparison of multivariate means: Suppose that two independent samples of random vectors $X_1, ..., X_{n_1}$ and $Y_1, ..., Y_{n_2}$ are given, both stemming from the d-dimensional cube $[-m, m]^d$, where $m > 0$ and $d \in \mathbb{N}$. Furthermore, we assume that both samples consist of independent identically distributed (i.i.d) observations. Conceptually, each vector corresponds to the data of one individual and we want to use these to test the "hypothesis-alternative"-pair

$$H_0 : \mu_X = \mu_Y , \qquad H_1 : \mu_X \neq \mu_Y , \qquad (2.1)$$

where $\mu_X := \mathbb{E}[X_1] \in \mathbb{R}^d, \mu_Y := \mathbb{E}[Y_1] \in \mathbb{R}^d$ denote the respective expectations. A standard way to test (2.1) is provided by *Hotelling's t^2-test*, which is based on the test statistic

$$t^2 = \frac{n_1 n_2}{n_1 + n_2} (\bar{X} - \bar{Y})^T \hat{\Sigma}^{-1} (\bar{X} - \bar{Y}) , \qquad (2.2)$$

where $\bar{X} = \frac{1}{n_1}\sum_{i=1}^{n_1} X_i$ and $\bar{Y} = \frac{1}{n_2}\sum_{i=1}^{n_2} Y_i$ denote the respective sample means and the pooled sample covariance is given by

$$\hat{\Sigma} = \frac{(n_1 - 1)\hat{\Sigma}_X + (n_2 - 1)\hat{\Sigma}_Y}{n_1 + n_2 - 2}.$$

Here, $\hat{\Sigma}_X = \frac{1}{n_1-1}\sum_{i=1}^{n_1}(X_i-\mu_X)(X_i-\mu_X)^\top$ and $\hat{\Sigma}_Y = \frac{1}{n_2-1}\sum_{i=1}^{n_2}(Y_i-\mu_Y)(Y_i-\mu_Y)^\top$ denote the sample covariance matrices of X_1 and Y_1, respectively. Assuming that $\Sigma_X = \Sigma_Y$ (a standard condition for Hotelling's t^2-test) $\hat{\Sigma}$ is a consistent estimator for the common covariance.

We briefly formulate a few observations regarding the t^2-statistic:

i) In the simple case of $d = 1$, the t^2-statistic collapses to the (squared) statistic of the better-known two sample t-test.
ii) We can rewrite

$$t^2 = \frac{n_1 n_2}{n_1 + n_2} \left\| \hat{\Sigma}^{-1/2}(\bar{X} - \bar{Y}) \right\|_2^2.$$

As a consequence, the t^2-statistic is non-negative and assumes high values if $\bar{X} - \bar{Y} \approx \mu_X - \mu_Y$ is large in the norm.
iii) The t^2-statistic is closely related to the Mahalanobis distance, which is a standard measure for multivariate mean comparisons (see [5]).

In order to formulate a statistical test based on the t^2-statistic, we consider its large sample behavior. Under the hypothesis $\sqrt{n_1 n_2/(n_1 + n_2)}\hat{\Sigma}^{-1/2}(\bar{X} - \bar{Y})$ follows (approximately) a d-dimensional, standard normal distribution, such that its squared norm (that is the t^2-statistic) is approximately χ_d^2 distributed (chi-squared with d degrees of freedom). Now if $q_{1-\alpha}$ denotes the upper α-quantile of the χ_d^2 distribution, the test decision "reject H_0 if $t^2 > q_{1-\alpha}$", yields a consistent, asymptotic level α-test for any $\alpha \in (0,1)$. For details on Hotelling's t^2-test we refer to [15].

2.3   Differential Privacy

Differential privacy (DP) has over the last decade become the de facto gold standard in privacy assessment of data disseminating procedures (see e.g. [9,14] or [18]). Intuitively, DP describes the difficulty of inferring individual inputs from the releases of a randomized algorithm. This notion is well suited to a statistical framework, where a trusted institution, like a hospital, publishes results of a study (algorithmic releases), but candidates would prefer to conceal participation (individual inputs). To make this notion mathematically rigorous, we consider databases $\mathbf{x}, \mathbf{x}' \in \mathcal{D}^n$, where $\mathcal{D}$ is some set, and call them *adjacent* or *neighboring*, if they differ in only one entry.

Definition 2.3.1. A randomized algorithm $A : \mathcal{D}^n \to \mathbb{R}$ is called ε-*differentially private* for some $\varepsilon > 0$, if for any measurable event $E \subset \mathbb{R}$ and any adjacent $\mathbf{x}, \mathbf{x}'$

$$\mathbb{P}(A(\mathbf{x}) \in E) \le e^\varepsilon \, \mathbb{P}(A(\mathbf{x}') \in E) \tag{2.3}$$

holds.

Condition (2.3) requires that the distribution of $A(\mathbf{x})$ does not change too much, if one entry of $\mathbf{x}$ is exchanged (where small ε correspond to less change and thus stronger privacy guarantees). In statistical applications, private algorithms are usually assembled modularly: They take as building blocks some well-known private algorithms (e.g., the Laplace or Exponential Mechanism), use them to privatize key variables (empirical mean, variance etc.) and aggregate the privatized statistic. This approach is justified by two stability properties of DP: Firstly, privacy preservation under post-processing, which ensures that if A satisfies ε-DP, so does any measurable transformation $h(A)$. Secondly, the composition theorem that maintains at least $\sum_{i=1}^{k} \varepsilon_i$-DP of a vector $(A_1, ..., A_k)$ of algorithms, where A_i are independent ε_i-differentially private algorithms. In the next section, we employ such a modular privatization of the Hotelling's t^2-statistic for private mean comparison. We conclude our discussion on privacy with a small remark on the role of the "trusted curator".

Remark 2.3.1. Discussions of (global) DP usually rely on the existence of some "trusted curator" who aggregates and privatizes data before publication. In reality this role could be filled by an automatized, cryptographic protocol (secure multi-party computation), which calculates and privatizes the statistic before publication without any party having access to the full data set (for details see [2,13]). This process has the positive side effect that it prevents a curator from re-privatizing if an output seems too outlandish (overturning privacy in the process).

3 Privatized Mean Comparison

In this section, we introduce a privatized version t^{DP} of Hotelling's t^2-statistic. Analogous to the traditional t^2-statistic, the rejection rule "$t^{DP} > q_{1-\alpha}$" yields in principle a consistent, asymptotic level-α test for H_0 (see Theorem 3.1.2). However, empirical rejection rates often exceed the prescribed nominal level α for a combination of low sample sizes and high privatization (see Example B). As a consequence, we devise a parametric bootstrap for a data-driven rejection rule. We validate this approach theoretically (Theorem 3.2.1) and demonstrate empirically a good approximation of the nominal level in Sect. 4.

3.1 Privatization of the t^2-Statistic

We begin this section by formulating the Assumptions of the following, theoretical results:

Assumption 3.1.1. (1) The samples $X_1, \ldots, X_{n_1}$ and $Y_1, \ldots, Y_{n_2}$ are independent, each consisting of i.i.d. observations and are both supported on the cube $[-m, m]^d$, for some known $m > 0$.
(2) The covariance matrices

$$\Sigma_X := \mathbb{E}[(X_1 - \mu_X)(X_1 - \mu_X)^T]; \quad \Sigma_Y := \mathbb{E}[(Y_1 - \mu_Y)(Y_1 - \mu_Y)^T].$$

are identical and invertible.

(3) The sample sizes n_1, n_2 are of the same order. That is with $n := n_1 + n_2$ we have

$$\lim_{n \to \infty} \frac{n_i}{n} = \xi_i \in (0,1) \qquad i = 1, 2.$$

We briefly comment on the Assumptions made.

Remark 3.1.1. (1): The assumption of independent observations is common in the literature on machine learning and justified in many instances. Boundedness of the data -with some known bound- is an important precondition for standard methods of privatization (such as the below discussed Laplace Mechanism or the ED algorithm). Generalization are usually possible (see e.g., [20]) but lie beyond the scope of this paper.

(2): Invertibility of the covariance matrices is necessary to define the Mahalanobis distance. If this assumption is violated, either using another distance measure (defining a different test) or a prior reduction of dimensions is advisable.

Equality of the matrices $\Sigma_X = \Sigma_Y$ is assumed for ease of presentation, but can be dropped, if the pooled estimate $\hat{\Sigma}$ is replaced by the re-weighted version

$$\hat{\Sigma}^{\neq} := \frac{n_2 \hat{\Sigma}_X + n_1 \hat{\Sigma}_Y}{n_1 + n_2}.$$

(3): We assume that asymptotically the size of each group is non-negligible. This assumption is standard in the analysis of two sample tests and implies that the noise in the estimates of both groups is of equal magnitude. If this was not the case and e.g. $\xi_1 = 0$ (in practice $n_1 << n_2$) it is more appropriate to model the situation as a one-sample test (as μ_Y is basically known).

Recall the definition of Hotelling's t^2-statistic in (2.2). By construction, we can express the t^2-statistic as a deterministic function of four data dependent entities: The sample means $\bar{X}, \bar{Y}$ and the sample covariance matrices $\hat{\Sigma}_X, \hat{\Sigma}_Y$. According to the *composition-* and *post-processing theorem* of DP (see Sect. 2.3) we can privatize the t^2-statistic by privatizing each of these inputs.

For the privatization of the sample means, we use the popular *Laplace Mechanism* (see [8], p.32): It is well-known that $\bar{X}^{DP} := \bar{X} + Z$ and $\bar{Y}^{DP} := \bar{Y} + Z'$ fulfill $\varepsilon/4$-DP, if $Z = (Z_1, \ldots, Z_d)^T$ and $Z' = (Z'_1, \ldots, Z'_d)^T$ consist of independent random variables $Z_k \sim Lap(0, \frac{2md}{n_1(\varepsilon/4)})$ and $Z'_k \sim Lap(0, \frac{2md}{n_2(\varepsilon/4)})$ for $k = 1, \ldots, d$.

For the privatization of the covariance matrices $\hat{\Sigma}_X, \hat{\Sigma}_Y$ we employ the *ED* Mechanism, specified in the Appendix (which is a simple adaption of the Algorithm proposed in [1]). We can thus define differentially private estimates $\hat{\Sigma}_X^{DP} := ED(\hat{\Sigma}_X, \varepsilon/4)$ and $\hat{\Sigma}_Y^{DP} := ED(\hat{\Sigma}_Y, \varepsilon/4)$, both satisfying $\varepsilon/4$-DP. We point out that the outputs of ED are always covariance matrices (positive semi-definite and symmetric). Therewith, we can define a privatized pooled sample covariance matrix as

$$\hat{\Sigma}^{DP} := \frac{(n_1 - 1)\hat{\Sigma}_X^{DP} + (n_2 - 1)\hat{\Sigma}_Y^{DP}}{n_1 + n_2 - 2} + diag(c_1 + c_2),$$

Algorithm 1. Privatized statistics (PS)

Input: means: $\bar{X}$, $\bar{Y}$, covariance matrices: $\hat{\Sigma}_X$, $\hat{\Sigma}_Y$,
        privacy level: ε

Output: $\bar{X}^{DP}$, $\bar{Y}^{DP}$, $\hat{\Sigma}_X^{DP}$, $\hat{\Sigma}_Y^{DP}$
1: **function** PS($\bar{X}$, $\bar{Y}$, $\hat{\Sigma}_X$, $\hat{\Sigma}_Y$, ε)
2:    **for** $i = 1, \ldots, d$ **do**
3:        Generate $Z_i \sim Lap(0, \frac{2md}{n_1\varepsilon/4})$
4:        Generate $Z_i' \sim Lap(0, \frac{2md}{n_2\varepsilon/4})$
5:    **end for**
6:    Set $\bar{X}^{DP} := \bar{X} + (Z_1, \ldots, Z_d)$, $\bar{Y}^{DP} := \bar{Y} + (Z_1', \ldots, Z_d')$
7:    Set $\hat{\Sigma}_X^{DP} = ED(\hat{\Sigma}_X, \varepsilon/4)$, $\hat{\Sigma}_Y^{DP} = ED(\hat{\Sigma}_Y, \varepsilon/4)$
8:    **return** $\bar{X}^{DP}, \bar{Y}^{DP}, \hat{\Sigma}_X^{DP}, \hat{\Sigma}_Y^{DP}$
9: **end function**

---

where $c_1 := 2(\frac{2md}{n_1(\varepsilon/4)})^2$, $c_2 := 2(\frac{2md}{n_2(\varepsilon/4)})^2$ are corrections accounting for variance increase, due to the mean privatizations. Finally, we can formulate a privatized version of the Hotelling's t^2-statistic as follows:

$$t^{DP} = \frac{n_1 n_2}{n_1 + n_2}(\bar{X}^{DP} - \bar{Y}^{DP})^T[\hat{\Sigma}^{DP}]^{-1}(\bar{X}^{DP} - \bar{Y}^{DP}) \qquad (3.1)$$

$$= \frac{n_1 n_2}{n_1 + n_2}\left\|[\hat{\Sigma}^{DP}]^{-1/2}(\bar{X}^{DP} - \bar{Y}^{DP})\right\|_2^2$$

Theorem 3.1.1. The privatized t^2-statistic t^{DP} is ε-differentially private.

In the one dimensional case, the covariance privatization by ED boils down to an application of the Laplace Mechanism and t^{DP} has a simple closed form.

Example 3.1.1. (Privatization in $d = 1$) Assume that $d = 1$. Then the data $X_1, \ldots, X_{n_1}$ and $Y_1, \ldots, Y_{n_2}$ originates from the interval $[-m, m]$ and we can write the privatized test statistic as

$$t^{DP} = \frac{n_1 n_2}{n_1 + n_2}\frac{(\bar{X}^{DP} - \bar{Y}^{DP})^2}{(\sigma^{DP})^2},$$

where

$$(\sigma^{DP})^2 := \frac{(n_1 - 1)(|\hat{\sigma}_X + L_1|) + (n_2 - 1)(|\hat{\sigma}_Y + L_2|)}{n_1 + n_2 - 2}$$
$$+ 2\left(\frac{2m}{n_1(\varepsilon/4)}\right)^2 + 2\left(\frac{2m}{n_2(\varepsilon/4)}\right)^2.$$

Here, L_1 and L_2 follow a centered Laplace distribution, with variance specified in the Appendix. Note that the privatization of $\hat{\sigma}_X^2$ and $\hat{\sigma}_Y^2$ is conforming with the privatization of Algorithm ED (see Appendix), since the first (and only) eigenvalue is the sample variance itself, while privatization of eigenvectors is a non-issue for $d = 1$.

As for the non-privatized t^2-statistic, we can prove under H_0 that t^{DP} approximates a χ_d^2-distribution as $n_1, n_2 \to \infty$. This means that (at least for large sample sizes) the perturbations introduced by the Laplace noise and the ED-algorithm are negligible.

Algorithm 2. Privatized Hotelling's t^2-test (PHT)

Input: means: $\bar{X}^{DP}, \bar{Y}^{DP}$, covariance matrices: $\hat{\Sigma}_X^{DP}, \hat{\Sigma}_Y^{DP}$, quantile: q

Output: *choice* $\in \{0, 1\}$ coding for acceptation (0) or rejection (1) of H_0
1: **function** PHT($\bar{X}^{DP}, \bar{Y}^{DP}, \hat{\Sigma}_X^{DP}, \hat{\Sigma}_Y^{DP}, q$)
2:      Compute t^{DP} (defined in 3.1)
3:      Define *choice* $= 0$
4:      **if** $t^{DP} > q$ **then**
5:          Set *choice* $= 1$
6:      **end if**
7:      **return** *choice*
8: **end function**

Theorem 3.1.2. The decision rule "reject if

$$t^{DP} > q_{1-\alpha} \tag{3.2}$$

(Algorithm 2)" where $q = q_{1-\alpha}$ is the $(1-\alpha)$-quantile of χ_d^2 distribution, yields a consistent, asymptotic level-α test for the hypotheses (2.1).

Theorem 3.1.2 underpins the assertion that "asymptotically, privatizations do not matter". Yet in practice, privatizations can have a dramatic impact on the (finite sample) performance of tests.

3.2   Bootstrap

In this section, we consider a modified rejection rule for H_0, based on t^{DP}, that circumvents the problem of inflated type-1-error (see Example B). Privatizations increase variance and therefore t^{DP} is less strongly concentrated than t^2, leading to excessive transgressions of the threshold $q_{1-\alpha}$. Consequently, to guarantee an accurate approximation of the nominal level, a different threshold is necessary.

Hypothetically, if we knew the true distribution of t^{DP} under H_0, we could analytically calculate the exact α-quantile $q_{1-\alpha}^{exact}$ and use the rejection rule "$t^{DP} > q_{1-\alpha}^{exact}$". Of course, in practice, these quantiles are not available, but we can use a *parametric bootstrap* to approximate $q_{1-\alpha}^{exact}$ by an empirical version $q_{1-\alpha}^*$ calculated from the data. In Algorithm 3 we describe the systematic derivation of $q_{1-\alpha}^*$.

Algorithm 3. Quantile Bootstrap (QB)

Input: Covariance matrices: $\hat{\Sigma}_X^{DP}$, $\hat{\Sigma}_Y^{DP}$, sample sizes: n_1, n_2, bootstrap iterations: B

Output: Empirical $1 - \alpha$ quantile of t^{DP}: $q_{1-\alpha}^*$.

1: **function** QB($\hat{\Sigma}_X^{DP}$, $\hat{\Sigma}_Y^{DP}$, n_1, n_2, B)
2:    **for** $i = 1, \ldots, B$ **do**
3:       Sample $\bar{X}^* \sim \mathcal{N}(0, \frac{\hat{\Sigma}_X^{DP}}{n_1})$ and $\bar{Y}^* \sim \mathcal{N}(0, \frac{\hat{\Sigma}_Y^{DP}}{n_2})$
4:       **for** $k = 1, \ldots, d$ **do**
5:          Generate $Z_k \sim Lap(0, \frac{2md}{n_1(\varepsilon/4)})$
6:          Generate $Z_k' \sim Lap(0, \frac{2md}{n_2(\varepsilon/4)})$
7:       **end for**
8:       Define $\bar{X}^{DP*} := \bar{X}^* + (Z_1, \ldots, Z_d)$
9:       Define $\bar{Y}^{DP*} := \bar{Y}^* + (Z_1', \ldots, Z_d')$
10:      Define $t_i^{DP*} := \frac{n_1 n_2}{n_1 + n_2} \left\| [\hat{\Sigma}^{DP}]^{-1/2} (\bar{X}^{DP*} - \bar{Y}^{DP*}) \right\|_2^2$
11:   **end for**
12:   Sort statistics in ascending order: $(t_{(1)}^{DP*}, \ldots, t_{(B)}^{DP*}) = sort((t_1^{DP*}, \ldots, t_B^{DP*}))$
13:   Define $q_{1-\alpha}^* := t_{(\lfloor (1-\alpha)B \rfloor)}^{DP*}$
14:   **return** $q_{1-\alpha}^*$
15: **end function**

Algorithm 3 creates B *bootstrap versions* $t_1^{DP*}, \ldots, t_B^{DP*}$, that mimic the behavior of t^{DP}. So, e.g., $\bar{X}^{DP*}$ (in t_i^{DP*}) has a distribution close to that of $\bar{X}^{DP}$ (in t^{DP}), which, if centered, is approximately normal with covariance matrix Σ_X/n_1. As a consequence of this parallel construction, the *empirical* $1 - \alpha$-*quantile* $q_{1-\alpha}^*$ is close to the true $(1 - \alpha)$-quantile of the distribution of t^{DP}, at least if the number B of bootstrap replications is sufficiently large. In practice, the choice of B depends on α (where small α require larger B), but our simulations suggest that for a few hundred iterations the results are already reasonable even for nominal levels as small as 1%.

Theorem 3.2.1. The decision rule "reject if

$$t^{DP} > q_{1-\alpha}^* \tag{3.3}$$

(Algorithm 2)", where $q_{1-\alpha}^*$ is chosen by Algorithm 3, yields a consistent, asymptotic level-α test in the sense that

$$\lim_{B \to \infty} \lim_{n_1, n_2 \to \infty} \mathbb{P}_{H_0}(t^{DP} > q_{1-\alpha}^*) = \alpha,$$

(level α) and

$$\lim_{n_1, n_2 \to \infty} \mathbb{P}_{H_1}(t^{DP} > q_{1-\alpha}^*) = 1$$

(consistency).

4   Simulation

In this section we investigate the empirical properties of our methodology by means of a small simulation study.

Data Generation: In the following, the first sample $X_1, ..., X_{n_1}$ is drawn from the uniform distribution on the d-dimensional cube $[-\sqrt{3}, \sqrt{3}]^d$, whereas the second sample $Y_1, ..., Y_{n_2}$ is uniformly drawn from the shifted cube $[-\sqrt{3} + a/\sqrt{d}, \sqrt{3} + a/\sqrt{d}]^d$. Here, $a \geq 0$ determines the mean difference of the two samples. In particular $a = 0$ corresponds to the hypothesis $\mu_X = \mu_Y = (0, ..., 0)^T$, whereas for $a > 0$, $\|\mu_X - \mu_Y\|_2 = a$. We also point out that both samples have the same covariance matrix $\Sigma_X = \Sigma_Y = Id_{d \times d}$. As a consequence, deviations in each component of $\mu_X - \mu_Y$ have equal influence on the rejection probability.

Parameter Settings: In the following we discuss various settings: We consider different group sizes n, between 10^2 and 10^5, privacy levels $\varepsilon = 1/10, 1/2, 1, 5$ and dimensions $d = 1, 10, 30$. The nominal level α is fixed at 5% and the number of bootstrap samples is consistently $B = 200$. All below results are based on 1000 simulation runs.

Empirical Type-1-Error: We begin by studying the behavior of our test decisions under the null hypothesis ($a = 0$). In Table 1 we report the empirical rejection probabilities for the bootstrap test (3.3) (top) and the asymptotic test (3.2) (bottom). The empirical findings confirm our theoretical results from the previous Section.

On the one hand, we observe that the bootstrap test approximates the nominal-level reasonably well (compare Theorem 3.2.1), even in scenarios with small sample size and high dimensions. In contrast, the validity of the asymptotic test (3.2) depends on the negligibility of privatization effects (see discussion of Theorem 3.1.2). Consequently, it works best for large ε and large sample sizes. However, for higher dimensions d, the asymptotic approach breaks down quickly, in the face of more noise by privatizations and thus stronger digressions from the limiting distribution.

Table 1. Empirical type-1-error

		$d = 1$				$d = 10$				$d = 30$			
	ε	$n_1 = n_2$											
		10^2	10^3	10^4	10^5	10^2	10^3	10^4	10^5	10^2	10^3	10^4	10^5
test (3.3)	0.1	0.052	0.046	0.051	0.048	0.058	0.05	0.068	0.063	0.04	0.057	0.056	0.062
	0.5	0.054	0.05	0.059	0.05	0.039	0.06	0.057	0.052	0.054	0.054	0.06	0.056
	1	0.053	0.05	0.054	0.053	0.048	0.061	0.038	0.069	0.048	0.063	0.056	0.054
	5	0.041	0.053	0.043	0.053	0.055	0.053	0.056	0.051	0.044	0.05	0.062	0.052
test (3.2)	0.1	0.738	0.676	0.328	0.093	1	1	1	1	1	1	1	1
	0.5	0.4	0.154	0.055	0.058	1	1	1	0.891	1	1	1	1
	1	0.24	0.063	0.057	0.044	1	1	0.993	0.428	1	1	1	1
	5	0.054	0.047	0.045	0.039	0.990	0.933	0.181	0.062	1	1	0.999	0.301

Empirical Power: Next we consider the power of our test. Given the poor performance of the asymptotic test (3.2) in higher dimensions (the key interest of this paper) we restrict our analysis to the bootstrap test (3.3) for the sake of brevity. In the following, we consider the alternative for $a = 1$. Recall that $\|\mu_X - \mu_Y\|_2 = a$ is independent of the dimension. However, we expect more power in low dimensions due to weaker privatization. In Fig. 1, we display a panel of empirical power curves, each graphic reflecting a different choice of the privacy parameter ($\varepsilon = 1/10, 1/2, 1, 5$) and each curve corresponding to a different dimension ($d = 1, 10, 30$). The group size is reported in logarithmic scale on the x-axis and the rejection probability on the y-axis. As might be expected, low dimensions and weak privatizations (i.e., large ε) are directly associated with a sharper increase of the power curves and smaller sample sizes to attain high power. For instance, moving from $\varepsilon = 1/2$ (high privatization) to the less demanding $\varepsilon = 5$ (low privatization) means that a power of 90% is attained with group sizes that are about an order of magnitude smaller. Similarly, increasing dimension translates into lower power: To attain for $\varepsilon = 0.1$ and $d = 30$, high power requires samples of a few ten thousand observations (see Fig. 1(a)). Even though such numbers are not in excess of those used in related studies (see e.g. [6]) nor of those raised by large tech cooperations, this trend indicates that comparing means of even higher dimensional populations might require (private) pre-processing to reduce dimensions.

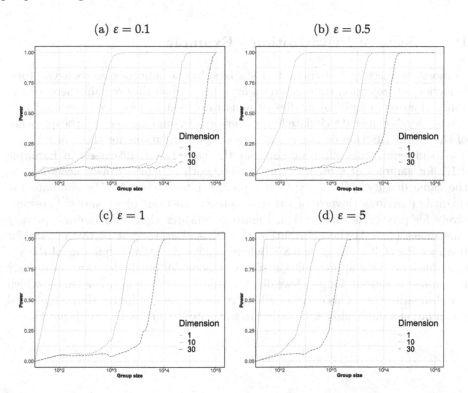

Fig. 1. Simulated power of the bootstrap test (3.3) under a uniform alternative for $\varepsilon = 0.1, 0.5, 1, 5$ and different group sizes.

5   Conclusion

In this paper, we have considered a new way to test multidimensional mean differences under the constraint of differential privacy. Our test employs a privatized version of the popular Hotelling's t^2-statistic, together with a bootstraped rejection rule. While strong privacy requirements always go hand in hand with a loss in power, the test presented in this paper respects the nominal level α with high precision, even for moderate sample sizes, high dimensions and strong privatizations. The empirical advantages are underpinned by theoretical guarantees for large samples. Given the easy implementation and reliable performance, the test can be used as an automatized part of larger analytical structures.

Acknowledgement. This work was partially funded by the DFG under Germany's Excellence Strategy - EXC 2092 CASA - 390781972.

A   Proofs

For the proofs, we refer to the arxiv version (see https://arxiv.org/abs/2110.07996).

B   Effects of Privatization - Example

In most instances, privatizing a test statistic has no influence on its asymptotic behavior, s.t. rejection rules based on asymptotic quantiles remain theoretically valid. However, empirical studies demonstrate, that in practice even moderate privacy levels can lead to inflated type-1-errors – in our case because the quantiles of the χ_d^2-distribution do not provide good approximations for those of t^{DP}.

To illustrate this effect we consider the case $d = 1$, discussed in Example 3.1.1 for samples of sizes $n_1 = n_2 = 500$, both of which drawn according to the same density, $f(t) \propto \exp(-2t^2)$ on the interval $[-1, 1]$. We simulate the quantile functions (inverse of the distribution function) of χ_1^2 and t^{DP} respectively for privacy levels $\varepsilon = 1, 4$. Figure 2 indicates that for moderate privacy guarantees ($\varepsilon = 4$) the distribution of t^{DP} is close to that of the χ_1^2, s.t. for instance $\mathbb{P}_{H_0}(t^{DP} > q_{0.95}) \approx 6.8\%$ (where again $q_{1-\alpha}$ is the α quantile of the χ_1^2-distribution). This approximation seems reasonable, but it deteriorates quickly for smaller ε. Indeed, for $\varepsilon = 1$ we observe that $\mathbb{P}_{H_0}(t^{DP} > q_{0.95}) \approx 18.9\%$, which is a dramatic error. This effect is still more pronounced in higher dimensions and much larger sample sizes are needed to mitigate it (for details see Table 1).

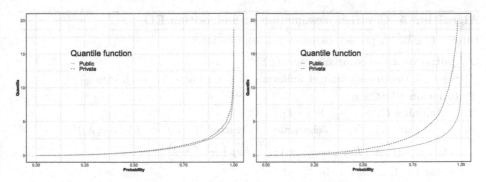

Fig. 2. Simulated quantile functions for χ_1^2 (red) and t^{DP} (blue) for privacy levels $\varepsilon = 4$ (left) and $\varepsilon = 1$ (right)

Summarizing this discussion, we recommend to use Hotelling's t^2-test (3.2) based on the privatized statistic t^{DP} with the standard (asymptotic) quantiles only in situations where sample sizes are large, the dimension is small and privatizations are weak. In all other cases, specifically for larger dimension and stronger privatization, the quantiles have to be adapted to avoid inflated rejection errors under the null hypothesis.

C   Algorithms

In the following, we will state two algorithms which describe the covariance privatization. Here, Algorithm 5 **ED** is used for the privatization, while Algorithm 4 describes the eigenvector sampling process. In Algorithm 5 **ED** the privatization budget is not supposed to be separated (for eigenvalues and eigenvectors) in the case $d = 1$ (as eigenvector privatization is unnecessary for $d = 1$). For more details, see [1].

Algorithm 4. Eigenvector sampling

Input: $\tilde{C} \in \mathbb{R}^{q \times q}$, privacy parameter ε

Output: Eigenvector u.

1: **function** SAMPLE($\tilde{C}, \varepsilon$)
2:     Define $A := -\frac{\varepsilon}{4}\tilde{C} + \frac{\varepsilon}{4}\hat{\lambda}_1 I_q$, where $\hat{\lambda}_1$ denotes the largest eigenvalue of C.
3:     Define $\Omega = I_q + 2A/b$, where b satisfies $\sum_{i=1}^{q} \frac{1}{b+2\lambda_i(A)} = 1$.
4:     Define $M := \exp(-(q - b)/2)(q/b)^{q/2}$.
5:     Set $ANS = 0$
6:     **while** $ANS = 0$ **do**
7:         Sample $X \sim \mathcal{N}_q\left(0, \Omega^{-1}\right)$ and set $u := z/\|z\|_2$.
8:         With probability $\frac{\exp(-u^T A u)}{M(u^T \Omega u)^{q/2}}$ $ANS = 1$
9:         **return** u.
10:     **end while**
11: **end function**

Algorithm 5. Covariance estimation with algorithm **ED**

Input: $\hat{C} \in \mathbb{R}^{d \times d}$, privacy parameter ε, sample size n

Output: Privatized covariance matrix $\hat{\Sigma}^{DP}$

1: Separate the privacy budget uniformly in $d+1$ parts, i.e. each step $\frac{\varepsilon}{d+1}$
2: **function** ED($\hat{C},\varepsilon,n$)
3:    Initialize $C_1 := \frac{n\hat{C}}{dm^2}$, $P_1 := I_d$.
4:    Privatize   the   eigenvalue   vector   by $(\bar{\lambda}_1, \ldots, \bar{\lambda}_d)^T$ $=$
   $\left| (\hat{\lambda}_1, \ldots, \hat{\lambda}_d)^T + \left(Lap\left(\frac{2}{(\varepsilon/(d+1))} \right), \ldots, Lap\left(\frac{2}{(\varepsilon/(d+1))} \right) \right)^T \right|$.
5:    **for** $i = 1, \ldots, d-1$ **do**
6:       Sample $\bar{u}_i \in S^{d-i}$ with $\bar{u}_i := Sample(\hat{C}, \frac{\varepsilon}{d+1})$ and let $\bar{v}_i := P_i^T \bar{u}_i$.
7:       Find an orthonormal basis $P_{i+1} \in \mathbb{R}^{(d-i) \times d}$ orthogonal to $\bar{v}_1, \ldots, \bar{v}_i$.
8:       Let $C_{i+1} := P_{i+1}\hat{C}P_{i+1}^T \in \mathbb{R}^{(d-i) \times (d-i)}$.
9:    **end for**
10:   Sample $\bar{u}_d \in S^0$ proportional to $f_{C_d}(u) = \exp\left((\frac{\varepsilon_i}{4})u^T C_d u \right)$ and let $\bar{v}_d := P_d^T \bar{u}_d$.
11:   $C^{ED} := \sum_{i=1}^{d} \bar{\lambda}_i \bar{v}_i \bar{v}_i^T$.
12:   **return** $\hat{\Sigma}^{DP} = \frac{1}{n} C^{ED}$
13: **end function**

References

1. Amin, K., Dick, T., Kulesza, A., Munoz, A., Vassilvitskii, S.: Differentially private covariance estimation. In: Wallach, H., Larochelle, H., Beygelzimer, A., d'Alché-Buc, F., Fox, E., Garnett, R. (eds.) Advances in Neural Information Processing Systems. vol. 32. Curran Associates, Inc. (2019). proceedings.neurips.cc/paper/2019/file/4158f6d19559955bae372bb00f6204e4-Paper.pdf
2. Bogetoft, P., et al.: Secure Multiparty Computation Goes Live. In: Dingledine, R., Golle, P. (eds.) FC 2009. LNCS, vol. 5628, pp. 325–343. Springer, Heidelberg (2009). https://doi.org/10.1007/978-3-642-03549-4_20
3. Canonne, C.L., Kamath, G., McMillan, A., Smith, A., Ullman, J.: The structure of optimal private tests for simple hypotheses. In: Proceedings of the 51st Annual ACM SIGACT Symposium on Theory of Computing, pp. 310–321 (2019)
4. Couch, S., Kazan, Z., Shi, K., Bray, A., Groce, A.: Differentially private nonparametric hypothesis testing. In: Proceedings of the 2019 ACM SIGSAC Conference on Computer and Communications Security, pp. 737–751 (2019)
5. De Maesschalck, R., Jouan-Rimbaud, D., Massart, D.: The mahalanobis distance. Chemomet. Intell. Lab. Syst.**50**(1), 1–18 (2000). https://doi.org/10.1016/S0169-7439(99)00047-7. www.sciencedirect.com/science/article/pii/S0169743999000477
6. Ding, B., Nori, H., Li, P., Allen, J.: Comparing population means under local differential privacy: with significance and power. In: Proceedings of the AAAI Conference on Artificial Intelligence, vol. 32 (2018)
7. Du, W., Foot, C., Moniot, M., Bray, A., Groce, A.: Differentially private confidence intervals. arXiv preprint arXiv:2001.02285 (2020)
8. Dwork, C., Roth, A.: The algorithmic foundations of differential privacy. Found. Trends Theor. Comput. Sci. **9**(3–4), 211–407 (2014). https://doi.org/10.1561/0400000042

9. Erlingsson, U., Pihur, V., Korolova, A.: Rappor: randomized aggregatable privacy-preserving ordinal response. In: Proceedings of the 2014 ACM SIGSAC Conference on Computer and Communications Security, pp. 1054–1067. CCS 2014, Association for Computing Machinery, New York, NY, USA (2014). https://doi.org/10.1145/2660267.2660348

10. Ferrando, C., Wang, S., Sheldon, D.: General-purpose differentially-private confidence intervals. arXiv preprint arXiv:2006.07749 (2020)

11. Gaboardi, M., Lim, H., Rogers, R., Vadhan, S.: Differentially private chi-squared hypothesis testing: goodness of fit and independence testing. In: International Conference on Machine Learning, pp. 2111–2120. PMLR (2016)

12. Karwa, V., Vadhan, S.: Finite sample differentially private confidence intervals. arXiv preprint arXiv:1711.03908 (2017)

13. Lindell, Y.: Secure multiparty computation for privacy preserving data mining. In: Encyclopedia of Data Warehousing and Mining, pp. 1005–1009. IGI Global (2005)

14. Machanavajjhala, A., Kifer, D., Abowd, J., Gehrke, J., Vilhuber, L.: Privacy: theory meets practice on the map, pp. 277–286, April 2008. https://doi.org/10.1109/ICDE.2008.4497436

15. Machanavajjhala, A., Kifer, D., Abowd, J., Gehrke, J., Vilhuber, L.: Privacy: theory meets practice on the map, pp. 277–286, April 2008. https://doi.org/10.1109/ICDE.2008.4497436

16. Raj, A., Law, H.C.L., Sejdinovic, D., Park, M.: A differentially private kernel two-sample test. In: Brefeld, U., Fromont, E., Hotho, A., Knobbe, A., Maathuis, M., Robardet, C. (eds.) ECML PKDD 2019. LNCS (LNAI), vol. 11906, pp. 697–724. Springer, Cham (2020). https://doi.org/10.1007/978-3-030-46150-8_41

17. Rogers, R., Kifer, D.: A new class of private chi-square hypothesis tests. In: Singh, A., Zhu, J. (eds.) Proceedings of the 20th International Conference on Artificial Intelligence and Statistics. Proceedings of Machine Learning Research, vol. 54, pp. 991–1000. PMLR, Fort Lauderdale, FL, USA 20–22 April 2017. http://proceedings.mlr.press/v54/rogers17a.html

18. Rogers, R., et al.: Linkedin's audience engagements api: A privacy preserving data analytics system at scale. arXiv preprint arXiv:2002.05839 (2020)

19. Sei, Y., Ohsuga, A.: Privacy-preserving chi-squared test of independence for small samples. BioData Mining **14**(1), 1–25 (2021)

20. Smith, A.: Privacy-preserving statistical estimation with optimal convergence rates. In: Proceedings of the Forty-Third Annual ACM Symposium on Theory of Computing, pp. 813–822 (2011)

21. Swanberg, M., Globus-Harris, I., Griffith, I., Ritz, A., Groce, A., Bray, A.: Improved differentially private analysis of variance. arXiv preprint arXiv:1903.00534 (2019)

22. Wang, Y., Lee, J., Kifer, D.: Revisiting differentially private hypothesis tests for categorical data. arXiv, Cryptography and Security (2015)

Asking the Proper Question: Adjusting Queries to Statistical Procedures Under Differential Privacy

Tomer Shoham[1](✉) and Yosef Rinott[2](✉)

[1] Department of Computer Science and Federmann Center for the Study of Rationality, The Hebrew University of Jerusalem, Jerusalem, Israel
Tomer.Shoham@mail.huji.ac.il
[2] Department of Statistics and Federmann Center for the Study of Rationality, The Hebrew University of Jerusalem, Jerusalem, Israel
Yosef.Rinott@mail.huji.ac.il

> "If I had an hour to solve a problem and my life depended on the solution, I would spend the first 55 minutes determining the proper question to ask".
>
> *Albert Einstein*

Abstract. We consider a dataset S held by an agency, and a vector query of interest, $f(S) \in \mathbb{R}^k$, to be posed by an analyst, which contains the information required for some planned statistical inference. The agency will release an answer to the queries with noise that guarantees a given level of Differential Privacy using the well-known Gaussian noise addition mechanism. The analyst can choose to pose the original vector query $f(S)$ or to transform the query and adjust it to improve the quality of inference of the planned statistical procedure, such as the volume of a confidence interval or the power of a given test of hypothesis. Previous transformation mechanisms that were studied focused on minimizing certain distance metrics between the original query and the one released without a specific statistical procedure in mind. Our analysis takes the Gaussian noise distribution into account, and it is non-asymptotic. In most of the literature that takes the noise distribution into account, a given query and a given statistic based on the query are considered and the statistic's asymptotic distribution is studied. In this paper we consider both non-random and random datasets, that is, samples, and our inference is on $f(S)$ itself, or on parameters of the data generating process when S is a random sample. Our main contribution is in proving that different statistical procedures can be strictly improved by applying different specific transformations to queries, and in providing explicit transformations for different procedures in some natural situations.

This work was supported in part by a gift to the McCourt School of Public Policy and Georgetown University, Simons Foundation Collaboration 733792, and Israel Science Foundation (ISF) grants 1044/16 and 2861/20, and by a grant from the Center for Interdisciplinary Data Science Research at the Hebrew University (CIDR).

J. Domingo-Ferrer and M. Laurent (Eds.): PSD 2022, LNCS 13463, pp. 46–61, 2022.
https://doi.org/10.1007/978-3-031-13945-1_4

Keywords: Gaussian mechanism · Differential privacy · Confidence region · Testing hypotheses · Statistical inference on noisy data

1 Introduction

1.1 Setting

Throughout the paper we consider a dataset S consisting of a whole population or a sample given as an $n \times d$ matrix, where each row pertains to an individual, with d variables that are measured for each of the n participant in the dataset. The dataset S is held by some agency and an analyst is interested in a vector function $f(S) = (f_1(S), \ldots, f_k(S)) \in \mathbb{R}^k$ of the data, to be called a query. Thus, a query consists of k functions of the data to be posed to the agency by the analyst. We consider throughout the case $k > 1$. We assume that in order to protect the privacy of individual data in S, the agency releases the response to the query $f(S)$ with noise, using a standard Gaussian mechanism that adds independent $N(0, \sigma^2)$ noise to each coordinate of $f(S)$. The distribution of the added noise is always assumed to be known to the analyst, a standard assumption in the differential privacy literature. Two datasets S and S' are said to be neighbors, denoted by $S \sim S'$, if they differ by a single individual, i.e., a single row. See, e.g., [12] for all needed details on Differential Privacy (henceforth DP). When we consider S and S' together we always assume that they are neighbors.

More generally, consider a noise mechanisms applied to S via a query $h(S) \in \mathbb{R}^k$ of the form $\mathcal{M}_h(S) = h(S) + U \in \mathbb{R}^k$, where U is a random vector. A mechanism $\mathcal{M}_h$ is said to be (ε, δ)–DP if for all (measurable) sets E we have

$$P(\mathcal{M}_h(S) \in E) \leq e^\varepsilon P(\mathcal{M}_h(S') \in E) + \delta \tag{1}$$

for all $S \sim S' \in \mathfrak{D}$, where the probability refers to the randomness of U, and $\mathfrak{D}$ is the universe of potential datasets. For example, if S is a sample of a given size n from some population, then $\mathfrak{D}$ is the universe of all samples that could have been drawn and considered for dissemination. The standard definition of DP takes $\mathfrak{D}$ to be a product C^n where C consists of all possible rows. Our results hold for any given $\varepsilon > 0$ and $\delta \in (0, 1)$, which we fix for the rest of this paper.

1.2 Our Contribution

We describe some simple examples where posing a linear transformation of the query $f(S)$, getting the agency's response via a mechanism that guarantees DP, and inverting the response to obtain the required information on $f(S)$ yields more information and better inference on $f(S)$ in the case where S is a given fixed dataset, or on the model that generates $f(S)$, when S is a random sample. More specifically, given a planned statistical procedure, we show that transforming the query can yield better results in terms of power of tests and size of confidence sets. We believe that the idea of transforming queries to improve efficiency of specific statistical procedure rather than to reduce some arbitrary distance between the posed query and the response is new, and we show that for different statistical procedures one should consider different suitable transformations. In this paper we provide simple formal mathematical evidence for the idea of adjusting queries towards statistical goals, and its practical value will be explored elsewhere.

The idea of transforming queries in order to minimize the effect of the noise while preserving privacy in terms of some metrics appears in [22] as the Matrix Mechanism. The difference between this line of work and ours is discussed next. In particular, our statistical goal is different, and our results are analytic and allow for continuous data rather than numerical and discrete.

1.3   Related Work

The principle of modifying queries is not new. The Matrix Mechanism (MM) was put forward in a line of work that started with [22]. Further literature includes [13, 23, 24, 26] and references therein. For given queries, MM linearly modifies the original data by applying a matrix that depends on the queries to be answered. The modified data is released with noise, and the answer to the original queries is estimated. The above literature studies numerical algorithms for finding optimal modifying matrices that minimizes the distance between the original queries and the mechanism's output relative to different metrics without regard to specific statistical purposes. Our work differs from the MM in that our goal is to improve the data for given statistical purposes, and in specific cases we propose analytic arguments. Criticism of the approach of minimizing some distance metrics between original and noisy statistics appears in [20]. Such criticism appears also in [3] in the context of optimizing the noise relative to a specific statistical goal.

In Sect. 3 we consider the problem of estimating and constructing a confidence region for the mean of a query having a multivariate Gaussian distribution, or testing hypothesis on it. This problem was studied with known and unknown covariance in the univariate case (that we do not discuss in this paper) in [9, 21, 33], and in the multivariate case, in [2, 6]; see also references therein. Some of the above results are involved because the query may be Gaussian, but with Laplace noise, or because the variance is unknown (although some bounds are sometimes assumed) and should be estimated, thus complicating the differentially private mechanism. Since our goal in this paper is to draw attention to the benefit of modifying queries for the statistical procedure, we do not compare those results to ours, and we confine ourselves to simple models.

When a dataset is randomly generated by some assumed distribution, it is well known that the analyst has to adjust the statistical procedure to the distribution of the observed data, taking the distribution of the added noise into account; see, e.g., [7, 15, 29, 33, 34] and references therein. Most of these results are asymptotic, and they do not consider transforming queries.

The idea of adjusting the noise mechanism to certain specific utility metrics and queries appears, e.g., in [16, 17], though most researchers consider simple mechanisms with a well-known distribution (e.g., addition of Laplace or

Gaussian i.i.d noise). Note again that we focus on adjusting queries using a given (Gaussian) noise mechanism, rather than adjusting the noise.

2 Fixed (Non-random) Datasets

Consider a dataset S held by an agency and an analyst who poses a query $f(S)$ in terms of measurement units of his choosing. For example, the components of $f(S)$ could be average age, average years of schooling, and median income in the dataset S. The observed response is given with noise through a privacy mechanism applied by the data-holding agency. The **analyst's goals** are to construct a confidence region for $f(S)$ and to test simple hypotheses about it. For any given level ε, δ of DP, we show that instead of posing the query $f(S)$, the analyst can obtain a smaller confidence region for $f(S)$ by computing it from a query of the form $f_\xi(S) = Diag(\xi)^{1/2} f(S)$ for a suitable $\xi \in \mathbb{R}^k_{\geq 0}$ (a vector having nonnegative coordinates), where $Diag(\xi)$ is a diagonal matrix whose diagonal elements form the vector ξ. For the goal of testing hypotheses, it turns out that a different choice of ξ maximizes the power of the standard likelihood-ratio test. Thus, the analyst can achieve better inference by adjusting his queries to the planned statistical procedure.

Consider a row $(x_1, \ldots, x_d)$ in the dataset S. For simplicity we assume that $x_i \in C_i$ for $i = 1, \ldots, d$ for suitable sets C_i. In this case each row is in the Cartesian product $C := C_1 \times \ldots \times C_d$ and we set $\mathfrak{D} := C^n$. We assume that the agency releases data under $(\varepsilon, \delta)-DP$ relative to this universe $\mathfrak{D}$, which is known to both the agency and the analyst.

In Sect. 2 we assume that the components f_i of the vector query $f = (f_1, \ldots, f_k)$ are functions of disjoint sets of columns of S. This assumption is not needed in Sect. 3. The quantity $\Delta(f) := \max_{S \sim S' \in \mathfrak{D}} \|f(S) - f(S')\|$, where $\| \cdot \|$ denotes the L_2 norm, is known as the sensitivity of f; higher sensitivity requires more noise for DP. Under simple assumptions on the functions f_i such as monotonicity, the agency can readily compute $\Delta(f)$, as well as

$$(\widetilde{S}, \widetilde{S}') := \operatorname*{argmax}_{S \sim S' \in \mathfrak{D}} \|f(S) - f(S')\|; \tag{2}$$

see Lemma 1, where it is shown that the maximization can be done separately for each coordinate of f. In general, the maximum in (2) is not unique, in which case arg max is a set of pairs.

The agency plans to release a response to the query $f(S)$ via a standard Gaussian mechanism; that is, the response is given by

$$\mathcal{M}(S) = f(S) + U \text{ where } U \sim N(0, \sigma^2 I)$$

and I is the $k \times k$ identity matrix. The variance σ^2 is the minimal variance such that the mechanism satisfies DP for given ε, δ; it can be determined by Lemma 2 below, which appears in [4]. This variance depends on $\Delta(f)$; however, here f is fixed and hence suppressed.

Consider a family of queries adjusted by $Diag(\xi)$:

$$f_\xi(S) := Diag(\xi)^{1/2} f(S) = \big(\xi_1^{1/2} f_1(S), \ldots, \xi_k^{1/2} f_k(S)\big).$$

In particular, for the vector ξ whose components are all equal to one we have $f_{\xi=1} = f$. Given a query from this family, the agency returns a perturbed response using a Gaussian mechanism $\mathcal{M}_\xi$ by adding to f_ξ a Gaussian vector $U \in \mathbb{R}^k$ where $U \sim N(0, \sigma^2 I)$, that is,

$$\mathcal{M}_\xi(S) = f_\xi(S) + U.$$

It is easy to see directly or from Lemma 2 that we can fix σ^2 and guarantee a given level of (ε, δ)–DP by normalizing ξ appropriately. Hence fixing σ^2 does not result in loss of generality. Below all mechanism $\mathcal{M}_\xi$ are assumed to have a common σ.

2.1   Confidence Regions

The following discussion concerns the choice of $\xi \in \mathbb{R}_{>0}^k$ such that the standard confidence region CR_ξ^t for $\mu^* := f(S)$ given in formula (3) below, which is based on the observed $\mathcal{M}_\xi(S)$, has the smallest volume.

The idea is simple: intuitively it appears efficient to add more noise to the more variable components of $f(S)$ rather than "waste noise" on components with low variability. Note that "more variable" depends on both the population being measured and the chosen units of measurement. Instead of asking the agency to adjust the noise to different components, we adjust the query, and thus the agency can use a standard Gaussian mechanism. This intuition, as the whole paper, is clearly relevant only for $k > 1$.

The analyst observes $\mathcal{M}_\xi(S) = f_\xi(S) + U$, where

$$\mathcal{M}_\xi(S) = \big(Diag(\xi)^{1/2} f(S) + U\big) \sim N(Diag(\xi)^{1/2} f(S), \sigma^2 I).$$

Thus, $Diag(\xi)^{-1/2} \mathcal{M}_\xi(S) \sim N(\mu^*, Diag(\xi)^{-1} \sigma^2)$.

The standard confidence region for μ_x based on $X \sim N(\mu_x, \Sigma)$ is $\{\mu : (X - \mu)^T \Sigma^{-1} (X - \mu) \le t\}$; see, e.g., [1], p. 79. Thus, the confidence region for $\mu^* = f(S)$ based on $Diag(\xi)^{-1/2} \mathcal{M}_\xi(S)$ becomes

$$CR_\xi^t = \{\mu \in \mathbb{R}^k : \big(Diag(\xi)^{-1/2} \mathcal{M}_\xi(S) - \mu\big)^T$$
$$\big(Diag(\xi)\sigma^{-2}\big)\big(Diag(\xi)^{-1/2} \mathcal{M}_\xi(S) - \mu\big) \le t\}. \quad (3)$$

For any $\xi \in \mathbb{R}_{>0}^k$ and any $\mu^* \in \mathbb{R}^k$, the coverage probability $P(\mu^* \in CR_\xi^t) = P(Y \le t)$ where $Y \sim \mathcal{X}_k^2$ (the chi-square distribution with k degrees of freedom), and thus all the regions CR_ξ^t have the same confidence (coverage) level. We denote the volume by $Vol(CR_\xi^t)$. For a discussion of the volume as a measure of utility of confidence regions see, e.g., [14]. We now need the notation

$$\psi := f(\widetilde{S}) - f(\widetilde{S}') = \big(f_1(\widetilde{S}) - f_1(\widetilde{S}'), \ldots, f_k(\widetilde{S}) - f_k(\widetilde{S}')\big),$$

where $(\widetilde{S}, \widetilde{S}')$ is any pair in the set defined in (2), and we assume that $\psi_i^2 = (f_i(\widetilde{S}) - f_i(\widetilde{S}'))^2 > 0$ for all i.

Theorem 1. (1) *For any fixed t, the confidence level of the regions CR_ξ^t defined in Eq. (3) is the same for all ξ, that is, the probability $P(\mu^* \in CR_\xi^t)$ depends only on t (and not on ξ).*

(2) *Assume all mechanism $\mathcal{M}_\xi$ below have a common σ. Set*

$$\Lambda(\xi) = \frac{\sqrt{\psi^T Diag(\xi)\psi}}{\sigma}.$$

If for two vectors ξ_a and ξ_b the mechanisms $\mathcal{M}_{\xi_a}$ and $\mathcal{M}_{\xi_b}$ have the same level of DP (that is, the same ε and δ) then $\Lambda(\xi_a) = \Lambda(\xi_b)$. In particular, for any $\mathcal{M}_\xi$ to have the same DP level as $\mathcal{M}_{\xi=1}$ we must have $\sqrt{\psi^T Diag(\xi)\psi} = \sqrt{\psi^T\psi}$.

(3) *The choice $\xi = \xi^* := c\,(1/\psi_1^2,\ldots,1/\psi_k^2)$ with $c = \|\psi\|^2/k$ minimizes $Vol(CR_\xi^t)$ for any $t > 0$ over all vectors $\xi \in \mathbb{R}_{>0}^k$ and associated mechanisms $\mathcal{M}_\xi$ having the same DP level. In particular,*

$$Vol(CR_{\xi^*}^t) \le Vol(CR_{\xi=1}^t),$$

with strict inequality when $max_i(\psi_i) \ne min_i(\psi_i)$. The right-hand side of the inequality pertains to the query f.

To prove Theorem 1 and others we use two lemmas given here.

Lemma 1. *For any $\xi \in \mathbb{R}_{>0}^k$.*

$$\Delta(f_\xi) \equiv \max_{S \sim S' \in \mathcal{D}} \|f_\xi(S) - f_\xi(S')\| = \|f_\xi(\widetilde{S}) - f_\xi(\widetilde{S}'))\|,$$

where the pair $(\widetilde{S}, \widetilde{S}')$ is defined in Eq. (2).

The proof of the Lemma is omitted due to space limitation, and can be found in the full version of this paper [32].

For an agency willing to release the query f, releasing f_ξ under the mechanism $\mathcal{M}_\xi$ with the same DP level does not add any complications. The agency needs to compute the sensitivity defined by $\Delta(f_\xi) \equiv \max_{S \sim S' \in \mathcal{D}} \|(f_\xi(S) - f_\xi(S'))\|$. By Lemma 1, this amounts to computing $\|f_\xi(\widetilde{S}) - f_\xi(\widetilde{S}'))\|$ using the pair $(\widetilde{S}, \widetilde{S}')$ from Eq. (2), which is needed to compute the sensitivity of f.

The next lemma can be obtained readily from the results of [4], which hold for any query f.

Lemma 2. *Let $\mathcal{M}(S) = f(S) + U$ be a Gaussian mechanism with $U \sim N(0, \sigma^2 I)$, and for given datasets S and S' set $D := D_{S,S'} = \|f(S) - f(S')\|$.*
(1) *If*

$$\Phi\left(\frac{D}{2\sigma} - \frac{\varepsilon\sigma}{D}\right) - e^\varepsilon \Phi\left(-\frac{D}{2\sigma} - \frac{\varepsilon\sigma}{D}\right) \le \delta, \tag{4}$$

then for all $E \subseteq \mathbb{R}^k$,

$$\mathbb{P}(\mathcal{M}(S) \in E) \le e^\varepsilon \mathbb{P}(\mathcal{M}(S') \in E) + \delta. \tag{5}$$

(2) *Setting $\widetilde{D} := \Delta(f) = \|f(\widetilde{S}) - f(\widetilde{S}'))\|$, with $(\widetilde{S}, \widetilde{S}')$ given in Eq. (2), Eq. (4) holds with D replaced by $\widetilde{D}$ if and only if the inequality (5) holds for all $S \sim S'$ and $E \subseteq \mathbb{R}^k$, that is, if and only if (ε, δ)–DP holds.*

Part (2) of Lemma 2 coincides with Theorem 8 of [4], and the first part follows from their method of proof.

Proof of Theorem 1. Part (1) follows from the fact mentioned above that all these regions have confidence level $P(Y \leq t)$ where $Y \sim \mathcal{X}_k^2$. Part (2) is obtained by replacing f of Part (2) of Lemma 2 by f_ξ; then $(\widetilde{D}/\sigma)$ becomes $\frac{\sqrt{\psi^T Diag(\xi)\psi}}{\sigma}$ and the result follows.

To prove Part (3), note that the confidence region for the adjusted query given in Eq. (3) is an ellipsoid whose volume is given by:

$$Vol(CR_\xi^t) = V_k \cdot (\sigma^2 t)^{k/2} (det[Diag(\xi)])^{-1/2}, \tag{6}$$

where V_k is the volume of the unit ball in k dimensions. By Part (2), we have to minimize the volume as a function of ξ subject to the constraint $\psi^T Diag(\xi)\psi = \psi^T \psi$, which we do by using Lagrange multipliers. See the Appendix for details.   □

Theorem 1 states that given a DP level, the volume is minimized by choosing ξ_i proportionally to $1/\psi_i^2$. Multiplying ξ by a suitable constant guarantees the desired DP level. It is easy to compute the ratio of the volumes of the optimal region and the one based on the original query f:

$$\frac{Vol(CR_{\xi*}^t)}{Vol(CR_{\xi=1}^t)} = \left(\frac{(\prod_{i=1}^k \psi_i^2)^{1/k}}{\frac{1}{k}\sum_{i=1}^k \psi_i^2} \right)^{k/2}.$$

Clearly the ratio is bounded by one, which can be seen again by the arithmetic-geometric mean inequality. Also, if one of the coordinates ψ_i tends to zero, so does the ratio, implying the possibility of a substantial reduction in the volume obtained by using the optimally adjusted query $f_{\xi*}$. We remark that the ratio is decreasing in the partial order of majorization applied to $(\psi_1^2, \ldots, \psi_k^2)$; see [25].

2.2   Testing Hypotheses: Likelihood-Ratio Test

As in Sect. 2.1, consider a query $f(S) \in \mathbb{R}^k$, which is observed with noise via a Gaussian mechanism. Now the analyst's goal is to test the simple hypotheses $H_0 : f(S) = 0$, $H_1 : f(S) = \eta$. The null hypothesis is set at zero without loss of generality by a straightforward translation. For any $\xi \in \mathbb{R}_{\geq 0}$ (a vector with nonnegative components), let $f_\xi(S) = Diag(\xi)^{1/2} f(S)$ and let $\mathcal{M}_\xi(S) = f_\xi(S) + U$, where $U \sim N(0, \sigma^2 I)$ and σ^2 is the smallest variance such that the Gaussian mechanism $\mathcal{M}_{\xi=1}(S)$ guarantees (ε, δ)–DP for the query f.

Let $h_{\xi i}$ denote the density of $\mathcal{M}_\xi(S)$ under the hypothesis H_i, $i = 0, 1$. The log-likelihood ratio based on the observed $\mathcal{M}_\xi(S)$, $\log\{\frac{h_{\xi 1}(\mathcal{M}_\xi(S))}{h_{\xi 0}(\mathcal{M}_\xi(S))}\}$, is proportional to $\frac{\mathcal{M}_\xi(S)^T Diag(\xi)^{1/2}\eta}{\sigma^2}$, which under H_0 has the $N(0, \frac{\eta^T Diag(\xi)\eta}{\sigma^2})$ distribution. The likelihood-ratio test (which by the Neyman–Pearson lemma has a well-known optimality property) rejects H_0 when the likelihood ratio is large. For a given significance level α, the rejection region has the form

$$R_\xi = \left\{ \mathcal{M}_\xi(S) : \frac{\mathcal{M}_\xi(S)^T Diag(\xi)^{1/2}\eta}{\sigma^2} > t \right\}, \tag{7}$$

where $t = \Phi^{-1}(1 - \alpha) \frac{\sqrt{\eta^T Diag(\xi)\eta}}{\sigma}$.

Let $\pi(R_\xi) := P_{H_1}(\mathcal{M}_\xi(\hat{S}) \in R_\xi)$ denote the power associated with R_ξ.

Theorem 2.

(1) *For any fixed α and for all $\xi \in \mathbb{R}_{\geq 0}$, the rejection regions R_ξ defined in* (7) *have significance level α, that is, $P_{H_0}(R_\xi) = \alpha$.*

(2) *Assume that for two vectors ξ_a and ξ_b the mechanisms $\mathcal{M}_{\xi_a}$ and $\mathcal{M}_{\xi_b}$ have the same level of DP (same ε and δ); then $\Lambda(\xi_a) = \Lambda(\xi_b)$, where $\Lambda(\xi) = \frac{\sqrt{\psi^T Diag(\xi)\psi}}{\sigma}$.*

(3) *Let $j^* = \arg\max_i(\eta_i^2/\psi_i^2)$, and define ξ^* by $\xi_{j^*}^* = ||\psi||^2/\psi_{j^*}^2$ and $\xi_i^* = 0 \;\; \forall i \neq j^*$; then the choice $\xi = \xi^*$ maximizes the power $\pi(R_\xi)$ over all vectors $\xi \in \mathbb{R}_{\geq 0}^k$ and the associated mechanisms $\mathcal{M}_\xi$ having the same DP level, and in particular $\pi(R_{\xi^*}) \geq \pi(R_{\xi=1})$, with strict inequality unless $\max_i(\eta_i^2/\psi_i^2) = \min_i(\eta_i^2/\psi_i^2)$.*

Note that $\pi(R_{\xi=1})$ is the power when the original query is posed. Theorem 2 states that the query of just one coordinate of f, the one having the largest ratio of (loosely speaking) signal (η_i^2) to noise (ψ_i^2), is the most informative for testing the hypothesis in question. Note the difference between the optimal query of Theorem 2 and that of Theorem 1, which uses all coordinates of f.

Proof of Theorem 2. Part (1) follows from (7) and the discussion preceding it with standard calculations; Part (2) is similar to that of Theorem 1. The proof of Part (3) is given in the Appendix.

3   Random, Normally Distributed Data

So far the dataset S was considered fixed, that is, nonrandom. Statisticians often view the data as random, construct a model for the data-generating process, and study the model's parameters. Accordingly, we now assume that the dataset, denoted by T, is randomly generated as follows: the rows of T, $T_1, \ldots, T_n$ are i.i.d, where each row $T_\ell \in \mathbb{R}^d$ represents d measurements of an individual in the random sample T. We also assume that f is a linear query, that is,

$$f(T) = (f_1(T), ..., f_k(T)) = \left(\frac{1}{n}\sum_{\ell=1}^{n} q_1(T_\ell), ..., \frac{1}{n}\sum_{\ell=1}^{n} q_k(T_\ell)\right)$$

for some functions $q_1, \ldots, q_k$. Set $q(T_\ell) := (q_1(T_\ell), \ldots, q_k(T_\ell))$. We assume that $q(T_\ell) \sim N(\mu^*, \Sigma)$ for some unknown μ^* and a known covariance matrix Σ. The normality assumption holds when the entries of T are themselves normal, and q_i are linear functions. Assuming normality, possibly after transformation of the data, and i.i.d observations is quite common in statistical analysis. It follows that $f(T) \sim N(\mu^*, \Sigma_n)$, where $\Sigma_n = \Sigma/n$. This may hold approximately by the central limit theorem even if normality of the dataset is not assumed. Here we assume that Σ is known. The case where it is obtained via a privatized query is beyond the scope of this paper.

Since the observed data will depend only on $q(T_\ell)$, we now redefine the dataset to be S, consisting of the n i.i.d rows $S_\ell := q(T_\ell)$, $\ell = 1, \ldots, n$. The assumption $q(T_\ell) \sim N(\mu^*, \Sigma)$ implies that these rows can take any value in $C := \mathbb{R}^k$. The universe of all such matrices S is $\mathfrak{D} := C^n = \mathbb{R}^{n \times k}$.

The **analyst's goal** is to construct a confidence region for the model parameter μ^* and test hypotheses about it. This can be done via the query $f(S) = \frac{1}{n} \sum_{\ell=1}^n S_\ell$ having the distribution $N(\mu^*, \Sigma_n)$; however, we show that posing the query $g(S) := \Sigma_n^{-1/2} f(S)$ under the same Random Differential Privacy parameters (RDP, to be defined below) results in smaller confidence regions. We also compare the powers of certain tests of hypotheses.

We say that a query f is invariant if $f(S)$ is invariant under permutations of the rows of S. This happens trivially when f is a linear query as defined above. If f is invariant then the distribution of the output of any mechanism that operates on f is obviously unchanged by permutations of rows. In this case it suffices to consider neighbors $S \sim S'$ of the form $S = (S_1, \ldots, S_{n-1}, S_n)$, $S' = (S_1, \ldots, S_{n-1}, S_{n+1})$. We assume that $S_1, \ldots, S_{n+1}$ are i.i.d rows having some distribution Q. The following definition is given in [18].

Definition 1. *A random perturbation mechanism $\mathcal{M}$ whose distribution is invariant under permutations of rows is said to be $(\varepsilon, \delta, \gamma)$-Randomly Differentially Private, denoted by $(\varepsilon, \delta, \gamma)$–RDP, if*

$$P_{S_1, \ldots, S_{n+1}} \left(\forall\, E \subseteq \mathbb{R}^k,\ P(\mathcal{M}(S) \in E | S) \leq e^\varepsilon \mathbb{P}(\mathcal{M}(S') \in E | S') + \delta \right) \geq 1 - \gamma,$$

where S and S' are neighbors as above, the probability $P_{S_1, \ldots, S_{n+1}}$ is with respect to $S_1, \ldots, S_{n+1} \overset{i.i.d}{\sim} Q$, and the probability $P(\mathcal{M}(S) \in E | S)$ refers to the noise U after conditioning on S.

In words, instead of requiring the condition of differential privacy to hold for all $S \sim S' \in \mathfrak{D}$, we require that there be a "privacy set" in which any two random neighboring datasets satisfy the DP condition, and its probability is bounded below by $1 - \gamma$.

On the Privacy Limitation of RDP. An objection to this notion may arise from the fact that under RDP "extreme" participants, who are indeed rare, are not protected, even though they may be the ones who need privacy protection the most. Since RDP is not in the worst-case analysis spirit of DP, we remark that DP can be obtained if, instead of ignoring worst cases having small probability as in RDP, the agency trims them by either removing them from the dataset or by projecting them to a given ball (that is independent of the dataset) which determines the sensitivity. Such trimming, if its probability is indeed small, corresponding to a small γ, will not overly harm the data analysis.

We define a "privacy set" H, which is a subset of $\mathfrak{D} \times \mathfrak{D}$ consisting of neighboring pairs (S, S'), and a mechanism $\mathcal{M}_h^H(S) = h(S) + U$ (see (1)) that satisfies $(\varepsilon, \delta, \gamma)$–RDP. Under the following two conditions

(A) $P((S, S') \in H) = 1 - \gamma$, where P is $P_{S_1, \ldots, S_{n+1}}$ of Definition 1,
(B) Equation (1) holds $\forall E$ and any pair of neighboring datasets $(S, S') \in H$,

it is easy to see that $\mathcal{M}_h^H(S)$ is $(\varepsilon, \delta, \gamma)$–RDP and we say that it is RDP with respect to the privacy set H and the query h.

To construct a suitable H satisfying condition (A) note that

$$f(S) - f(S') = \frac{1}{n}[q(S_n) - q(S_{n+1})] \sim N(0, 2\Sigma/n^2), \text{ and} \qquad (8)$$

$$\|g(S) - g(S')\|^2 = \|\Sigma_n^{-1/2}[f(S) - f(S')]\|^2$$

$$= \|\Sigma^{-1/2}[q(S_n) - q(S_{n+1})]\|^2 \sim \frac{2}{n}\mathcal{X}_k^2.$$

Thus, if $Y \sim \mathcal{X}_k^2$ satisfies $P(Y \leq r^2) = 1 - \gamma$ then $P(\|g(S) - g(S')\|^2 \leq 2r^2/n) = 1 - \gamma$, and we can choose the set H to be

$$H_g := \{(S, S') \in \mathfrak{D} \times \mathfrak{D} : \|g(S) - g(S')\|^2 \leq 2r^2/n\}.$$

Another privacy set we consider is given by

$$H_f := \{(S, S') \in \mathfrak{D} \times \mathfrak{D} : \|f(S) - f(S')\|^2 \leq C^2\},$$

where C is such that $P((S, S') \in H_f) = 1 - \gamma$, and by (8) the constant C depends on Σ and n.

We consider three Gaussian mechanisms:

$$\mathcal{M}_g^{H_g}(S) = g(S) + U, \text{ where } U \sim N(0, \sigma_g^2 I),$$

$$\mathcal{M}_f^{H_g}(S) = f(S) + U \text{ with } U \sim N(0, \sigma_{fg}^2 I),$$

$$\mathcal{M}_f^{H_f}(S) = f(S) + U \text{ with } U \sim N(0, \sigma_f^2 I),$$

where the first two are with respect to the privacy set H_g, and the third is with respect to H_f. For each of the three, an appropriate noise variance σ_g^2, σ_{fg}^2, and σ_f^2 has to be computed, given the privacy set and the RDP parameters, so that condition (B) above holds. To determine the noise variance we have to compute the sensitivity of the query g on the set H_g and the sensitivity of f on both H_g and H_f.

Define the sensitivity of f and g on H_g, denoted by $D(fg)$ and $D(g)$, respectively, and the sensitivity of f on H_f, denoted by $D(f)$, as follows:

$$D(fg) := \max_{(S,S') \in H_g} \|f(S) - f(S')\|,$$

$$D(g) := \max_{(S,S') \in H_g} \|g(S) - g(S')\| = \sqrt{2}\,r/\sqrt{n}, \qquad (9)$$

$$D(f) := \max_{(S,S') \in H_f} \|f(S) - f(S')\| = C.$$

We compare the above three mechanisms with the same RDP level in terms of the volume of confidence regions and the power of tests of hypotheses for the model parameter μ^*, computed from data given by these mechanisms.

By the definition of RDP, the mechanism $\mathcal{M}_f^{H_g}(S)$ satisfies $(\varepsilon, \delta, \gamma)$–RDP when (4) holds with $D = D(fg)$ and $\sigma = \sigma_{fg}$, as does the mechanisms $\mathcal{M}_g^{H_g}(S)$ with $D = D(g)$ and $\sigma = \sigma_g$, and likewise the mechanism $\mathcal{M}_f^{H_f}(S)$ with $D = D(f)$ and $\sigma = \sigma_f$.

Lemma 3. *If $\mathcal{M}_g^{H_g}$, $\mathcal{M}_f^{H_g}$, and $\mathcal{M}_f^{H_f}$ have the same RDP, then $D(g)/\sigma_g = D(fg)/\sigma_{fg} = D(f)/\sigma_f$. The first equality is equivalent to $\sigma_{fg}^2 = \lambda_{max}(\Sigma_n)\sigma_g^2$, where $\lambda_{max}(\Sigma_n)$ denotes the largest eigenvalue of Σ_n.*

The proof is omitted due to lack of space and can be found in [32].

3.1   Confidence Regions

We have

$$\Sigma_n^{1/2}\mathcal{M}_g^{H_g}(S) \sim N\left(\mu^*, \Sigma_n(1 + \sigma_g^2)\right),$$
$$\mathcal{M}_f^{H_g}(S) \sim N(\mu^*, \Sigma_n + \sigma_{fg}^2 I),$$
$$\mathcal{M}_f^{H_f}(S) \sim N(\mu^*, \Sigma_n + \sigma_f^2 I).$$

The standard confidence regions for $\mu^* := E[f(S)]$ based on $\mathcal{M}_g^{H_g}(S)$, $\mathcal{M}_f^{H_g}(S)$, and $\mathcal{M}_f^{H_f}(S)$ are

$$CR_g^t = \left\{\mu \in \mathbb{R}^k : (\Sigma_n^{1/2}\mathcal{M}_g^{H_g}(S) - \mu)^T (\Sigma_n(1 + \sigma_g^2))^{-1}(\Sigma_n^{1/2}\mathcal{M}_g^{H_g}(S) - \mu) \le t\right\},$$
$$CR_{fg}^t = \left\{\mu \in \mathbb{R}^k : (\mathcal{M}_f^{H_g}(S) - \mu)^T (\Sigma_n + \sigma_{fg}^2 I)^{-1}(\mathcal{M}_f^{H_g}(S) - \mu) \le t\right\},$$
$$CR_f^t = \left\{\mu \in \mathbb{R}^k : (\mathcal{M}_f^{H_f}(S) - \mu)^T (\Sigma_n + \sigma_f^2 I)^{-1}(\mathcal{M}_f^{H_f}(S) - \mu) \le t\right\}.$$

The next theorem shows that confidence regions based on $\mathcal{M}_g^{H_g}$ have a smaller volume than those based on $\mathcal{M}_f^{H_g}$, and, for γ sufficiently small, also than those based on $\mathcal{M}_f^{H_f}$. Thus, of the three natural candidates we consider, $\mathcal{M}_g^{H_g}$ is the best mechanism for small γ.

Theorem 3.
  (**1**) *For any fixed t, the confidence regions CR_g^t, CR_{fg}^t, and CR_f^t have the same confidence level; that is, for any μ^* we have $P(\mu^* \in CR_g^t) = P(\mu^* \in CR_{fg}^t) = P(\mu^* \in CR_f^t)$.*
  (**2**) *If the mechanisms $\mathcal{M}_g^{H_g}$, $\mathcal{M}_f^{H_g}$, and $\mathcal{M}_f^{H_f}$ have the same level of $(\varepsilon, \delta, \gamma)$–RDP then $D(g)/\sigma_g = D(fg)/\sigma_{fg} = D(f)/\sigma_f$.*
  (**3**) *$Vol(CR_g^t) \le Vol(CR_{fg}^t)$, with strict inequality unless all the eigenvalues of Σ_n are equal.*
  (**4**) *For sufficiently small γ, $Vol(CR_g^t) \le Vol(CR_f^t)$, with strict inequality, unless all the eigenvalues of Σ_n are equal.*

Proof. Part (1) holds as in Theorem 1, and Part (2) holds by Lemma 3. The proof of Part (3), given in the Appendix, uses the relation $\sigma_{fg}^2 = \lambda_{max}(\Sigma_n)\sigma_g^2$ of Lemma 3, and a straightforward eigenvalue comparison. The proof of Part (4) is somewhat more involved. It uses a comparison of distribution functions of weighted sums of independent gamma random variables and a majorization argument. Details and references are given in the Appendix.

3.2 Testing Hypotheses: Likelihood-Ratio Test

With $E[f(S)] = \mu^*$ we consider the hypotheses $H_0 : \mu^* = 0$ and $H_1 : \mu^* = \eta$ and the mechanisms $\mathcal{M}_f^{H_g}(S)$ and $\mathcal{M}_g^{H_g}(S)$ defined above. If $\mathcal{M}_f^{H_g}(S)$ is observed then the rejection region R_{fg} of the likelihood-ratio test with significance level α has the form

$$R_{fg} = \{\mathcal{M}_f^{H_g}(S) : \mathcal{M}_f^{H_g}(S)^T(\Sigma_n + \sigma_{fg}^2 I)^{-1}\eta > t\},$$

where $t = \Phi^{-1}(1 - \alpha)\sqrt{\eta^T(\Sigma_n + \sigma_{fg}^2 I)^{-1}\eta}$.

If $\mathcal{M}_g^{H_g}(S)$ is observed then the testing problem becomes $H_0 : \mu^* = 0$ vs. $H_1 : \mu^* = \Sigma_n^{-1/2}\eta$, and the rejection region R_g of the likelihood-ratio test with significance level α has the form

$$R_g = \{\mathcal{M}_g^{H_g}(S) : \mathcal{M}_g^{H_g}(S)^T[(1+\sigma_g^2)I]^{-1}\Sigma_n^{-1/2}\eta > t\},$$

where $t = \Phi^{-1}(1 - \alpha)\sqrt{\frac{\eta^T \Sigma_n^{-1}\eta}{\sigma_g^2+1}}$.

Theorem 4.

    (**1**) *The rejection regions R_{fg} and R_g have the same significance level α.*

    (**2**) *If both mechanisms $\mathcal{M}_g^{H_g}$ and $\mathcal{M}_f^{H_g}$ have the same level of $(\varepsilon, \delta, \gamma)$–RDP then $D(g)/\sigma_g = D(fg)/\sigma_{fg}$.*

    (**3**) *Let $\pi(R_g)$ and $\pi(R_{fg})$ denote the power associated with the rejection regions R_g and R_{fg}, respectively; then $\pi(R_g) \geq \pi(R_{fg})$ with strict inequality, unless all the eigenvalues of Σ_n are equal.*

Proof. Part (1) is similar to Part (1) of Theorem 2. Part (2) is already given in Theorem 3. The proof of Part (3), given in the Appendix, involves a simultaneous diagonalization argument and a comparison of eigenvalues using $\sigma_{fg}^2 = \lambda_{max}(\Sigma_n)\sigma_g^2$.

4 Numerical Example

A numerical example can be found in the longer version of this paper, [32]. Space limitation prevent its presentation here.

Acknowledgements. We are grateful to Katrina Ligett and Moshe Shenfeld for very useful discussions and suggestions.

5   Appendix

Proof of **Theorem** 1 Part (3). The confidence region for the adjusted query given in (3) is an ellipsoid whose volume is given by

$$Vol(CR_\xi^t) = V_k \cdot (\sigma^2 t)^{k/2} (det[Diag(\xi)])^{-1/2},$$

where V_k is the volume of the unit ball in k dimensions. In view of Part (2) we minimize the log of the volume as a function of ξ subject to the constraint $\psi^T Diag(\xi)\psi = \psi^T \psi$. We consider the Lagrangian

$$\mathcal{L}(\xi_1, \ldots \xi_k, \lambda) = -\sum_{i=1}^k \log(\xi_i) - \lambda \Big[\sum_{i=1}^k \psi_i^2 \xi_i - \sum_{i=1}^k \psi_i^2 \Big].$$

We are minimizing a strictly convex function subject to a linear constraint. Differentiating and setting the Lagrangian to zero we readily obtain the unique minimum when ξ_i is proportional to $1/\psi_i^2$. The constraint $\psi^T Diag(\xi)\psi = \psi^T \psi$, which by Part (2) guarantees the same DP level, implies that $\xi^* = c (1/\psi_1^2, \ldots, 1/\psi_k^2)$ with $c = ||\psi||^2/k$.   □

Proof of **Theorem** 2 **Part (3)**. Note first that for $\mathcal{M}_\xi$ and $\mathcal{M}_{\xi=1}$ to have the same DP(ε, δ) level we must have

$$\Lambda(\xi) = \frac{\sqrt{\psi^T Diag(\xi)\psi}}{\sigma} = \Lambda(1) = \frac{\sqrt{\psi^T \psi}}{\sigma}.$$

The power of the rejection region R_ξ is

$$\pi(R_\xi) = P_{H_1}\left(\frac{\mathcal{M}_\xi(S)^T \xi^{1/2} \eta}{\sigma^2} > \Phi^{-1}(1-\alpha) \frac{\sqrt{\eta^T Diag(\xi)\eta}}{\sigma} \right)$$

$$= 1 - \Phi\left(\Phi^{-1}(1-\alpha) - \frac{\sqrt{\eta^T Diag(\xi)\eta}}{\sigma} \right),$$

which is increasing in $\eta^T Diag(\xi)\eta$. Thus in order to maximize the power we have to maximize $\eta^T Diag(\xi)\eta$ over ξ subject to $\psi^T Diag(\xi)\psi = \psi^T \psi$. Defining $v_i = \xi_i \psi_i^2$ the problem now is to maximize $\sum_i v_i \frac{\eta_i^2}{\psi_i^2}$ over $v_i \geq 0$, subject to $\sum_i v_i = ||\psi||^2$. Clearly the maximum is attained when $v_{j^*} = ||\psi||^2$, where $j^* = \arg\max_i(\eta_i^2/\psi_i^2)$, and $v_i = 0$ for $i \neq j^*$, completing the proof.   □

Proof of **Theorem** 3. The proof of Part (3) uses the fact that

$$Vol(CR_{fg}^t) = b_k \sqrt{det[\Sigma_n + \sigma_{fg}^2 I]} \quad \text{and} \quad Vol(CR_g^t) = b_k \sqrt{det[\Sigma_n(1 + \sigma_g^2)]},$$

where $b_k = t^{k/2} V_k$ and V_k is the volume of the k-dimensional unit ball, and the relation $\sigma_{fg}^2 = \lambda_{max}(\Sigma_n)\sigma_g^2$. The required inequality follows from the relations

$$det\left(\Sigma_n(\sigma_g^2 + 1)\right) = \prod_{i=1}^k \left[\frac{\lambda_i}{\lambda_{max}(\Sigma_n)} \sigma_{fg}^2 + \lambda_i \right] \leq \prod_{i=1}^k \left[\sigma_{fg}^2 + \lambda_i \right] = det\left(\Sigma_n + \sigma_{fg}^2 I\right),$$

where λ_i denote the eigenvalues of Σ_n.

To prove Part (4), we need the following fact, which is a special case of a result stated in [10], Proposition 2.7 and Equation (10). The last part is given by [30], p. 999, and [31], Theorem 2.2.

Fact. *Let $X_i \sim \mathcal{X}_1^2$ be i.i.d. Without loss of generality assume that $\lambda_1, \ldots, \lambda_k$ with $\lambda_i > 0$ satisfy $\overline{\lambda} := \sum_{i=1}^{k} \lambda_i / k = 1$. Define $F_\lambda(x) = P(\sum_{i=1}^{k} \lambda_i X_i \leq x)$ and let $F(x)$ denote the distribution function of $\mathcal{X}_k^2$. Then for sufficiently large x we have $F_\lambda(x) \leq F(x)$.*

More specifically, the latter inequality holds for $x > 2k$. The latter lower bound, given by [30,31], is far from being tight, as suggested by numerical computations.

Recall that $C = D(f)$ satisfies $P((2/n)\sum_{i=1}^{k} \lambda_i X_i \leq C^2) = 1 - \gamma$. For the rest of the proof set $D = D(g)$; then D is defined by $P((2/n)Y \leq D^2) = 1 - \gamma$, where $Y \sim \mathcal{X}_k^2$; see Eq. (9). For $\overline{\lambda} = 1$, which can be assumed without loss of generality, the above Fact immediately implies that $C^2 \geq D^2$ for sufficiently small γ. By the last part of the above fact, for $k = 6, 10, 20$, and 30, sufficiently small means $\gamma \leq 1 - P(Y < 12) = 0.062$ and $\gamma \leq 0.03, 0.005$, and 0.001, respectively.

***Proof* of Part (4).** As in the proof of Part (3), by Lemma 3 and then for sufficiently small γ such that $C^2 \geq D^2 \overline{\lambda}$ (where $\overline{\lambda} = 1$), we have $det\left(\Sigma_n(\sigma_g^2 + 1)\right) = \prod_{i=1}^{k}\left[\lambda_i + \frac{\lambda_i D^2}{C^2}\sigma_f^2\right] \leq \prod_{i=1}^{k}\left[\lambda_i + \lambda_i \sigma_f^2/\overline{\lambda}\right]$ and now it remains to show that the latter product is bounded above by $\prod_{i=1}^{k}[\lambda_i + \sigma_f^2] = det\left(\Sigma_n + \sigma_f^2 I\right)$. Dividing by $\prod_{i=1}^{k}\lambda_i$ and taking log, we see that the required bound is equivalent to $\sum_{i=1}^{k}\log\left[1 + \sigma_f^2/\overline{\lambda}\right] \leq \sum_{i=1}^{k}\log[1 + \sigma_f^2/\lambda_i]$. This follows from the fact that $\log[1 + \sigma_f^2/\lambda]$ is convex in λ and therefore $\sum_{i=1}^{k}\log[1 + \sigma_f^2/\lambda_i]$ is a Schur convex function; see [25]. $\qquad\square$

***Proof* of Theorem 4 Part (3).** The power of the rejection region R_{fg} is given by

$$\pi(R_{fg}) = P_{H_1}\left(\mathcal{M}_f^{H_g}(S)^T(\Sigma_n + I\sigma_{fg}^2)^{-1}\eta > \Phi^{-1}(1 - \alpha)\sqrt{\eta^T(\Sigma_n + I\sigma_{fg}^2)^{-1}\eta}\,\right)$$

$$= 1 - \Phi\left(\Phi^{-1}(1 - \alpha) - \sqrt{\eta^T(\Sigma_n + I\sigma_{fg}^2)^{-1}\eta}\right) \qquad (10)$$

$$= 1 - \Phi\left(\Phi^{-1}(1 - \alpha) - \sqrt{\eta^T(\Sigma_n + I\lambda_{max}(\Sigma_n)\sigma_g^2)^{-1}\eta}\right).$$

Likewise $\pi(R_g) = 1 - \Phi\left(\Phi^{-1}(1 - \alpha) - \sqrt{\frac{\eta^T\Sigma_n^{-1}\eta}{\sigma_g^2 + 1}}\right)$. Therefore, $\pi(R_g) \geq \pi(R_{fg})$ if and only if

$$\frac{\eta^T\Sigma_n^{-1}\eta}{\sigma_g^2 + 1} \geq \eta^T(\Sigma_n + I\lambda_{max}(\Sigma_n)\sigma_g^2)^{-1}\eta.$$

Diagonalizing the two matrices $(1 + \sigma_g^2)\Sigma_n$ and $\Sigma_n + I\lambda_{max}(\Sigma_n)\sigma_g^2$ by the common orthogonal matrix of their eigenvectors, we see that the diagonal terms, that is, the eigenvalues $(\sigma_g^2 + 1)\lambda_i$ of the first matrix are less than or equal to those of the second one, $\lambda_i + \lambda_{max}\sigma_g^2$. It follows that $\frac{\Sigma_n^{-1}}{\sigma_g^2+1} \succeq (\Sigma_n + I\lambda_{max}(\Sigma_n)\sigma_g^2)^{-1}$, where $A \succeq B$ means that $A - B$ is nonnegative definite; see [19], Chapter 4. The result follows.

References

1. Anderson, T.W.: An Introduction to Multivariate Statistical Analysis. Wiley, New York (2003)
2. Ashtiani, H., Liaw, C.: Private and polynomial time algorithms for learning Gaussians and beyond. arXiv preprint arXiv:2111.11320 (2021)
3. Awan, J., Slavkovć, A.: Differentially private uniformly most powerful tests for binomial data. In: Advances in Neural Information Processing Systems, vol. 31 (2018)
4. Balle, B., Wang, Y.X.: Improving the Gaussian mechanism for differential privacy: analytical calibration and optimal denoising. In: International Conference on Machine Learning, pp. 394–403. PMLR, July 2018
5. Blum, C.B., Dell, R.B., Palmer, R.H., Ramakrishnan, R., Seplowitz, A.H., Goodman, D.S.: Relationship of the parameters of body cholesterol metabolism with plasma levels of HDL cholesterol and the major HDL apoproteins. J. Lipid Res. **26**(9), 1079–1088 (1985)
6. Brown, G., Gaboardi, M., Smith, A., Ullman, J., Zakynthinou, L.: Covariance-aware private mean estimation without private covariance estimation. In: Advances in Neural Information Processing Systems, vol. 34 (2013)
7. Canonne, C.L., Kamath, G., McMillan, A., Smith, A., Ullman, J.: The structure of optimal private tests for simple hypotheses. In: Proceedings of the 51st Annual ACM SIGACT Symposium on Theory of Computing, pp. 310–321 (2019)
8. Castelli, W.P., et al.: Distribution of triglyceride and total, LDL and HDL cholesterol in several populations: a cooperative lipoprotein phenotyping study. J. Chronic Dis. **30**(3), 147–169 (1977)
9. Degue, K.H., Le Ny, J.: On differentially private Gaussian hypothesis testing. In:2 018 56th Annual Allerton Conference on Communication, Control, and Computing (Allerton), pp. 842–847. IEEE, October 2018
10. Diaconis, P., Perlman, M.D.: Bounds for tail probabilities of weighted sums of independent gamma random variables. Lecture Notes-Monograph Series, pp. 147–166 (1990)
11. Dwork, C., McSherry, F., Nissim, K., Smith, A.: Calibrating noise to sensitivity in private data analysis. In: Halevi, S., Rabin, T. (eds.) TCC 2006. LNCS, vol. 3876, pp. 265–284. Springer, Heidelberg (2006). https://doi.org/10.1007/11681878_14
12. Dwork, C., Roth, A.: The algorithmic foundations of differential privacy. Found. Trends Theor. Comput. Sci. **9**(3–4), 211–407 (2014)
13. Edmonds, A., Nikolov, A., Ullman, J.: The power of factorization mechanisms in local and central differential privacy. In: Proceedings of the 52nd Annual ACM SIGACT Symposium on Theory of Computing, pp. 425–438 (2020)
14. Efron, B.: Minimum volume confidence regions for a multivariate normal mean vector. J. R. Stat. Soc. Ser. B (Stat. Methodol.) **68**(4), 655–670 (2006)
15. Gaboardi, M., Lim, H., Rogers, R., Vadhan, S.: Differentially private chi-squared hypothesis testing: goodness of fit and independence testing. In: International Conference on Machine Learning, pp. 2111–2120. PMLR, June 2016

16. Ghosh, A., Roughgarden, T., Sundararajan, M.: Universally utility-maximizing privacy mechanisms. SIAM J. Comput. **41**(6), 1673–1693 (2012)
17. Gupte, M., Sundararajan, M.: Universally optimal privacy mechanisms for minimax agents. In: Proceedings of the Twenty-Ninth ACM SIGMOD-SIGACT-SIGART Symposium on Principles of Database Systems, pp. 135–146 (2010)
18. Hall, R., Rinaldo, A., Wasserman, L.: Random differential privacy. J. Privacy Confident. **4**(2), 43–59 (2012)
19. Horn, R.A., Johnson, C.R.: Matrix Analysis. Cambridge University Press, Cambridge (2012)
20. Karwa, V., Slavković, A.: Inference using noisy degrees: differentially private β-model and synthetic graphs. Ann. Stat. **44**(1), 87–112 (2016)
21. Karwa, V., Vadhan, S.: Finite sample differentially private confidence intervals. In: 9th Innovations in Theoretical Computer Science Conference, vol. 94, p. 44. Schloss Dagstuhl-Leibniz-Zentrum fuer Informatik (2018)
22. Li, C., Hay, M., Rastogi, V., Miklau, G., McGregor, A.: Optimizing histogram queries under differential privacy. arXiv preprint arXiv:0912.4742 (2009)
23. Li, C., Miklau, G.: An adaptive mechanism for accurate query answering under differential privacy. Proc. VLDB Endow. **5**(6) (2002)
24. Li, C., Miklau, G., Hay, M., McGregor, A., Rastogi, V.: The matrix mechanism: Optimizing linear counting queries under differential privacy. VLDB J. **24**(6), 757–781 (2015)
25. Marshall, A.W., Ingram Olkin, B.C.: Arnold Inequalities: Theory of Majorization and Its Applications. Springer, New York (2011). https://doi.org/10.1007/978-0-387-68276
26. McKenna, R., Miklau, G., Hay, M., Machanavajjhala, A.: Optimizing error of high-dimensional statistical queries under differential privacy. Proc. VLDB Endow.-**11**(10), 1206–1219 (2018)
27. Mohammady, M., et al.: R2DP: a universal and automated approach to optimizing the randomization mechanisms of differential privacy for utility metrics with no known optimal distributions. In: Proceedings of the 2020 ACM SIGSAC Conference on Computer and Communications Security, pp. 677–696, October 2020
28. Qureshi, N.A., Suthar, V., Magsi, H., Sheikh, M.J., Pathan, M., Qureshi, B.: Application of principal component analysis (PCA) to medical data. Ind. J. Sci. Technol. **10**(20), 1–9 (2017)
29. Rogers, R., Kifer, D.: A new class of private chi-square hypothesis tests. In: Artificial Intelligence and Statistics, pp. 991–1000. PMLR (2017)
30. Roosta-Khorasani, F., Székely, G.J.: Schur properties of convolutions of gamma random variables. Metrika **78**(8), 997–1014 (2015). https://doi.org/10.1007/s00184-015-0537-9
31. Roosta-Khorasani, F., Széekely, G.J., Ascher, U.M.: Assessing stochastic algorithms for large scale nonlinear least squares problems using extremal probabilities of linear combinations of gamma random variables. SIAM/ASA J. Uncert. Quant. **3**(1), 61–90 (2015)
32. Shoham, T., Rinott, Y.: Adjusting queries to statistical procedures under differential privacy. arXiv preprint arXiv:2110.05895 (2021)
33. Solea, E. Differentially private hypothesis testing for normal random variables (2014)
34. Wang, Y., Kifer, D., Lee, J., Karwa, V.: Statistical approximating distributions under differential privacy. J. Privacy Confident. **8**(1) (2018)

Towards Integrally Private Clustering: Overlapping Clusters for High Privacy Guarantees

Vicenç Torra[(✉)]

Department of Computing Science, Umeå University, Umeå, Sweden
vtorra@ieee.org

Abstract. Privacy for re-identification, k-anonymity, and differential privacy are the main privacy models considered in the literature on data privacy. We introduced an alternative privacy model called integral privacy, that can be seen as a model for computations avoiding membership inference attacks, as well as other inferences, on aggregates and computations from data (e.g., machine learning models and statistics). In previous papers we have shown how we can compute integrally private statistics (e.g., means and variance), decision trees, and regression.

In this paper we introduce clustering with overlapping clusters. The goal is to produce integrally private clusters. We formulate the problem in terms of an optimization problem, and provide a (sub-optimal) solution based on genetic algorithms.

Keywords: Data privacy · Statistical disclosure control · Clustering · Integral privacy · Overlapping clustering

1 Introduction

Differential privacy [4] has been established as a convenient privacy model when privacy is related to computations from databases. That is, we have a database X and we need to compute $f(X)$ for a given function f. Then, we usually provide a function K_f instead of f that is differentially private. Functions K_f satisfying differential privacy roughly correspond to functions where adding or removing a record from a database do not change much the output. A large number of differential privacy mechanisms have been defined for different families of functions f. The definition of differential privacy includes a parameter ϵ that controls the privacy level. The larger the ϵ, the lower the privacy level.

Another well known privacy mechanism is k-anonymity [8]. It provides anonymity for databases. The scenario is as follow. We have a database X that needs to be shared with a third party. In this case, we provide this third party with a sanitized or anonymized database X'. A database X' is k-anonymous when for each row (record) in X', there are other $k-1$ indistinguishable rows (records). Properly speaking, indistinguishability is only for quasi-identifiers.

© Springer Nature Switzerland AG 2022
J. Domingo-Ferrer and M. Laurent (Eds.): PSD 2022, LNCS 13463, pp. 62–73, 2022.
https://doi.org/10.1007/978-3-031-13945-1_5

That is, the set of attributes that can make at least a record unique and can be at the hands of intruders. In k-anonymity, k is a parameter of the model. The larger the k, the larger the privacy level.

Different privacy models have their pros and cons. For example, k-anonymity requires assumptions on the attributes that intruders can access to, and the side information they can use to attack X'. Differential privacy does not have this latter drawback, but this is usually at the cost of less quality on the output. Both privacy models need to take into account whether records in the database are independent or there are dependencies. k-Anonymity is for database sharing and, thus, any computation f can be computed or estimated from X'. In contrast, differential privacy requires knowledge on the function f to be computed. To improve the quality of computations using differential privacy, some relaxations have been defined that provide less privacy guarantees. E.g., (ϵ, δ)-differential privacy [3]. In addition, some companies implement differential privacy with large values of the parameter ϵ, so that computations have large utility, which can lead to some high levels of disclosure.

Integral privacy [10,11] was introduced as a new privacy model with a goal similar to the one of differential privacy. That is, privacy for computations $f(X)$. This privacy model ressembles k-anonymity in the sense that a function $f(X)$ is integrally private (k-anonymous integral privacy) if there is a set of at least k different databases that can generate $f(X)$. In other words, if $y = f(X)$, we consider the set of generators of y. I.e., possible databases X_i that can produce y. That is, $y = f(X_i)$. The value y is private enough if there are at least k such databases. The definition is more strict than that, because it is also required that the databases do not share records to avoid e.g. membership attacks. To achieve integral privacy we can proceed as follows [9]. First, we sample a database multiple times, and produce multiple subsets of this database to generate a set of models. From these models, then we select the ones that are recurrent and satisfy the privacy constraints.

In this paper, we provide a completely different approach. Our goal is to achieve integral privacy by design. We focus on clustering (i.e., we consider f to be a clustering algorithm). Then, we formalize an optimization problem for clustering which, when solved, provides independent sets of records to define the same clusters. This is to implement privacy by design. Nevertheless, as we discuss later, the approach does not always guarantees this for an arbitrary data set. In this paper, we describe a way to solve our clustering algorithm using a combination of a clustering algorithm and genetic algorithms.

The structure of the paper is as follows. In Sect. 2 we present some concepts we need later. Then, in Sect. 3 we formalize the clustering problem. We call it κ-centroid c-means. Then we describe in Sect. 4 the experiments and results. The paper finishes with some conclusions.

2 Preliminaries

In this section we review the main topics we need in this paper.

2.1 Integral Privacy

Integral privacy is defined for a function or algorithm f to be computed from a database. The application of f to X returns a computation or model G. This result G is integrally private if there are enough databases X_i that generate the same computation or model G. Given a population P, and some background knowledge S^* an intruder has on P, we define $Gen^*(G, S^*) = \{S' \setminus S^* | S^* \subseteq S' \subseteq P, f(S') = G\}$ as the set of generators of G. That is, $Gen^*(G, S^*)$ is the set of possible databases that can produce G when f is applied to them. k-Anonymous integral privacy holds when there are at least k databases and they do not share any common record. Note that a common record would imply that a membership inference attack can be successful.

Definition 1. *Let P represent the data of a population, let f be a function to compute from databases $S \subseteq P$ into $\mathcal{G}$. Let $G \in \mathcal{G}$, let $S^* \subseteq P$ be some background knowledge on the data set used to compute G, let $Gen(G, S^*)$ represent the possible databases that can generate G and are consistent with the background knowledge S^*. Then, k-anonymous integral privacy is satisfied when the set $Gen(G, S^*)$ contains at least k-elements and*

$$\cap_{S \in Gen^*(G, S^*)} S = \emptyset.$$

2.2 k-Anonymity, Microaggregation, and MDAV

It is well known that for a database X, k-anonymity holds when for each combination of quasi-identifiers there are at least k-indistinguishable records. Recall that a quasi-identifier is a set of attributes that makes at least a record unique. There are two main families of methods to produce a k-anonymous database from an original unmasked database. They are the methods based on generalization and the methods based on clustering (microaggregation).

Microaggregation [1,2] methods build a set of clusters, each one with at least k records (and at most $2k$ records), and then replace the values of these records by the centroids or cluster centers. These cluster centers are a kind of aggregation of the original data (e.g., the mean of the records associated to a cluster). In contrast, generalization methods change the level of detail of these records, replacing their original value by a value that represents the information of the records in the cluster. E.g., an interval for a set of numerical data, a disjunction of terms for a set of categorical data or a more general term if exists (e.g. replacing town by county or region).

Several heuristic algorithms have been constructed for microaggregation. Heuristic methods usually build heuristically the clusters so that the condition of having at least k records hold, and, then, they produce the protected database

X' from X replacing the records by the arithmetic mean of the records associated to the cluster. In this paper we will use MDAV [2], a well known algorithm extensively used in the literature. We will use its definition for numerical data, as in our experiments we will use only databases with numerical data.

2.3   Genetic Algorithms

Genetic algorithms [6,7] is a meta-heuristic search method. It is suitable method for large search spaces. They are inspired in evolution, and terminology is based on the one in biology. Genetic algorithms are defined as an iterative process taking a number of epochs (i.e., iterations). In each epoch, we consider a population (i.e., a set of possible solutions) of chromosomes (i.e., a codification of a particular solution). In each epoch we evaluate each chromosome in terms of a fitness function (i.e., a proxy of the objective function and of the constraints, if any) and we select the chromosomes that better fit and then create a new population by means of genetic operators on these chromosomes that better fit. Here genetic operators are functions that given a codification of one or two possible solutions builds a new one. Mutation and cross-over are the two most common operators. Mutation is about changing the value in one of the gens (e.g., the value of one of the variables of a possible solution) and cross-over is about mixing two possible solutions (e.g., taking the values of some of the variables in one possible solution and the values of the other variables in the other possible solution).

3 κ-Centroid c-Means

Our goal is to cluster data into a set of $c \times \kappa$ clusters. We will use the terms macro-cluster and micro-cluster to distinguish two types of clusters. We have c macro-clusters each with κ micro-clusters. Each micro-cluster is represented by κ centroids. We want these micro-clusters to overlap as much as possible and macro-clusters be as indistinguishable as possible.

For a given macro-cluster, all κ centroids should be similar enough, and they are computed by means of a set of elements that fall into their area of influence. That is, they are computed by the objects associated to the micro-cluster. Because of that, objects in X will be assigned to a micro-cluster (in the same way as for the c-means). We call the area of influence or the set of objects attached to a particular micro-cluster, a region. All centroids associated to a macro-cluster are similar one another (i.e., their distance is bounded) and the number of elements in the area of influence of all centroids of the same cluster should be similar. In this way, we can observe each cluster as a set of strongly overlapping clusters, whose centroids are strongly similar. In other words, we want that the regions associated to all micro-clusters of the same macro-cluster are basically the same region.

3.1 Formalization

Let us consider objects x in the reference set X. We build c macro-clusters, each with micro-clusters with their corresponding κ centroids. Let v_{jk} for $j = 1, \ldots, c$ and $k = 1, \ldots, \kappa$ be the kth centroid of the jth cluster. Let $\mu_{jk}(x)$ represent the membership of x to the kth centroid of the jth cluster. We assume $\mu_{jk} \in \{0, 1\}$.

To formalize the requirement that centroids of the same cluster are similar, we use δ as the maximum distance between centroids of the same cluster. That is, v_{jk_1} and v_{jk_2} are at most at a distance δ. In addition, we expect that all centroids of the same cluster have a similar number of objects. This is to force the area of influence to be of similar size (where size is the cardinality of elements). We use A as the maximum difference allowed between the number of objects associated to a centroid.

Taking into account this notation, we formalize the κ-centroid c-means as follows:

$$\min J(\mu, v) = \sum_{j=1}^{c} \sum_{k=1}^{\kappa} \sum_{x \in X} \mu_{jk}(x) \|x - v_{jk}\|^2$$

$$\text{subject to } \sum_{j=1}^{c} \sum_{k=1}^{\kappa} \mu_{jk}(x) = 1 \text{ for all } x \in X$$

$$\left| \sum_{x \in X} \mu_{jk_1}(x) - \sum_{x \in X} \mu_{jk_2}(x) \right| \leq A$$
$$\text{for all } j \in \{1, \ldots, c\}, \ k_1 \neq k_2 \in \{1, \ldots, \kappa\}$$

$$\|v_{jk_1} - v_{jk_2}\|^2 \leq \delta$$
$$\text{for all } j \in \{1, \ldots, c\}, \ k_1 \neq k_2 \in \{1, \ldots, \kappa\}$$

$$\mu_{jk}(x) \in \{0, 1\}$$
$$\text{for all } j \in \{1, \ldots, c\}, \ k \in \{1, \ldots, \kappa\}, \text{ and } x \in X$$

In this definition, the function to minimize is naturally the distance of objects to cluster centroids. The summations are over all cluster centroids (j and k) and over all objects (x). The constraints included in the problem are (i) that each element belongs to exactly one cluster region (so, membership for all clusters and regions adds to one), (ii) the number of elements in each region of a macro-cluster is similar and at most with a difference of A (this is, of course, for all macro-clusters j and for all micro-clusters k), (iii) cluster centroids belonging to the same cluster are at a distance of at most δ (this is also, of course, for all clusters j and for all regions k), (iv) membership of objects to clusters is binary.

3.2 Properties

The goal of the problem above is to construct c clusters, and to provide integral privacy by means of having at least κ alternative sets of records that return these same clusters. These sets of records need to be disjoint to satisfy κ-anonymous integral privacy.

As we have κ records for each of the c clusters, we can easily build $c \times \kappa$ different partitions that, in principle, produce the same clustering results. That is, we can take one of the clusters for c_1, another cluster for c_2, ..., and another for c_c. The union of the corresponding records produce a dataset X_i. As there are k different and disjoint sets for each macro-cluster, appropriate selection of sets produce disjoint X_i.

Let k' for $k' = 1, \ldots, \kappa$ denote one of the κ selections and $X_{k'}$ the resulting set. This selection needs to choose one of the micro-clusters for each of the macro-clusters.

Let $k'_{k',j'}$ for $j' = 1, \ldots, c$ denotes the micro-cluster for macro-cluster j' for the k'th selection. Naturally, $k'_{k',j'} \in \{1, \ldots, \kappa\}$.

In order that two selections do not take the same micro-cluster, we need that for any cluster j' we have that the micro-clusters differ. I.e., for any cluster j' we have $k'_{k',j'} \neq k'_{k'',j'}$ for $k' \neq k''$. This is easy to implement, we can just define $k'_{k'',j'}$ for $j' = 1, \ldots, c$ as a permutation of $1, \ldots, c$. Then, we define a partition of X into κ parts in which the $X_{k'}$ part is the union of the records associated to the c clusters denoted by $k'_{k',j'}$ for $j' = 1, \ldots, c$.

This construction satisfies the constraints of integral privacy, in the sense that there are κ parts and the intersection of these parts is empty.

Unfortunately, while the construction satisfies the constraints of integral privacy in the sense that we have κ disjoint sets, we cannot mathematically prove that the (optimal) solution of the optimization problem will satisfy integral privacy. This is so because of the following.

- The optimization problem formulated above with $\kappa \neq 1$ and the full dataset X does not necessarily lead to the same solution as the *reduced* problem with any of the sets $X_{k'}$ for $k' = 1, \ldots, \kappa$ setting their own $\kappa = 1$. For example, if we solve the problem with $\kappa = 5$, and then define X_1 as the union of 5 different micro-clusters and solve the optimization problem without overlapping, we may not get the same clusters and centroids.

Some considerations apply here. First, separated enough clusters can provide these privacy guarantees, as optimal solutions for both $X_{k'}$ and X will be the same. It is easy to see that we can construct artificial data sets with these properties. Second, computational solutions for clustering algorithms do not usually provide global optima but local optima. Similarly, our solution for the optimization problem formalized in Sect. 3.1 is not optimal. Therefore, it may be enough to find that the solution of the κ-centroid c-means can produce $X_{k'}$ that are local optima of the corresponding clustering algorithm. These considerations need to be further studied and discussed.

4   Experiments

We have used a combination of standard clustering and genetic algorithms to solve the optimization problem described above. In this section we describe our implementation, and the experiments we have performed.

4.1 Solving the Optimization Problem

Our main tool to solve the optimization problem has been genetic algorithms [6,7]. This requires the representation of the solution using a population of chromosomes, the definition of operations on the chromosomes, and a fitness function. We also need a strategy for the initialization. We detail these elements below.

- Chromosomes. Each chromosome represents the membership of each $x \in X$ into its assigned cluster. Then, a population represents a set of possible assignments. Our representation is based on a list of integers (i.e., we have an integer for each $x \in X$), where an integer represents a micro-cluster.
- Operations. In relation to the operations on chromosomes to create a new population, we have considered three operations: structural mutation, cluster swapping, and structural crossover. Structural mutation moves an element from one cluster to another one. Cluster swapping takes the chromosome and two clusters, and then swaps elements x_i from cluster c_i to cluster c_j with a given probability. Structural crossover takes two chromosomes c_1 and c_2, and incorporates a given cluster c present in c_2 in c_1. This requires disassembling the original cluster in c_1 and relocating the elements (at random to other clusters). In addition, some chromosomes from the previous population are transferred to the new population without change.
- Fitness function. Three main metrics computed from a chromosome have been used to evaluate its quality. They are: (1) the error of the objective function, (2) the difference on the cardinalities in the same macro-cluster, (3) the difference on the norms of the centroids in the same macro-cluster. We have tested several fitness functions. In the experiments reported below we have used the following has permitted to solve some of the problems considered. It is based on a scoring system, where the score increases when some conditions are fulfilled. We have scores for the cardinalities ((2) above) and for the norms ((3) above). For the cardinalities we use the following:
  - scoreCard = 100 - min(mean(Δc),100)
  - if ($mean(\Delta C) \leq 2A$), then scoreCard = scoreCard+100
  - if ($mean(\Delta C) \leq A$), then scoreCard = scoreCard+100
  - if ($max(\Delta C) \leq A$), then scoreCard = scoreCard+100

  Similarly, for the norm we have:
  - scoreNorm = 100 - min(mean(Δ Norms), 100)
  - if ($mean(\Delta DNorms) <= 2 * \Delta$), then scoreNorm = scoreNorm + 100
  - if ($mean(\Delta DNorms) <= \Delta$), then scoreNorm = scoreNorm + 100
  - if ($max(\Delta DNorms) <= \Delta$), then scoreNorm = scoreNorm + 100

Table 1. Values of the objective function at initialization for different values of κ and c using (i) random initialization of records to clusters and (ii) initialization based on MDAV. Parameters of the optimization problem $\delta = 0.0005$, $A = 5$.

κ	c	OF		δ'	
		Random	MDAV	Random	MDAV
2	3	5.76	5.56	0.000495	0.000294
3	5	15.08	14.02	0.001851	0.000962
4	10	13.52	13.64	0.020078	0.003188

We combine the three values as follows. First, as the error of the objective function is to minimize and the other scores are to maximize, we compute $(m_1 - error)$, where $m_1 = 100$ and then we add the three terms. The exact way to combine these three elements was done by trial and error.

- Initialization. We have considered two different strategies for assigning an initial population. A random initialization of the population lead to extremely bad results, which were not converging to a solution satisfying the constraints (except for unacceptable values A and Δ). Because of that we developed a different initialization strategy. It is based on k-anonymous clustering. In particular, we first run MDAV on the data which produces an assignment of elements to clusters (say macro-clusters). Then, we assign each record in a given macro-cluster to any of its corresponding micro-clusters following a uniform distribution. Different assignments to micro-clusters are used in each chromosome of the population. This initialization leads to better results in the sense that the objective function is better and the constraints are sooner satisfied. Using MDAV in the initialization instead of using another clustering algorithm we have that at least in the initialization, all clusters have a similar number of records.

4.2 Datasets

We have considered the dataset `Concrete` [12,13] from the UCI repository, and the CASC dataset [5].

The datasets have been normalized (i.e., column mean has been substracted to each value and then divided by standard deviation). Our clustering approach has been applied using different parameters κ and c.

4.3 Parameters

We apply our algorithm considering the following parameters: $\delta = 0.0005$, $A = 5$. We consider different number of macro-clusters and micro-clusters. That is, different values for c and κ. In particular, we consider the pairs listed in Table 1.

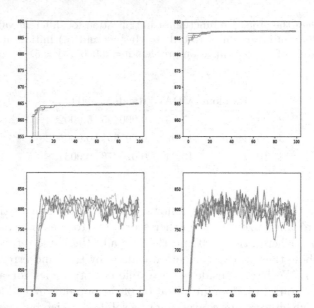

Fig. 1. Case of $c = 2$ and $\kappa = 3$ for Concrete dataset. Best (top) and mean (bottom) fitness functions for random initialization and for MDAV initialization (5 independent executions, 100 epochs).

We apply genetic algorithms with the following parameters. Probability of structural mutation $prSM = 0.4$, probability of structural crossover $prSC = 0.2$, probability of cluster swapping $prCS = 0.2$, probability of creating a copy of the best chromosome $prBC = 0.1$, number of epochs 100. For cluster swapping we need an additional probability of actually doing a swapping of the elements in the selected chromosome. This is $prCSe = 0.5$. Selection of the clusters to swap is done by uniform distribution on all possible clusters.

4.4 Results

Table 1 show the effects of the two initialization approaches on both the objective function (columns OF) and difference on cluster centers (columns δ'). We can see that for both indices the MDAV initialization leads to better results. The improvement on the objective function is not so relevant (e.g., from 5.76 to 5.56, or even slightly worsening in the case of the pair $(4, 10)$ from 13.52 to 13.64. Nevertheless, the change on the difference between cluster centers is significant as initialization using MDAV reduces it to about a half.

Figure 1 display the results of the fitness function for the case of two macro-clusters $(c = 2)$ each with three micro-clusters $(\kappa = 3)$. We have 5 executions of the genetic algorithms for each initialization strategy. After 100 epochs, in the case of using MDAV, we have that all 5 executions lead to a feasible solution for $c = 2$ and $\kappa = 5$ with $A = 5$ and $\delta = 0.0005$. In contrast, only two of the five executions using random initialization lead to a feasible solution. Table 2

Table 2. Best values of A and δ achieved when looking for the best solution of $c = 2$, $\kappa = 5$, $A = 5$, $\delta = 0.0005$.

Best A random	Best A MDAV	δ random	δ MDAV
3.0	1.0	0.000595	0.000272
12.0	3.0	0.000412	0.000324
4.0	3.0	0.001112	0.000285
4.0	5.0	0.000374	0.000158
8.0	5.0	0.000094	0.000329

displays the solutions obtained. We can observe that we can achieve (with random) solutions that violate the constraint related to δ and the one related to A independently.

Figure 1 shows the results for random initialization (left) and MDAV initialization (right). We can observe the difference on the objective functions, better for MDAV. We also display the value of the fitness function of the best chromosome found so far, and the mean value of the fitness function. The value of the best fitness function is naturally, always non-decreasing, and it does not change much after the first initial epochs.

We have tried to solve the same problem with $c = 3$ and $A = 5$, but the best solution obtained was with $A' = 8$ and $\delta' = 0.000933$ after 4000 epochs. For $c = 3$ and $A = 4$ we also obtain (with 4000 epochs) a solution with $A' = 8$ and in this case δ satisfies the constraint and is, in addition, very small $\delta' = 3.9568 \cdot 10^{-08}$. A still larger problem is $c = 4$ and $\kappa = 10$. The best solution is with $A' = 11$ and $\delta' = 0.00542$, but no optimal solution is found in 100 epochs (Fig. 2 show the fitness functions of 5 executions).

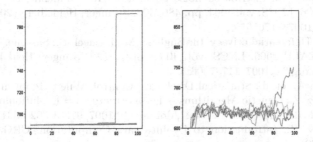

Fig. 2. Case of $c = 4$ and $\kappa = 10$ for Concrete dataset. Best (left) and mean (right) fitness functions for MDAV initialization (5 independent executions, 100 epochs).

5   Conclusions

In this paper we have introduced κ-centroid c-means as a way to provide integral privacy. We have formalized the problem as an optimization problem with constraints related to the distance among cluster centroids (of micro-cluster) in a macro-cluster, and related to the number of records associated to a cluster. We have discussed that the solution of this optimization problem for clearly separated clusters will provide a way to define integrally private clusters. For non-clearly separated data, this is not necessarily the case. As both our implementation of the approach and usual methods for standard clustering produce sub-optimal solutions, we consider that further work is needed to evaluate these solutions. Our algorithm produces solutions that are feasible for small number of clusters. We have applied the algorithm to larger number of clusters, but then we need much more number of epochs and the solutions have major difference on the number of records. i.e., parameters A in the optimization problem. We plan to study this problem better to solve the optimization problem for larger datasets and larger number of micro-clusters and macro-clusters.

Acknowledgements. This work was partially supported by the Wallenberg AI, Autonomous Systems and Software Program (WASP) funded by the Knut and Alice Wallenberg Foundation

References

1. Domingo-Ferrer, J., Mateo-Sanz, J.M.: Practical data-oriented microaggregation for statistical disclosure control. IEEE Trans. Knowl. Data Eng. **14**(1), 189–201 (2002)
2. Domingo-Ferrer, J., Torra, V.: Ordinal, continuous and heterogeneous k-anonymity through microaggregation. Data Mining Knowl. Disc. **11**(2), 195–212 (2005)
3. Dwork, C., Kenthapadi, K., McSherry, F., Mironov, I., Naor, M.: Our data, ourselves: privacy via distributed noise generation. In: Vaudenay, S. (ed.) EUROCRYPT 2006. LNCS, vol. 4004, pp. 486–503. Springer, Heidelberg (2006). https://doi.org/10.1007/11761679_29
4. Dwork, C.: Differential privacy. In: Bugliesi, M., Preneel, B., Sassone, V., Wegener, I. (eds.) ICALP 2006. LNCS, vol. 4052, pp. 1–12. Springer, Heidelberg (2006). https://doi.org/10.1007/11787006_1
5. Hundepool, A., et al.: Statistical Disclosure Control. Wiley, Hoboken (2012)
6. Michalewicz, Z.: Genetic Algorithms + Data Structures = Evolutionary Programs. Springer, Heidelberg (2000). https://doi.org/10.1007/978-3-662-03315-9
7. Michalewicz, Z.: Introduction to Evolutionary Computation. CRC Press, Boca Raton (2000)
8. Samarati, P.: Protecting respondents' identities in microdata release. IEEE Trans. Knowl. Data Eng. **13**(6), 1010–1027 (2001)
9. Senavirathne, N., Torra, V.: Integrally private model selection for decision trees. Comput. Secur. **83**, 167–181 (2019)
10. Torra, V., Navarro-Arribas, G.: Integral privacy. In: Foresti, S., Persiano, G. (eds.) CANS 2016. LNCS, vol. 10052, pp. 661–669. Springer, Cham (2016). https://doi.org/10.1007/978-3-319-48965-0_44

11. Torra, V., Navarro-Arribas, G., Galván, E.: Explaining recurrent machine learning models: integral privacy revisited. In: Domingo-Ferrer, J., Muralidhar, K. (eds.) PSD 2020. LNCS, vol. 12276, pp. 62–73. Springer, Cham (2020). https://doi.org/10.1007/978-3-030-57521-2_5
12. Yeh, I.-C.: Modeling of strength of high performance concrete using artificial neural networks. Cement Concr. Res. **28**(12), 1797–1808 (1998)
13. Yeh, I.-C.: Analysis of strength of concrete using design of experiments and neural networks. J. Mater. Civil Eng. ASCE **18**(4), 597–604 (2006)

Tabular Data

Perspectives for Tabular Data Protection – How About Synthetic Data?

Felix Geyer, Reinhard Tent, Michel Reiffert, and Sarah Giessing[✉]

Federal Statistical Office of Germany, 65180 Wiesbaden, Germany
{Felix.Geyer,Reinhard.Tent,Michel.Reiffert,
Sarah.Giessing}@destatis.de

Abstract. The advance of small-scale analysis and user demand driven tabulations bring cell suppression based SDC to its limits. Perturbative SDC strategies could be flexible alternatives. The present paper compares results obtained by first experiments with synthetic data generated by the Synthpop package [13] on one hand to those of a more traditional input perturbation approach, i.e. targeted record swapping [19], and on the other hand to those of the Cell Key Method [6, 8, 12] which adds noise directly to tabular outputs.

1 Introduction

Developing SDC strategies allowing to release detailed geographic grid level aggregates for an administrative data set involving a large variety of highly sensitive, partially skewed continuous variables is a challenge. The challenge becomes especially demanding when the same aggregates shall be released also at various administrative and non-administrative geographies, like at municipality and at grid level. This holds in particular in cases where releases involve a huge variety of mostly user demand driven tabulations. When those tables may present aggregates of sensitive variables they must be protected properly. But when tables can be defined by crossings of some of perhaps numerous discrete variables of a dataset, maybe even along with crossings defined by discretized versions of the continuous variables, the "space" of potential output to protected in a consistent way can become extremely complex.

In such a situation, SDC strategies relying solely on primary and secondary cell suppression are especially prone to differencing and linking risks, because a fully controlled coordination of secondary cell suppression would be enormously complex and in fact practically impossible.

In this paper we therefore look at perturbative methods as alternative to, or maybe even to be used along with cell suppressions within a hybrid SDC concept. We wonder, if synthetic data methods as introduced by [14, 18], or [16] can be a realistic alternative as disclosure control concept. This means, a disclosure concept not for public or scientific use microdata sets, but when those data shall be used as basis for producing the official outputs of NSIs. To that aim and with inspiration from [2] (and a related workshop), we used the Synthpop package [13] for some first experiments. We also tried a hybrid approach mixing original with partially synthetic (c.f. [10, 15]) records generated with the

© Springer Nature Switzerland AG 2022
J. Domingo-Ferrer and M. Laurent (Eds.): PSD 2022, LNCS 13463, pp. 77–91, 2022.
https://doi.org/10.1007/978-3-031-13945-1_6

CART approach [1, 17] of the Synthpop package [13]. We compare the results to those obtained by a targeted record swapping method [9, 19] on one hand, and to the output perturbation approach of the Cell Key Method [6, 8, 12] on the other hand. Notably, it is clear from the beginning that output perturbation cannot generally be outperformed by input perturbation, because with output perturbation it is much easier to select parameters yielding acceptable risk/utility balance. On the other hand, output perturbation has certain conceptual disadvantages (like non-additivity) which is why for example under organisation aspects and in particular from IT-perspective input perturbation is sometimes considered as tempting SDC alternative.

In Sect. 2 we introduce the methods under consideration. Section 3 describes our test data and test scenarios, also proposing ideas for partial or hybrid implementations. Information loss and disclosure risk measures for evaluation will be introduced in Sect. 4. After presentation and discussion of test results (Sect. 5) the paper finishes with a conclusive summary.

2   Recalling the Methods under Consideration

This paper compares results obtained by synthetic data generation, c.f. Sect. 2.1, and by Targeted Record Swapping (Sect. 2.2) to random noise implemented by a Cell Key Method (CKM) (Sect. 2.3). Technical details of the setup of those methods used in the present study are described in Sects. 3.2, 3.3 and 3.4.

2.1   Synthetic Data

The idea of synthetic data was first introduced in [18], followed up and further developed by [14–16]. Using an approach building on concepts of [21] leads to a flexible method that does not rely on the assumption of multivariate normality. It specifies for each variable to be synthesized a conditional distribution, conditional on the other variables. The outcome will then of course depend on the order of those variables. The Classification and Regression Tree (CART) machine learning method developed by [1] offers a non-parametric approach (see also [17] and [3]). The synthpop package [13] integrates algorithms implementing those concepts.

2.2   Targeted Record Swapping (TRS)

Data Swapping is a common pre-tabular method of SDC, especially for cases when the released outputs are typically frequency tables, as in the case of a population Census. Pairs of records are determined within strata defined by control variables. With a TRS Implementation such as that of [9] the swapping is targeted to ensure that certain units ("records") that are considered to be most of risk for disclosure will be selected for the swapping with higher probabilities [19].

After identifying the units with the highest disclosure risks the algorithm determines which pairs of units to swap. In the TRS implementation we use in the present context, swapping means to swap the geography variables of the records relating to such a pair of units.

In order to use the TRS tool, a user must select those variables of the dataset that will be considered the "geography related" ones, referred to as "hierarchy variable(s)", which will be swapped. The user must also decide which variables will be used to determine disclosure risk of a record, those are called "risk variable(s)". In order to control information loss, the most important user choice is that of so called "similar variables", i.e. the control variables used to determine the strata. Notably, any tabulations of the swapped data that are defined solely by cross-combinations of "similar variables" will exactly match the respective results obtained with the original data, because those variables will be identical for all the record pairs with swapped entries.

Another parameter with strong effect on the impact of the method is the swap rate. The swap rate defined by the user specifies how many records should ideally be swapped in the process. Note, that this swap rate will usually not be reached exactly depending on the data structure and choice of the parameters mentioned above. The swap rate will be exceeded, if there are more critical units in the data set than the swap rate would imply and all of those units have suitable swapping partners under the restrictions given by the "similar variable(s)". Just as well, the real swap rate can turn out to be below the rate defined by the parameter, if there are no more suitable swapping partners left in the data set before reaching that limit.

2.3 CKM Noise Design for Tabulations of Continuous Variables

The Cell Key Method (CKM) for statistical disclosure limitation by random noise is implemented for example in the package τ-Argus and as separate R package cellKey, c.f. [12], relying on the R-package $p$table [4] to compute noise random distributions by maximizing entropy [7, 11].

When generated to protect tables of counts, perturbation tables typically define noise distributions with constant variance σ^2 and fixed maximum perturbation, specified by a parameter D. For tabulations of continuous (magnitude) variables, τ-Argus and cellKey offer to compute the noise by adding the sum of $topK$ random variables $\widehat{X}_j (j = 1, .., topK)$. In the present paper we generally assume the simplest case, $topK = 1$.

A perturbed value for original value $x = \sum_{i=1}^{n} y_i$ will then be computed according to (1) in [5] as

(1) $\hat{x} := x + x_\delta \cdot V$, with $x_\delta = x_1 m(x_1)$ for $x_1 > z_f$.

The function $m(z)$ decreases monotonously, such that for user defined parameters z_f, σ_0, σ_1, where $\sigma_0 < \sigma_1$, we have $x_\delta \cong \sigma_0 x_1$ for "large" x_1, and $x_\delta = \sigma_1 x_1$, for "small" $x \le z_f$, and some extra provisions to ensure a minimum noise variance also for x_1 tending to zero. V refers to a random variable defining standardized noise with a noise variance of 1.

3 Study Design

In this study we examine the usability of synthetic microdata as basis for tables released by national statistics offices as "official" figures. To this end we use one large, regionally detailed dataset as test data and apply the input perturbation methods of Sects. 2.1 and 2.2 to the dataset. In the next step, we tabulate the original and perturbed data at different levels of geographic detail. We also apply the Cell Key Method of Sect. 2.3 to the tabulated (original) data, thus generating a noisy version of the tabulated data, protected by output perturbation. We then compare the publication tables derived on basis of the original data to those derived from the synthetic data and their version after application of CKM, in terms of utility and disclosure risk.

3.1   Test Data

The test data set involves geographic information at several levels of a hierarchical administrative geography plus information relating to a grid geography. Each record contains a set of continuous variables we denote by $X_i (i = 0, 1, \ldots, 12)$. All X_i provide sensitive income by source information. The records relate to natural persons. Notably, those variables are not statistically independent, some of them are related by simple linear models that might be denoted by, say, $X_{i(m)} = \left(\sum_{j \in J_m} X_j \right) + \varepsilon_m$, $i(m) \notin J_m$, where ε_m denote a small error term for model m, and J_m the set of indices of its explanatory variables. But not all variables which would appear in those models are actually present in the data. In particular, there are three cases corresponding to a model such as $X_k = X_{k_1} + X_{k_2} + \varepsilon$, with many zero entries in the X_{k_2} component but X_{k_1} not included in the data. Such variable index pairs (k, k_2) are (1,2), (3,4) and (5,6).

Notably, one of the models could be regarded as "master model" with, say, $m = 0$ and $i(0) = 0$. X_0 as well as all X_j where all $j \in J_0$ are present in the test data. Typical tabulations of the dataset present aggregates of the X_j variables $(j = 0, 1, \ldots, 12)$ by size of X_0. In the following, we refer to this classification variable as $SCL(X_0)$. The master model can also be used to create another classification of the records. We define for each record an index variable $XMAX$. $XMAX$ characterizes a record by the explanatory variable in the master model with the largest value such that for each record $X_{XMAX} = max(X_j)_{j \in J_0}$.

We should mention that those test data are only a small excerpt (the most popular variables) of a highly important, huge administrative dataset with many more continuous and discrete variables tabulated for publication in various ways. Pre-planned tables make up for only a minor part of dissemination. Most of the released outputs are user demand driven or could be regarded as research outputs.

For the risk utility analysis in our small study, we only rely on a large tabulation with results for each of the thirteen X_j variables of our test data set. The tabulation is defined by crossing the geographic information (at all levels) with classification $SCL(X_0)$.

3.2   Application Settings for Synthetic Data Generation

For data synthesis in our study we used the synthpop package [13], synthesizing the data sequentially, one variable at a time. The algorithm predicts each variable using CART with the previous variables used as predictors. So, the order of variables has an impact. The order we chose for the continuous variables X_0 to X_{12} of our dataset was

$X_1, X_2, \ldots, X_{12}, X_0$. This means in particular, the other variables were used as predictors for X_0, but not the other way round.

Following ideas of [10] and [15] on partially synthetic data, and assuming the sensitive information of our data set relating to its continuous variables in the first place, we decided only to synthesize those. This means, the resulting dataset is only partially synthetic. All discrete variables of the dataset keep their original values and are used as predictors, in particular the "similar variables" of the TRS setting (3.2), and a variable relating to the grid geography. Regarding the "risk variable" of the TRS setting, $SCL(X_0)$, we tested two scenarios: with and without including it (as predictor) into the synthesis. In summary, this means, synthpop has synthesized the X_i variables conditional on the joint distribution of the original data of all the discrete variables of the data set (excluding $SCL(X_0)$ in one scenario). Apart from those settings, the package was used with default parameters.

Now, considering that our test scenario assumes that the synthetic data are not meant for direct publication, but only as basis to release aggregate statistics, and recalling moreover, that in the record swapping approach not all the records are swapped, but only those considered as the ones with the highest disclosure risks, it might suffice, if we do not replaces all the records of the dataset by partially synthetic ones, but only a sample of them:

Mixing Partially Synthetic Records with Original Records – Targeted Sample
Assuming it would be enough to mix just as many synthesized records with original records as swapped in the TRS approach, we identify those records of the dataset swapped by TRS. We select those records, that is, their original versions, and apply synthpop to only this subset of records, ending up with a (again partially) synthetic version of the subset. We then generate a "full" dataset by adding the remaining (original) records of the original dataset to the synthesized subset. To improve this approach with respect to utility, we also tested a variant where we randomly selected only 50% of the records swapped by TRS and synthesised only those.

Mixing Partially Synthetic Records with Original Records – Random Sample
However, concentrating synthesis on the swapped, risky records leads to relatively strong effects. From an information loss perspective, it might be better to just randomly select the records subjected to synthesis: This is the approach we follow in the final synthetization experiment. Using again the Swap Rate of the TRS experiment, we this time randomly select 10% of the original records to generate partially synthetic data from this subset and add the synthesized subset to the remaining set of original records.

3.3 Application Settings for Targeted Record Swapping

In the first phase of the study, Targeted Record Swapping as outlined in Sect. 2.2 was applied to the dataset. For this application we have used the following parameters and settings:

$SCL(X_0)$, as introduced in Sect. 3.1, was chosen as only risk variable. At low geographic levels it carries highly sensitive information and is also used as a classification variable in typical tabulations of our data set. The choice of the "similar variables" was not that straightforward. In our test application, finally three "similar variables" were selected from the set of discrete variables, with personal information on the individuals relating to the records of the data set. The index variable $XMAX$ (c.f. Sect. 3.1) is one of them.

The hierarchy is defined by the underlying administrative geography in conjunction with information from the grid geography.

To define risky records, the concept of k-anonymity is applied, with $k = 3$ as a parameter. The swap rate was set to a rather large value of $s = 0.1$, i.e. if necessary (and possible) 10% of the records will be swapped with suitable partners.

In the resulting swapped data set we actually observe a slightly higher swap rate of 13.2%. As mentioned in Sect. 2.2, it is possible for the swap rate to be higher when there are more critical units in the data set than the swap rate would imply, of course, always assuming the availability of a sufficient number of suitable swapping partners. In our study, the high swap rate is caused by the very small-scale evaluation of the data on grid level, which implies a high number of risky records.

3.4 Application Settings for the Cell Key Method

For application to the continuous variables, we used three different settings regarding the noise coefficient parameters σ_0 and σ_1, e.g. $\sigma_0 = 0.05, 0.1$ and 0.15, and $\sigma_1 = 2 \, \sigma_0$. However, the plots in Sect. 4. present only the "middle" variant with $\sigma_0 = 0.1$ because the three variants always score closer to each other than to any of the other methods in the comparison.

For application to counts data, we only choose one p-table with noise variance $\sigma^2 = 2$. For the continuous variable, we also produced mean-before-sum adjusted variants, i.e. variants where the noisy result $\widehat{X}$ is multiplied by $(n+j)/n$, where n denote the original frequency and $n + j$ the frequency after perturbation. This way, users of the adjusted data estimating a mean by dividing by the published frequency $n + j$ they get the estimate $\widehat{X}/n$ which is usually closer to the original mean, c.f. [20].

For input perturbation methods, such an adjustment should not be necessary, because they do not perturb counts and magnitudes independent of each other. Therefore, even though the tabulations of our test scenario do not present means (they present only sums), including the adjusted version in the comparison makes some sense: Though of course indirectly only, the information loss of the adjusted sums in a way "represents" the increased information loss of means computed from not adjusted noisy magnitudes caused by "no adjustment".

4 Measuring Utility and Disclosure Risk

For the risk utility analysis in our study we rely on tabulation of the X_j variable data, defined by crossing the geographic information (at all levels) with the size-classes $SCL(X_0)$, c.f. Sect. 3.1., re-generated on basis of the perturbed version of X_0 in case of the synthetic data.

For comparison of disclosure risk avoidance capacities of the different methods, we decided to concentrate on the *singletons* in our publication tables: we focus on table cells which relate to only a single record in the microdata. Singleton cells are typically considered as "risky", because they may disclose a confidential attribute of a single unit with a non-zero X_{k_2} (c.f. Sect. 3.1) which might be identifiable (especially at grid geography or municipality level). And even though the corresponding attribute (i.e. cell value) may differ from the "true" attribute as reported for this unit, due to use of noisy, swapped or synthetic data, there still remains a risk of perceived disclosure: a unit appearing as single unit in a tabulation of the perturbed data assumes itself identifiable and associated with a value and may feel exposed, no matter (or maybe even more so), if that associated value is false. In our study, we use R_{pert}, the share of disseminated singletons (units *appearing* to be single units after perturbation) that correspond to real singletons, representing the probability that a published one leads to exposure of a real singleton[1].

For utility loss, regarding count tables, we use the Hellinger distance. Regarding the magnitude tables, we use a simple absolute distance measure. To this end, for each table assumed in the study as "to be published", we compare the original (true) cell values and the altered values and sum up the absolute differences. Subsequently, for reasons of scaling and comparability, we divide this value by the sum of all absolute cell values of the original table. This is actually the ℓ^1norm of the differences divided by the ℓ^1norm of the original cell values and can be expressed as

$$(2) \quad U := \frac{\|\hat{x} - x\|_1}{\|x\|_1} = \frac{\sum_{i \in \{cells\ to\ be\ published\}} |\hat{x}_i - x_i|}{\sum_{j \in \{cells\ to\ be\ published\}} |x_j|},$$

where x_i denote the original value of cell i, whereas $\hat{x}_i$ is the value of that same cell in the altered table. The measure can be interpreted as mean deviation of the perturbed cells in terms of the mean absolute (original) cell value.

We also compute this measure for the subgroup of singleton table cells, i.e. cells that are singletons in both, the table before, and after disclosure control is applied: U_{single}. Then 1-min $(1; U_{single})$ could be regarded as indicator for the risk of proximity of a published "one" to the original value of this singleton.

5 Results

To give a first impression, Figs. 1a and 1b present risk-utility maps comparing the performance of the various methods and their variants at the lowest geographic levels, using the risk and utility measures of Sect. 4, i.e. for utility loss we use indicator U, and for risk we use the combined indicator

$$R := \left(1 - \min(1; U_{single})\right) R_{pert}$$

The R-U maps plot medians of utility loss indicator U and risk indicator R, observed for the 13 variables of our test instance. Clearly, at low level geography, TRS (Swap)

---

[1] This indicator is different from R_{true}, the share of original singletons remaining singletons after disclosure control is applied, reflecting the probability for an identifiable unit to stay identifiable. R_{true} is a parameter of CKM – for the other methods we did not observe any relevant differences between R_{true} and R_{pert}.

performs worst regarding utility with much higher utility loss compared to the other methods, but best regarding risk. According to the indicators, targeting "risky" records for synthesizing (Synth (Swap)) performs similar to the Cell Key Method with mean-before-sum adjustment (CKM(MBS)) regarding both, risk and utility. Whereas without adjustment CKM scores lowest compared to all other methods regarding utility loss, with still the same risk score. Synthesizing only half of the "risky" records (Synth (Half)) clearly reduces utility loss, but also increases risk. Compared to that, synthesizing randomly selected records (Synth (Rand)) leads to much less utility loss, but also to the highest risk score.

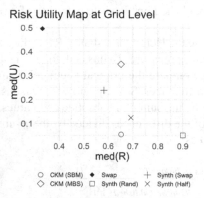

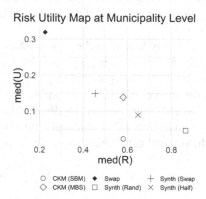

Fig. 1a. Risk vs. utility loss at grid level

Fig. 1b. Risk vs. utility loss at municipality level

5.1 Comparing Utility Loss

To enhance comparison and presentation of the results, we first introduce an ordering sequence into the 13 continuous variables of our test data set, ranking the variables by size of the utility indicator (c.f. (2) in Sect. 4) when applied to the (not mean-before-sum adjusted) CKM results. So, we have $U_{CKM}\left(X_{(1)}\right) \leq U_{CKM}\left(X_{(2)}\right) \leq \cdots \leq U_{CKM}\left(X_{(13)}\right)$. Notably, $X_{(1)}$ is a fairly even distributed variable, $X_{(3)} = X_0$, i.e. it is the dependent variable in the master model mentioned in Sect. 3.1, and $X_{(2)}$ is the dependent variable of a similar model, also involving several of the other variables as explanatory variables. Typical for the variables at the end of the sequence are very large proportions of zero entries.

Indeed, for CKM without adjustment, we observe for the most detailed versions of our test tabulation at grid level utility losses close to or even slightly above the σ_0 parameter of the respective CKM variant. We observe this for the five "difficult" variables at the end of the utility loss ranking, $X_{(9)}$ to $X_{(13)}$. This is not unexpected, see Appendix A.1 for some theoretical reasoning.

For the variables at the other end of the ranked list, utility loss indicator values are much lower. Utility loss for $X_{(1)}$ is then less than 10% compared to that obtained for $X_{(13)}$.

Figures 2a and 2b compare for the methods we tested utility loss results for $X_{(1)}$ to $X_{(13)}$, skipping $X_{(10)}$ to $X_{(12)}$, because for those variables the synthetic data utility loss indicator results are way too far outlying for the scale of the plot[2]. For similar reasons plots present only synthetic data results for our second scenario, i.e. including the size class variable $SCL(X_0)$ as predictor in the synthesis. For most variables, including this predictor improves utility losses tremendously. For example, for the "targeted" 'Synth (Swap)' variant without the predictor in the synthesis, for $X_{(1)}$ to $X_{(4)}$ utility loss U scores between 6 to 15 times higher compared to the variant computed including the predictor.

Comparing Figs. 2a to 2b shows that moving from grid to municipality level slightly improves the scores for all variants and variables, but not very much. Which is probably due to still many small and zero cells in our test tabulation on municipality level as well.

At both geography levels, the best scoring methods are CKM without mean-before-sum adjustment (CKM (SBM)), closely followed by synthetic data variant (Synth (Rand)) which only synthesizes randomly selected 10% of the records. At grid level, for $X_{(1)}$ to $X_{(8)}$, the next best scoring methods are the synthetic data variants which either synthesize all records (variant "Synth", not included in the plot), or only the records selected by targeted record swapping (Synth (Swap)), or at least half of them (Synth (Half)). For these variables they all outperform the adjusted mean-before-sum CKM variant (for the base variant "Synth", the latter is true only for $X_{(1)}$ to $X_{(4)}$).

With exception of $X_{(8)}$, for all variables in the plots utility loss scores are highest for the targeted record swapping (Swap). However, for the three variables skipped in the plot, the extremely poor results of Synth, Synth (Swap) and Synth (Half) are worse. For $X_{(9)}$ and $X_{(13)}$ synthesizing all records (Synth) performs similar to record swapping and much worse compared to synthesis of selected records only.

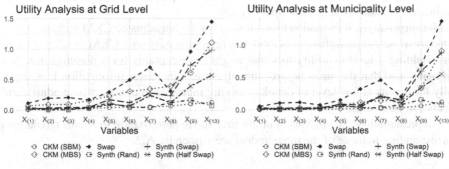

Fig. 2a. Utility loss U at grid level   **Fig. 2b.** Utility loss U at municipality level

Moving to higher levels of aggregation in the test tabulation generally reduces utility losses. Moving from grid to municipality level, they decrease by about approx. 20% to 50% for most variables and methods. From municipality to district level, utility losses shrink further, and even more so when moving to the state level. On that top level of our test geography, utility loss is of course zero for the record swapping because records are only swapped on the levels below (Figs. 2c and 2d).

---

[2] Notably, $X_{(10)}$ to $X_{(12)}$ relate to the k_2 variables of the special (k, k_2) index pairs mentioned in Sect. 3.1: X_2, X_4 and X_6.

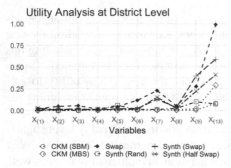

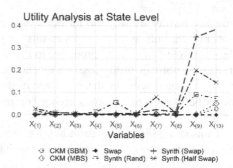

Fig. 2c. Utility loss U at district level   **Fig. 2d.** Utility loss U at state level

To summarize findings regarding utility of the perturbed magnitude data: for variables $X_{(1)}$ to $X_{(8)}$, except for $X_{(7)}$, it can be imagined that the synthetic data might actually be sufficiently close to the real data, i.e. close enough for publishing aggregates derived on basis of such data at the levels considered here. Especially when we consider that at grid and municipality level not all these variables should be published in combination with a fine size-classification such as the one considered in the tests, i.e. there would be some further aggregation anyway.

With exception of the state level, for most variables targeted record swapping leads to the highest utility losses, though for some of the variables, especially those where the "similar variables" of the setting are relatively efficient, utility losses are somewhat less extreme. However, a mean deviation of about 11% of the mean cell value for the most important variable can probably not be accepted in an official publication of administrative data, even at municipality level.

Not surprisingly, information loss is lowest for the not adjusted CKM variant: Due to its nature as output perturbation method, it is relatively easy with CKM to select parameters yielding a level of utility loss that might be acceptable for a disseminator. Not surprising also that the mean-before-sum adjustment increases information loss, especially for results at low level of detail, involving many small counts, where adjustment affects can be strong.

For presentation and analysis of utility loss due to perturbation in counts tabulations (corresponding to those of the magnitudes) see Appendix A.2.

6   Summary, Open Issues, Conclusions

The question we have started to address in this paper is, if synthetic data methods can be a realistic alternative as disclosure control concept not for public or scientific use microdata sets, but when partially synthesized data sets shall be used as basis for producing the official outputs of NSIs. To that aim we have compared several variants of synthetic data generation implemented using the CART approach of the synthpop package to results obtained by an "established" input perturbation method (though typically used with datasets mainly consisting of discrete variables) on one hand, and to the output perturbation approach of the Cell Key Method on the other hand.

Results of the paper have been obtained in a rather small study, looking at just one tabulation of several continuous variables computed for different geographies, and considering only two different modelling approaches for data synthesis. For analysing utility, we mainly rely on a simple, easy to interpret indicator which, however, does for example not provide information on the tails of tabulation cell level distributions of the perturbation. Another important open issue not yet addressed in this study is inspecting tabulation cell level distributions of the perturbation resulting for ratios of pairs of variables.

Regarding risk, we focus on the share of singleton cells in the test tabulation, also considering how much their cell values change.

As expected, we observe that the performance of data synthesis strongly depends on the predictor information included in the model. It is quite encouraging that for some of the variables even the simple predictors in our model seem to work quite well. On the other hand, for the other variables, suitable predictors would probably have to be generated variable by variable which may become a challenge for data sets involving a large variety of variables. If such efforts could be worthwhile should be clarified by more detailed follow up studies.

Appendix

A.1 Approximate Behavior of Utility Loss Indicator U for CKM in Extremely Detailed Tabulations

As explained in Sect. 3.4, a lower bound for the noise added by CKM to cell value x is approximately $\sigma_0 x_1 V$, where V is a realization of a discrete, symmetric random variable V. It is statistically independent of the cell values x, with VAR $(V) = 1$, and $E(|V|) := \vartheta$. In our setting, $\vartheta \approx 0.8$.

Typically, realizations of V will be $v \in \left\{ -D, \frac{1}{I} - D, \frac{2}{I} - D, \ldots, D - \frac{1}{I}, D \right\}$, obtained by a "lookup" in a suitable perturbation table (with user defined "step-width" $\frac{1}{I}$) using the cell key of x. This perturbation table defines probabilities $(p_{D,k})_{k=0, 1, \ldots, 2 \cdot I \cdot D}$ for an original value $x \geq D$ to be perturbed by $\frac{k}{I} - D$.

Consider now the extreme case of a very detailed tabulation where all cells to be published are either empty, or consist of only a single contribution, i.e. $x_i = x_{i_1}$. Then we have $U_{CKM} \approx \dfrac{\sum_{i \in \{cells\ to\ be\ published\}} \sigma_0 |x_{i_1}| \left| \frac{k_i}{I} - D \right|}{\sum_{j \in \{cells\ to\ be\ published\}} |x_{j_1}|}$. Because of the definition of V, and because of its independence of the x_i data, we can further approximate

$$U_{CKM} \approx \frac{\sigma_0 \sum_{k=0,1,\ldots,2 \cdot I \cdot D} p_{D,k} \left| \frac{k}{I} - D \right| \left(\sum_{i \in \{cells\ to\ be\ published\}} |x_{i_1}| \right)}{\sum_{j \in \{cells\ to\ be\ published\}} |x_{j_1}|} = \sigma_0 E(|V|) = \sigma_0 v.$$

Appendix A.2

In terms of which variables are affected most by the perturbation, and which least, the distances for the counts data do not really change the picture observed for the continuous variables. Looking at grid level results (Table 1) we find that except for the variant synthesizing all records, all other methods tend to perform better than record swapping (Swap), and for $X_{(1)}$ to $X_{(8)}$ they also do better as CKM. This impression is especially striking for the "best" three variables, $X_{(1)}$ to $X_{(3)}$.

Appendix A.3

Table 2 presents results for risk indicator R_{pert}, i.e. the probability for a published count of 1 to relate to a real singleton. As we see, TRS (Swap) is the only method for which this probability is below 2/3 for all variables. In contrast to that, for CKM, R_{pert} is larger than 2/3 for each of the 13 variables. The same holds for the synthetic data variants where only a random 10% sample of all records is synthesized, or only one half of the swapped records, respectively. For these variants, a large proportion (for $X_{(1)}$ to $X_{(3)}$ almost all, actually) of the published singletons coincide with real singletons and so might be identifiable. It is therefore important to check how close the associated magnitudes are to the original ones, c.f. Table 3, further below.

Table 1. Hellinger distances at grid Level

Variable	Swap	CKM	Synthetic data variants			
			SwapHalf	*Synth*	*Swap*	*Rand*
$X_{(1)}$	0,15	0,10	0,08	0,03	0,02	0,02
$X_{(2)}$	0,15	0,10	0,02	0,02	0,01	0,01
$X_{(3)}$	0,15	0,09	0,02	0,00	0,00	0,00
$X_{(4)}$	0,20	0,15	0,07	0,13	0,10	0,07
$X_{(5)}$	0,27	0,20	0,14	0,29	0,20	0,13
$X_{(6)}$	0,31	0,24	0,15	0,34	0,21	0,11
$X_{(7)}$	0,29	0,21	0,14	0,25	0,17	0,08
$X_{(8)}$	0,31	0,27	0,19	0,47	0,28	0,15
$X_{(9)}$	0,58	0,38	0,34	0,67	0,48	0,22
$X_{(10)}$	0,57	0,46	0,38	0,93	0,57	0,28
$X_{(11)}$	0,51	0,46	0,37	0,89	0,47	0,28
$X_{(12)}$	0,63	0,51	0,49	0,99	0,69	0,24
$X_{(13)}$	0,75	0,51	0,46	0,92	0,65	0,21

Table 2. Risk indicator R_{pert} at grid level

Variable	Swap	CKM	Synthetic data variants			
			SwapHalf	Synth	Swap	Rand
$X_{(1)}$	0,30	0,70	0,71	0,96	0,98	0,98
$X_{(2)}$	0,28	0,70	0,97	0,98	0,99	1,00
$X_{(3)}$	0,27	0,69	0,98	1,00	1,00	1,00
$X_{(4)}$	0,37	0,74	0,88	0,69	0,80	0,90
$X_{(5)}$	0,37	0,73	0,75	0,40	0,58	0,80
$X_{(6)}$	0,44	0,76	0,82	0,41	0,69	0,89
$X_{(7)}$	0,37	0,73	0,77	0,55	0,71	0,93
$X_{(8)}$	0,54	0,78	0,78	0,28	0,62	0,86
$X_{(9)}$	0,38	0,82	0,67	0,37	0,49	0,85
$X_{(10)}$	0,64	0,97	0,83	0,11	0,62	0,90
$X_{(11)}$	0,64	0,92	0,80	0,11	0,70	0,88
$X_{(12)}$	0,59	1,00	0,69	0,01	0,44	0,95
$X_{(13)}$	0,39	0,96	0,74	0,12	0,53	0,98

Table 3 shows the utility loss indicator results considering only the single units (after perturbation), U_{single}, for the grid level. Although an indicator for data utility, applied to only those cells with counts of 1 before and after perturbation, it can also be interpreted as risk measure. Singletons are especially prone to risk of identification. So, protection of such cells benefits from a larger deviation. With CKM, the amount of perturbation of the largest contributor is a parameter. It is therefore not surprising that we observe a relatively homogenous mean perturbation of about 15% for those cases where largest contribution and cell value always coincide. Again, synthesis of only a random subsample of 10% of the records results in the smallest deviations - not a desirable feature in this context, as explained before.

Table 3. Utility loss indicator for single units (after perturbation), U_{single} at Grid Level

Variable	Swap	CKM	Synthetic data variants			
			SwapHalf	Synth	Swap	Rand
$X_{(1)}$	0,32	0,15	0,19	0,18	0,17	0,03
$X_{(2)}$	0,29	0,13	0,22	0,34	0,34	0,02

(continued)

Table 3. (*continued*)

Variable	Swap	CKM	Synthetic data variants			
			SwapHalf	*Synth*	*Swap*	*Rand*
$X_{(3)}$	0,28	0,13	0,32	0,34	0,35	0,02
$X_{(4)}$	0,12	0,16	0,06	0,16	0,11	0,03
$X_{(5)}$	0,18	0,16	0,08	0,33	0,20	0,01
$X_{(6)}$	0,18	0,15	0,08	0,22	0,16	0,02
$X_{(7)}$	0,44	0,14	0,46	0,38	0,49	0,03
$X_{(8)}$	0,04	0,16	0,04	0,23	0,06	0,02
$X_{(9)}$	0,12	0,15	0,10	0,35	0,24	0,02
$X_{(10)}$	0,01	0,15	0,01	0,43	0,03	0,00
$X_{(11)}$	0,08	0,13	0,01	0,47	0,78	0,00
$X_{(12)}$	0,00	0,14	0,00	0,35	0,00	0,00
$X_{(13)}$	0,04	0,17	0,07	0,82	0,37	0,00

References

1. Breiman, L., Friedman, J., Stone, C.J., Olshen, R.A.: Classification and Regression Trees. Wadsworth International Group (1984)
2. Burnett-Isaacs, K., et al.: Synthetic data for national statistical organization: a starter guide, version 2 (2021). https://statswiki.unece.org/download/attachments/282330193/HLG-MOS%20Synthetic%20Data%20Guide.docx?version=2&modificationDate=164244 9524045&api=v2. Accessed 20 May 2022
3. Drechsler, J., Reiter, J.P.: Disclosure risk and data utility for partially synthetic data: an empirical study using the German IAB establishment survey. J. Official Statist. **25**(4), 589–603 (2009)
4. Enderle, T., Giessing, S.: Implementation of a 'p-table generator' as separate R-package. Deliverable D3.2 of Work Package 3 "Prototypical implementation of the cell key/seed method" within the Specific Grant Agreement "Open Source tools for perturbative confidentiality methods" (2018b)
5. Enderle, T., Giessing, S., Tent, R.: Calculation of risk probabilities for the cell key method. In: Domingo-Ferrer, J., Muralidhar, K. (eds.) Privacy in Statistical Databases. LNCS, vol. 12276, pp. 151–165. Springer, Cham (2020). https://doi.org/10.1007/978-3-030-57521-2_11
6. Fraser, B., Wooton, J.: A proposed method for confidentialising tabular output to protect against differencing. In: Monographs of Official Statistics. Work session on Statistical Data Confidentiality, Eurostat-Office for Official Publications of the European Communities, Luxembourg, pp. 299–302 (2006)
7. Giessing, S.: Computational issues in the design of transition probabilities and disclosure risk estimation for additive noise. In: DomingoFerrer, J., PejićBach, M. (eds.) Privacy in Statistical Databases. LNCS, vol. 9867, pp. 237–251. Springer, Cham (2016). https://doi.org/10.1007/978-3-319-45381-1_18

8. Giessing, S., Tent, R.: Concepts for generalising tools implementing the cell key method to the case of continuous variables. In: Joint UNECE/Eurostat Work Session on Statistical Data Confidentiality, The Hague, 29–31 October 2019. http://www.unece.org/fileadmin/DAM/stats/documents/ece/ces/ge.46/2019/mtg1/SDC2019_S2_Germany_Giessing_Tent_AD.pdf

9. Gussenbauer, J.: Record swapping: Package vignette (2019). https://github.com/sdcTools/recordSwapping

10. Little, R.J.A.: Statistical Analysis of Masked Data (1993)

11. Marley, J.K., Leaver, V.L.: A method for confidentialising user-defined tables: statistical properties and a risk-utility analysis. In: Proceedings of 58th World Statistical Congress, pp. 1072–1081 (2011)

12. Meindl, B., Kowaric, A., De Wolf, P.P.: Prototype implementation of the cell key method, including test results and a description on the use for census data. Deliverable D3.1 of Work Package 3 "Prototypical implementation of the cell key/seed method" within the Specific Grant Agreement "Open Source tools for perturbative confidentiality methods" (2018)

13. Nowok, B., Raab, G., Dibben, C.: synthpop: Bespoke creation of synthetic data in R. J. Statist. Softw. **74**, 1–26 (2016). https://doi.org/10.18637/jss.v074.i11, https://www.jstatsoft.org/article/view/v074i11

14. Raghunathan, T.E., Reiter, J.P., Rubin, D.B.: Multiple Imputation for statistical disclosure limitation. J. Official Statist. **19**(1), 1–16 (2003)

15. Reiter, J.P.: Inference for partially synthetic, public use microdata sets. Surv. Methodol. **29**(2), 181–188 (2003)

16. Reiter, J.P.: Releasing multiply imputed, synthetic public use microdata: an illustration and empirical study. J. R. Statist. Soc. Ser. A. Statist. Soc. **168**(1), 185–205 (2003b). https://doi.org/10.1111/j.1467-985X.2004.00343.x

17. Reiter, J.: Using CART to generate partially synthetic public use microdata. J. Official Statist. **21**(3), 441–462 (2005)

18. Rubin, D.B.: Statistical disclosure limitation. J. Official Statist. **9**(2), 461–468 (1993)

19. Shlomo, N., Tudor, C., Groom, P.: Data swapping for protecting census tables. In: Domingo-Ferrer, J., Magkos, E. (eds.) Privacy in Statistical Databases. LNCS, vol. 6344, pp. 41–51. Springer, Heidelberg (2010). https://doi.org/10.1007/978-3-642-15838-4_4

20. Thompson, G., Broadfoot, S., Elazar, D.: Methodology for the automatic confidentialisation of statistical outputs from remote servers at the Australian Bureau of statistics. In: Joint UNECE/Eurostat Work Session on Statistical Data Confidentiality, Ottawa, 28–30 Oktober 2013. http://www.unece.org/fileadmin/DAM/stats/documents/ece/ces/ge.46/2013/Topic_1_ABS.pdf

21. Van Buuren, S., Brand, J.P.L., Groothuis-Oudshoorn, C.G.M., Rubin, D.B.: Fully conditional specification in multivariate imputation. J. Stat. Comput. Simul. **76**(12), 1049–1064 (2006)

On Privacy of Multidimensional Data Against Aggregate Knowledge Attacks

Ala Eddine Laouir[(✉)] and Abdessamad Imine

Lorraine University, Cnrs, Inria, 54506 Vandœuvre-lès-Nancy, France
{ala-eddine.laouir,abdessamad.imine}@loria.fr

Abstract. In this paper, we explore the privacy problem of individuals in publishing data cubes using SUM queries, where a malicious user is expected to have an aggregate knowledge (e.g., average information) over the data ranges. We propose an efficient solution that maximizes the utility of SUM queries while mitigating inference attacks from aggregate knowledge. Our solution combines cube compression (i.e., suppression of data cells) and data perturbation. First, we give a formal statement for the privacy of aggregate knowledge based on data suppression. Next, we develop a Linear Programming (LP) model to determine the number of data cells to be removed and a heuristic method to effectively suppress data cells. To overcome the limitation of data suppression, we complement it with suitable data perturbation. Through empirical evaluation on benchmark data cubes, we show that our solution gives best performance in terms of utility and privacy.

Keywords: Data cubes · Privacy preservation · Cell suppression · Cell perturbation · Cube compression

1 Introduction

Multidimensional data (or data cubes) are widely used in many fields to store all collected data, as these data structures are optimized for Online Analytical Processing (or OLAP) [2,8,9]. For business or research purposes, the data collected is made available to external parties (e.g., analysts, and organizations) to enable them to query and analyze trends and patterns necessary for decision making. Although most external parties have legitimate usage interests and data is anonymized before publication or query, there are situations where a malicious user can mine this data in order to endanger the privacy of individuals, such as leaking medical records. Because of this privacy risk, many research works have addressed this issue and different models have been proposed [5,12]. For example, aggregation of a single cell value is not allowed, and query set size and access controls are deployed to provide additional security (for more details see [15]). However, most of the proposed techniques focus on the privacy of individual data (or cells of data cubes) and neglect insights that can be gained by simply

---

This work is funded by DigiTrust (http://lue.univ-lorraine.fr/fr/article/digitrust/).

J. Domingo-Ferrer and M. Laurent (Eds.): PSD 2022, LNCS 13463, pp. 92–104, 2022.
https://doi.org/10.1007/978-3-031-13945-1_7

analyzing aggregate patterns (such as average information) [3]. These aggregate patterns may not be allowed for security reasons, but they can be easily inferred in many cases. Even for multidimensional data where only SUM is allowed, knowledge of the average could be a valuable indicator to know the trend of the data (e.g., whether or not the average salary is above the minimum wage).

Illustrative Example. Consider that a large retail company allows access to sales from all its stores during the year through range sum queries (here and in the rest of the paper, we consider a data cube that contains only positive values). Due to business and analytical needs, our retail company publishes a view (or part) of the data cube and makes it accessible only through SUM range queries (see Fig. 1). Even though the COUNT information is not explicitly available, a malicious user can get it easily: either by knowing some metadata of the published cube and queries [3], or by using sophisticated approaches such as the Volume Leakage attack [6]. The result of the aggregate average AVG can then be inferred and exploited to violate the privacy of individual cells.

Suppose our malicious user knows that two stores had similar sales on certain days (weekend, holidays, events, etc.). Using the result of SUM query and the knowledge of COUNT, he can deduce the result of AVG. For the region defined by the range $\{(Shop2 : Shop3), (Day2 : Day3)\}$ in Fig. 1, the AVG is equal to $129.5k$. Based on this AVG, the attacker can now estimate the sum of the range $\{(Shop2 : Shop4), (Day2 : Day2)\}$ as $(129.5 + 129.5 + X)$ and deduce that X is $172k$, which gives the cell value with high precision. The attack can proceed in the same way to disclose the rest of the values. Note that the attacker can also assume that the negative or small scale results are the empty cells, allowing him to reconstruct the entire region.

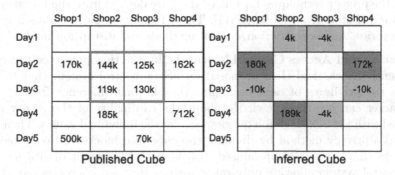

Fig. 1. Data cube for retail company.

Contributions. To disturb the AVG and prevent these attacks, cell suppression techniques are better suited since they simultaneously modify the COUNT (by -1) and the SUM (by *minus* the value of the cell), for each cell suppressed [3]. Cell suppression is very similar to a well-known technique for speeding up query responses in OLAP, namely *cube compression*.

In this paper, we present an efficient method that maximizes the utility of SUM queries while mitigating inference attacks from AVG aggregate knowledge.

Our method combines cube compression via cell suppression and cell perturbation to ensure better utility and privacy. Our contributions are as follows:

1. We develop a Linear Programming (LP) model for finding the optimal allocation needed to provide the best utility/privacy results.
2. We design a cell suppression technique that maximizes privacy against attacks based on AVG, and can be extended to other aggregate operators but this will be the subject of future work.
3. To overcome the limitation of cell suppression, in the case where the values are close to each other, we complement it with suitable cell perturbation to ensure a higher level of privacy while maintaining utility.
4. All of our contributions are validated by extensive experiments, in which our techniques have outperformed the state-of-the-art [3] approach by ensuring both utility and privacy against attacks that use AVG aggregate inferred from data cubes.

Paper Organization. In Sect. 2, we review related work. In Sect. 3, we introduce the notation used and the problem statement of the aggregate knowledge privacy based on data suppression. We present our privacy-preserving method in Sect. 4. We provide an experimental evaluation of our method in Sect. 5 and discuss the limitations of our solution in Sect. 6. Finally, we conclude in Section 7 by presenting some future works.

2   Related Work

Most of the privacy techniques for OLAP data are derived from the literature on the privacy of statistical database [11]. These techniques can be classified into two categories: Access restriction/control methods and disruption methods.

Restriction and Access Control Methods. In [13], they presented two types of inference attacks and then proposed an access control system that further restricts the privileges of each user until they become inference free. In [16], the attacker can use the knowledge about the cardinality of the empty cells, combined with the SUM singleton queries, to infer the individual values. Then, [16] proposed a privacy method by dividing the cube into blocks and only keeping the blocks that are not compromised. [14] is an enhancement of [16] with an audit control system allowing only range queries that are inference-free. [17] is another query auditing method that uses information theory and only responds to user queries if the user's prior knowledge (represented by his previous queries) and new knowledge do not compromise the cells targeted by the query.

Perturbation Methods. Another way to provide privacy is to add noise to the cube cells, in such a way any inference of a cell's value will yield a perturbed value and reconstruction will not give the original data. [1] presented an approach to add noise to cells, where each cell is kept as it is with probability p or noised (using a noise sampled from the normal distribution) with probability $1 - p$.

In [11], they present another method where the cube is considered as a group of blocks that will be perturbed individually. In each block, the sum of the noise added to each cell in a row (as well as for the cells in a column) is equal to 0 so that it can provide accurate range sum queries. Their results show that they were able to change the values significantly, while providing accurate answers to queries. In [3], they present another perturbation approach that uses cell deletion (also called sampling). Based on the average aggregate knowledge, their algorithm applies data suppression to modify not only the response of the queries but also the aggregate patterns inferred from the queries. They compared to [11], and the results show that adding noise alone cannot prevent this type of inference without the loss of utility and cell suppression is a better suited solution. Our work targets the same aspects of inference and privacy as defined in [3], and we have provided a better algorithm for cell suppression and perturbation.

Another perturbation method is the Differential Privacy (DP) [5], considered the gold standard. However, DP cannot be applied to all the possible scenarios [4]. In our work, we considered SUM queries, and one of the main problems in applying DP is to define the global sensitivity for the SUM functions [10]. Also it requires the addition of a lot of noise to significantly disrupt the AVG, resulting in poor utility. For these reasons, DP is out of the scope of this work and given other privacy considerations, we will investigate the application of differential privacy to secure high-dimensional data (multidimensional cubes) in future work. Another privacy model is k-anonymity [12], which was used by [8] to avoid re-identification attacks on the dimensions of a data cube. This is also outside the scope of this paper.

3 Problem Statement

In this section, we give the notation used throughout the paper and a formulation of the problem under consideration.

3.1 Preliminaries

Data Cube. A *data cube* C is a multidimensional data over a relational table for a set of *dimensions* $D = \{d_1, d_2, \ldots, d_n\}$, where each dimension d_i corresponds to an attribute and each cell contains the result of an aggregated *measure*. Figure 1 illustrates a 2-dimensional cube where attributes *Shops* and *Days* are dimensions, and each cell contains the *measure* which is the result of the aggregate operator SUM on the amount of sales. In this work, we consider only SUM, as this aggregate operator is (i) extensively used in many multi-dimensional frameworks [15], and (ii) used to compute other aggregate operators such as AVG.

We consider here a special class of queries called *continuous range queries* in such a way that SUM is performed over a range query $R = \{r_1, r_2, \ldots, r_n\}$, noted by SUM($R$), where r_i is a continuous range on dimension d_i specified by the start and end positions. For instance, in Fig. 1, SUM(R) results in $1004k$ with $R = \{(Shop1 : Shop3), (Day3 : Day5)\}$. Let $\mathcal{R} = \{R_1, R_2, \ldots, R_m\}$ be a *query workload* used by the data publisher and the end user as a contract that defines the view to be published (via *range queries*) instead of the whole data cube. Let $|R|$ be the size (i.e., the number of non-empty cells) of range query R. In Fig. 1, the size of $R = \{(Shop1 : Shop3), (Day3 : Day5)\}$ is 5.

Metrics. Our goal is to suppress (and possibly perturb) some cell values from each range R to get R'. Any privacy-preserving data publishing method for data cubes is evaluated on two criterias: *utility* and *privacy*. As utility objective, we consider boosting the accuracy of SUM queries in our solution. For that, we compute the *relative error of accuracy* $\mathcal{A}_e$ which shows how different the SUM answer on the exact range R and the altered one R':

$$\mathcal{A}_e(R, R') = \frac{|\text{SUM}(R) - \text{SUM}(R')|}{\text{SUM}(R)} \tag{1}$$

It is clear that the smaller the relative error of accuracy, the better the utility.

As for privacy issues, our objective is to prevent inference attacks like those presented in Sect. 1. Let us recall that in these attacks the information on the average was deduced from the data cube. Therefore, we compute the *inference error* $\mathcal{I}_e$ which shows how well the solution we propose is able to disrupt this inferred average and thus mitigate successful attacks [3,7]:

$$\mathcal{I}_e(R, R') = \frac{|\text{AVG}(R) - \text{AVG}(R')|}{\text{AVG}(R)} \tag{2}$$

Note that the higher the inference error, the better the privacy of the data cube. Given both metrics, a good privacy-preserving data publishing method should provide the best utility-privacy tradeoff.

3.2 Problem Definition

Let C be a data cube with a query workload $\mathcal{R} = \{R_1, R_2, \ldots, R_m\}$. We consider attackers whose knowledge is limited to the average information and/or the distribution of some cells (as described in Sect. 1). The attackers can discover the exact and/or approximate values of other cells, and accordingly infer sensitive attribute information. To prevent these attacks, we propose to suppress some values from data cube C (only the region defined by $\mathcal{R}$) while preserving the accuracy of SUM queries and mitigating the inference due to the use of the average operator. More precisely, for each R_i in $\mathcal{R}$, we create a non empty replica R_i' that contains a *minimal subset* of the cells in R_i and the others are left null. Our privacy-preserving solution can be defined as a multi-objective optimization problem:

$$\text{maximize } \sum_{k=1}^{m} |R_i| - |R_i'|$$

$$\text{minimize } \frac{1}{m} \sum_{k=1}^{m} \mathcal{A}_e(R_i, R_i')$$

$$\text{maximize } \frac{1}{m} \sum_{k=1}^{m} \mathcal{I}_e(R_i, R_i') \tag{3}$$

$$\text{subject to } R_i' \subset R_i \text{ for each } R_i \in \mathcal{R}$$

The first objective function of Eq. 3 maximizes the difference between R_i and R_i' to meet the requirement of cube compression. In real world scenarios, the sizes of data cube and its query workload are very huge. Finding an optimum solution to multi-optimization problem 3 is hard and intractable task. This type of resolution falls under the broad category of multi-criteria, or vector optimization problems. Unlike single-objective problems, it is not always possible to find an optimal solution that satisfies all the function objectives under consideration. Moreover, even several solutions may not meet the expectations of the data publisher in terms of utility and privacy requirements. Therefore, it would be more beneficial if the data publisher had the ability to select and verify their own levels of utility and privacy.

In the next section, we propose a heuristic to problem 3 combining cell suppression and perturbation and allowing us to find plausible solutions.

4   Privacy-Preserving Method

We present a method to publish a view of a given data cube while ensuring good utility of SUM queries and protecting cells privacy against average-based inference attacks. The view is built using two operations on non-empty cells: suppression and perturbation of data cells. Our method proceeds in three steps: (i) splitting query ranges at finer grid granularity in order to increase the accuracy of potential SUM queries; (ii) defining and solving a Linear Programming (LP) problem to determine how many cells are needed from $R_i \in \mathcal{R}$ and its sub-ranges to prepare a range view given a storage space $\mathcal{B}$; (iii) preparing range views by deleting some data cells and perturbing others as necessary as possible.

4.1   Preprocessing Step

Each R_i in the query workload $\mathcal{R}$ represents a region of the data cube that we want to publish while ensuring utility and privacy. If the range of R is large, then dealing directly with this range would be a coarse granularity to the problem, thereby penalizing small user queries after the data cube view is published. Indeed these queries will not be efficient (i.e., null result) since they target regions where cells have been deleted unnecessarily due to coarse granularity. Accordingly, to ensure that the retained/removed cells from region R_i are evenly distributed across all parts, we consider a smaller unit on which to apply cell suppression. We call this unit a sub-query (or sub-region), so each R_i will be divided (logically by indices only) into smaller sub-regions. We build a view for each sub-region of R_i, which together constitute the global view of R_i. Splitting data into smaller units is a common preprocessing step, either on the data cube directly [11] or on a query workload [3].

4.2   Space Allocation Step

Unlike multi-objective optimization problem 3 (where the decision variables are the cells values of each R_i), we define a simple LP problem to compute the view size of ranges as well as their sub-ranges (resulting from the preprocessing step). Given an allocated space $\mathcal{B}$ in which we want to build a view of the query workload $\mathcal{R} = \{R_1, R_2, \ldots, R_m\}$. The objective is to distribute this space in such a way to guarantee a minimum $\mathcal{A}_e$ for all R_i:

$$\text{minimize } \frac{1}{m} \sum_{i=1}^{m} \frac{|\text{SUM}(R_i) - \text{SUM}(R_i')|}{\text{SUM}(R_i)} \tag{4}$$

The view R_i' is a subset of R_i to be published. The size and the data cells of R_i' are unknown at this stage. As the data cells of R_i are known, we can compute the average value $\text{AVG}(R_i) = \text{SUM}(R_i)/|R_i|$. Instead of determining which data cells of R_i to include in R_i', we consider a single a decision variable b_i, the size of R_i', and replace the data cells of R_i' by a single cell $\text{AVG}(R_i)$. Thus, Eq. 4 is redefined as follows:

$$\text{minimize } \frac{1}{m} \sum_{i=1}^{m} \frac{|\text{SUM}(R_i) - b_i \times \text{AVG}(R_i)|}{\text{SUM}(R_i)} = |1 - \sum_{i=1}^{m} \frac{1}{m \times |R_i|} \times b_i| \tag{5}$$

From Eq. 5, our allocation problem needs only the size of each R_i in $\mathcal{R}$. Therefore, our LP problem is stated as follows:

$$\begin{aligned} &\textbf{minimize } |1 - \sum_{i=1}^{m} \frac{1}{m \times |R_i|} \times b_i| \\ &\textbf{subject to } \sum_{i=1}^{m} b_i <= \mathcal{B} \\ &\text{minimum}_{space} <= b_i < |R_i| \text{ for each } i \in \{1, \ldots, m\} \end{aligned} \tag{6}$$

The first constraint of Eq. 6 ensures that the space allocated to all range queries does not exceed the given space $\mathcal{B}$. As for minimum_{space}, it is an input parameter passed to the algorithm to ensure that each query gets a minimum space allocation. The solution of Eq. 6 will give us the allocated space b_i for each R_i, and since each R_i consists of a group of subqueries, we need to divide the b_i and give each subquery its appropriate allocation. To do this, we reuse our LP problem to calculate the space allocation for each subquery of R_i but with $\mathcal{B}$ equal to the space found for R_i.

4.3   View Creation Step

After the allocation step, each query (and its subqueries) in $\mathcal{R}$ would be allocated a space (in number of cells). To prepare the data cube view for publishing, the next step is to find the cells to keep in each region R (working on each of its subregions) relative to $\mathcal{A}_e$ and $\mathcal{I}_e$. To this end, we have designed two algorithms, the first based solely on cell suppression and the second being a perturbation-based approach.

Cell Suppression Algorithm. To get the optimal $\mathcal{A}_e$ and $\mathcal{I}_e$, we have to try all possible combinations of cells (that fit in the allocated space b), which is inconvenient and computationally expensive. So an approximate solution might be a better way to solve this problem. In [3], they used a heuristic that considers first the utility (by selecting outliers or the largest values), and privacy in second order.

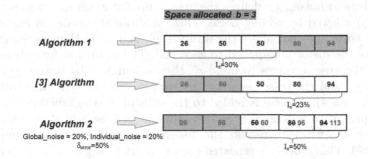

Fig. 2. Output comparison between Algorithm 1, Algorithm 2 and [3].

Algorithm 1. Cells suppression **Algorithm 2.** Cell perturbation

```
Inputs: r - subquery region
        b - allocated space
Output: r' - view of r
  r' ← []
  T ← sort in asc. order r by values
  (index, best_index, best_Ie) ← (0, 0, 0)
  while index + b < |r| do
    r' ← get b values of T
         from position index
    if Ie(r, r') >= best_Ie then
      best_Ie ← Ie(r, r')
      best_index ← index
    end if
    index ← index + 1
    r' ← []
  end while
  r' ← get b values of T
       from position best_index
  return r'
```

```
Inputs: r - subquery region
        b - allocated space
        noise_global - maximum distortion
        noise_individual - distortion rate
        δerror - reachable level of Ie
Output: perturbed_r' - noised view of r
  r' ← get b largest values from r
  noise_budget ← SUM(r) × (1 + noise_global) − SUM(r')
  noise_step ← 1
  perturbed_r' ← r'
  noise ← 0
  while (Ie(r, perturbed_r') < δerror) and
        (noise < noise_budget) and
        (noise_step < noise_individual) do
    perturbed_r', noise ← add_noise(r', noise_step)
    if noise > noise_budget then
      perturbed_r', noise ← add_noise(r', noise_step − 1)
      return perturbed_r'
    end if
    noise_step ← noise_step + 1
  end while
  perturbed_r', noise ← add_noise(r', noise_step)
  return perturbed_r'
```

Relegating $\mathcal{I}_e$ to the second order of priority does not guarantee the best results in all cases, so we propose Algorithm 1 that optimizes $\mathcal{I}_e$ in priority.

Given a sub-region r of a region R and space allocation b, the algorithm first sorts the cells by values (each cell contains a value and a location in the cube) in ascending order. It computes iteratively the maximal value of $\mathcal{I}_e$ by considering successive and overlapping b values. At the end, Algorithm 1 constructs r' with the subset of cells that gave the best results (R' is composed of all r'). Figure 2 shows by an example the difference between the results provided by our heuristic and [3]. In our solution, we give priority to $\mathcal{I}_e$, but without neglecting $\mathcal{A}_e$. If the first and last b values yield the best (and similar) $\mathcal{I}_e$, our solution will choose the last b values. Since the values are ordered, this set of cells offers the least (best) $\mathcal{A}_e$.

Cell Perturbation Algorithm. When the values are close to each other, cell suppression alone will not be able to provide good $\mathcal{I}_e$. For these cases, we have designed Algorithm 2, which is based on cell suppression and perturbation and can be triggered when Algorithm 1 does not perform well. In addition to r and b, it takes three other parameters: δ_{error} represents the level of $\mathcal{I}_e$ we want to reach, $noise_{individual}$ limits the noise (distortion rate) added to each cell individually, and $noise_{global}$ defines the maximum distortion rate between r and r' (i.e., $\mathcal{A}_e$) allowed by adding noise. These additional thresholds ensure that the added noise does not affect the utility (i.e., $\mathcal{A}_e$) more than necessary, and also control the distortion applied to each cell. The algorithm first chooses the cells with the largest values to build r', then calculates the $noise_{budget}$ based on $noise_{global}$ and the noise generated by suppressing the cells (discarding the most small values). A noise is added to the cells of r' (the function add_noise returns a perturbed version of r' and the amount of noise added) incrementally using $noise\_step$ which represents the distortion rate (e.g. in Algorithm 2 it is equal to 1%). This process is repeated each time with bigger $noise\_step$ as long as: (i) $noise_{global}$ is not exceeded (ii) δ_{error} is not satisfied, and (iii) $noise\_step$ is smaller than $noise_{individual}$. Figure 2 shows that the perturbation method provides better utility and privacy than other suppression-based algorithms.

Using these different thresholds allows us to better control the balance between utility and privacy. For example, by allowing more noise (accuracy limit defined by $noise_{global}$), we can ensure a higher δ_{error}.

5 Experimental Evaluation

For experimental evaluation, we implemented[1] our method and [3] approach (as this one is closest to our work) in C# to facilitate compatibility with SSAS[2] used to create data cubes from TPCDS[3] and AdventureWork2012[4] (see Table 1 for the details on the cubes). For solving our LP problem in allocation step, we used Google OrTools with SCIP solver[5]. To measure the performance of our method, we conducted several experiments to observe how $\mathcal{A}_e$ and $\mathcal{I}_e$ vary depending on two parameters. The first parameter is the allocated space $\mathcal{B}$ which we varied relatively (e.g., 50%) from the original size of $\mathcal{R}$ in the data cubes in order to check the performance of $\mathcal{A}_e$ and $\mathcal{I}_e$ during the allocation step. The second parameter is the selectivity of each range query $R = \{r_1, r_2, \ldots, r_n\}$ defined as $\|R\| = (|r_1| * |r_2| * \ldots * |r_n|)$ where $|r_i|$ represents the length of the range on the i_{th} dimension. In the experiments, we create a random query workload $\mathcal{R}$, with initial selectivity for all queries. Next, we modify (multiply by 2, 3, ...) the initial selectivity to observe how the performance changes if the size of the input data increases (with $\mathcal{B}$ between 50% and 60%). Our experiments are divided into two parts: the first part compares our method to [3] in terms of cell suppression and

---

Table 1. Data cubes used in our experiments.

Databases	Dimensions					Measures	Total
TPCDS	Product 18000	Household 7200	Store 6	Date 1824	Promo 300	Net paid Type reel	~4.25 E14 cells
Adventure works	Product 158	Date 1124	Customer 18484			InternetSales Type reel	~36 M cells

space allocation. The second part is devoted to our method in order to compare the techniques of suppression and perturbation of cells.

In the first part, we used the TPCDS data cube to compare both methods in terms of space allocation using the cell suppression technique of [3] (**Experiment (A)**) and our cell suppression technique (**Experiment (B)**). In Fig. 3, we find the expected behavior of our method in both experiments (**A**) and (**B**). Indeed, both $\mathcal{A}_e$ and $\mathcal{I}_e$ converge to 0 when the given allocation is too large. This is something that [3] cannot replicate, because their allocation scheme is based on the data distribution (i.e., it relies heavily on variance analysis). If the data in the query workload is not balanced according, their allocation scheme fails to distribute the allocated space. Unlike our method, [3] gives more space than necessary to some regions and neglects others (for example, this is visible when the allocation is given at 90%), [3] still fails to get a $\mathcal{A}_e$ minimal. As for selectivity in both experiments (**A**) and (**B**), Fig. 3 shows that our method is able to provide better results in $\mathcal{A}_e$ and $\mathcal{I}_e$ when the query size increases. Despite an evolution of the results in the right way for the experiment (**A**), [3] nevertheless presents worse results in the experiment (**B**) when the size of the queries increases. From Fig. 3, we can also see that our view creation step provides better results than the suppression heuristic of [3] as shown in experiments (**A**) and (**B**). From this first part of the experiments, we conclude that our method (allocation and view creation steps) outperforms [3].

For the second part of our experiments, we seek to compare the effectiveness of the perturbation algorithm Algorithm 2 and the cell suppression algorithm Algorithm 1. In addition to TPCDS, we used the Adventure Work data cube because it contains cell values that are close to each other, which allows us to better highlight the performance of both algorithms. To perform these experiments, we used Algorithm 2 with the following parameters: $noise_{global} = 20\%$, $noise_{individual} = 27\%$, $\delta_{error} = 50\%$. Applying both algorithms to the TPCDS data cube, the results for $\mathcal{A}_e$ and $\mathcal{I}_e$ are similar in all cases as illustrated in Fig. 4. Whether by allocation or selectivity, our perturbation technique did not add significant noise. On the other hand, in the case of the adventure work data cube, we find that cell suppression (see Algorithm 1) alone does not provide any level of privacy. This is due to the closeness of the cell values giving poor results for $\mathcal{I}_e$ as explained in Sect. 4.3. However, we find that the perturbation technique (see Algorithm 2) is far superior in terms of performance for $\mathcal{A}_e$ and $\mathcal{I}_e$. From the allocation, the perturbation provides less $\mathcal{A}_e$ with much more $\mathcal{I}_e$. We notice that Algorithm 1 is only able to provide 3% of $\mathcal{I}_e$, while Algorithm 2 is able to

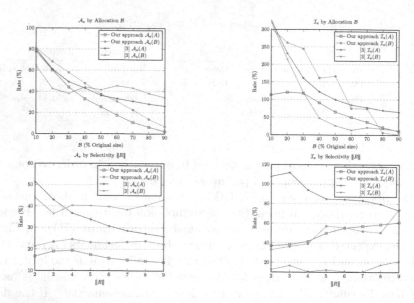

Fig. 3. Comparaison with [3] in terms of selectivity and allocation.

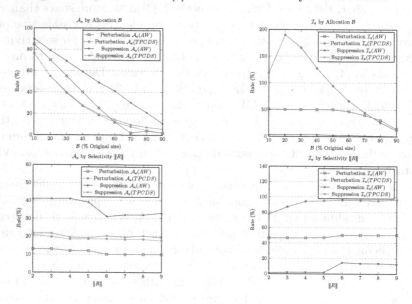

Fig. 4. Comparaison between suppression and perturbation

achieve the desired 50% of $\mathcal{I}_e$. Indeed, Algorithm 2 guarantees a better value of $\mathcal{A}_e$ because it adds noise to cells kept in view, which reduces the effect of deleted cells on $\mathcal{A}_e$. Since cell perturbation is limited by the $noise_{global}$ parameter, it cannot increase the value of $\mathcal{A}_e$ beyond what is allowed. The results also show that Algorithm 2 is able to provide the required δ_{error} passed to the parameters and respects the utility constraint. To sum up, Algorithm 1 may be enough to provide a good balance between Utility/Privacy when the data distribution is uniform. Otherwise, in the case of close values, Algorithm 2 is suitable and gives better results.

6   Discussion

Our method incurs some computational costs and presents some limitations. Despite a cost due to the resolution of our LP problem (see Eq. 6), it should be noted that all computations are performed in offline mode and without impact on the end user. However, comparing our method and [3] in term of computation time, we can see that our allocation algorithm is faster in all experiments because it only requires the count ($|R|$) of each region compared to multiple scans of cells needed for [3]. For view creation, the cell deletion algorithm proposed by [3] takes the least computation time compared to Algorithm 1 and Algorithm 2, due to the fact that their algorithm does not test many subsets to choose the best one for privacy.

We have proposed a LP problem to divide a given space $\mathcal{B}$ over the query workload $\mathcal{R}$ to have minimal utility loss, but we can transform it into a multi-objective optimization problem by including the objective function minimizing $\mathcal{B}$. As said in Sect. 3.2, finding an optimum solution to this kind of optimization problem is hard and intractable task. In addition, to minimize the number of decision variables, we replaced the data cells of the view R'_i by the average value $\mathtt{AVG}(R_i)$ where R_i is the original range (see Eq. 6). Using this simplification, it will eliminate the maximization of the inference error $\mathcal{I}_e$ (see Eq. 2) as $\mathtt{AVG}(R'_i)$ will be replaced by $\mathtt{AVG}(R_i)$. For this reason, we considered privacy only in the third step (see Sect. 4.3) of our method.

7   Conclusion

In this paper, we have proposed a privacy-preserving method for creating sanitized view of a data cubes. Our approach is based on data cube compression (by cell suppression), using a LP model that allows for the best cell deletion while maintaining maximum utility. Given a set of parameters, we also proposed a perturbation algorithm that is able to balance utility and privacy. We conducted extensive experimental tests to evaluate our approach, which was found to give better performances in terms utility and privacy.

In future work, we plan to explore other aspects of privacy to further develop our approach and compare it to well-known standards such as Differential Privacy.

References

1. Agrawal, R., Srikant, R., Thomas, D.: Privacy preserving OLAP. In: Proceedings of the 2005 ACM SIGMOD International Conference on Management of Data, pp. 251–262 (2005)

2. Chatenoux, B., et al.: The Swiss data cube, analysis ready data archive using earth observations of Switzerland. Sci. Data **8**(1), 1–11 (2021)
3. Cuzzocrea, A., Saccà, D.: A theoretically-sound accuracy/privacy-constrained framework for computing privacy preserving data cubes in OLAP environments. In: Meersman, R., et al. (eds.) OTM 2012. LNCS, vol. 7566, pp. 527–548. Springer, Heidelberg (2012). https://doi.org/10.1007/978-3-642-33615-7_6
4. Domingo-Ferrer, J., Sánchez, D., Blanco-Justicia, A.: The limits of differential privacy (and its misuse in data release and machine learning). Commun. ACM **64**(7), 33–35 (2021)
5. Dwork, C., Roth, A., et al.: The algorithmic foundations of differential privacy. Found. Trends Theor. Comput. Sci. **9**(3–4), 211–407 (2014)
6. Grubbs, P., Lacharité, M.-S., Minaud, B., Paterson, K.G.: Pump up the volume: practical database reconstruction from volume leakage on range queries. In: Proceedings of the 2018 ACM SIGSAC Conference on Computer and Communications Security, pp. 315–331 (2018)
7. Hylkema, M.: A survey of database inference attack prevention methods. Educational Technology Research (2009)
8. Kim, S., Lee, H., Chung, Y.D.: Privacy-preserving data cube for electronic medical records: an experimental evaluation. Int. J. Med. Inform. **97**, 33–42 (2017)
9. Nativi, S., Mazzetti, P., Craglia, M.: A view-based model of data-cube to support big earth data systems interoperability. Big Earth Data **1**(1–2), 75–99 (2017)
10. Sarathy, R., Muralidhar, K.: Evaluating Laplace noise addition to satisfy differential privacy for numeric data. Trans. Data Priv. **4**(1), 1–17 (2011)
11. Sung, S.Y., Liu, Y., Xiong, H., Ng, P.A.: Privacy preservation for data cubes. Knowl. Inf. Syst. **9**(1), 38–61 (2006). https://doi.org/10.1007/s10115-004-0193-2
12. Sweeney, L.: k-anonymity: a model for protecting privacy. Int. J. uncertainty Fuzziness Knowl.-Based Syst. **10**(05), 557–570 (2002)
13. Wang, L., Jajodia, S., Wijesekera, D.: Securing OLAP data cubes against privacy breaches. In: IEEE Symposium on Security and Privacy, Proceedings 2004, pp. 161–175. IEEE (2004)
14. Wang, L., Jajodia, S., Wijesekera, D.: Parity-based inference control for range queries. In: Wang, L., Jajodia, S., Wijesekera, D. (eds.) Preserving Privacy in On-Line Analytical Processing (OLAP). ADIS, vol. 29, pp. 91–117. Springer, Boston (2007). https://doi.org/10.1007/978-0-387-46274-5_6
15. Wang, L., Jajodia, S., Wijesekera, D.: Preserving Privacy in On-Line Analytical Processing (OLAP), vol. 29. Springer, New York (2007). https://doi.org/10.1007/978-0-387-46274-5
16. Wang, L., Wijesekera, D., Jajodia, S.: Cardinality-based inference control in data cubes. J. Comput. Secur. **12**(5), 655–692 (2004)
17. Zhang, N., Zhao, W.: Privacy-preserving OLAP: an information-theoretic approach. IEEE Trans. Knowl. Data Eng. **23**(1), 122–138 (2010)

Synthetic Decimal Numbers as a Flexible Tool for Suppression of Post-published Tabular Data

Øyvind Langsrud(✉) [ID] and Hege Marie Bøvelstad [ID]

Statistics Norway, P.O. Box 8131 Dep., 0033 Oslo, Norway
{Oyvind.Langsrud,Hege.Bovelstad}@ssb.no

Abstract. Cell suppression is a widely used statistical disclosure control method for tabular data. Commonly, several linked tables are suppressed simultaneously. After publication, additional tables may be requested. In many contexts, new tables mean new ways of grouping and aggregating data that has already been published. The suppression of the new tables must be coordinated with the tables that have already been disseminated. A certain type of synthetic decimal numbers has proven to be very useful for this purpose. Based on the aggregation of these decimal numbers, one can decide whether a cell should be suppressed or not. An aggregation summing up to a whole number means the same as non-suppression. This article describes the theoretical basis for such decimal numbers. This is based on standard methodology from ordinary linear regression. The method is illustrated by a small example. In addition, two practical applications at Statistics Norway are presented, where one involves large hierarchical and linked tables where more than 50000 unique cells were primarily suppressed.

Keywords: Statistical disclosure control · Confidentiality · Tabular data · Cell suppression · Synthetic data · Linear regression · Official statistics

1 Introduction

Cell suppression is a common statistical disclosure control method for tabular data [5,7]. Then the attacker cannot disclose the value of the sensitive cells exactly, but intervals can still be computed. Releasing intervals instead of suppression is also a method of disclosure control in itself [3]. Another possibility is to release synthetic values as replacements for the suppressed values, and leave unsuppressed values unchanged. Langsrud [8] described such a method as a special application of information preserving statistical obfuscation (IPSO) [2]. Special attention was given to frequency tables. Then, the synthetic values generated are decimal numbers. The method ensures that table additivity is preserved.

© Springer Nature Switzerland AG 2022
J. Domingo-Ferrer and M. Laurent (Eds.): PSD 2022, LNCS 13463, pp. 105–115, 2022.
https://doi.org/10.1007/978-3-031-13945-1_8

A very interesting feature of this type of synthetic data is that any sum of cell values resulting in a whole number corresponds to an unsynthesized true value that does not require protection and is thus publishable. This has proven to be very useful for practical applications, especially for the situation where the new tables requested are linked to already published tables. Then, the main goal of the synthetic data is to use them as a tool for suppression. In particular, this is a possible way to deal with certain frozen cell problems [6]. The present paper explores this method, and compared to [8], the method is slightly reformulated. The description below is based on standard tools from ordinary linear regression combined with random draws from the normal distribution. In this paper, we have chosen to use the term *whole number* instead of *integer*. The paper can be read as if these are synonyms, but in the applications considered, negative integers will not be relevant.

This paper is outlined as follows. In Sect. 2 a motivating example is given, before the theory behind the synthetic decimal numbers is described in Sect. 3. Various technical aspects related to application of the method are discussed in Sect. 4. Finally, Sect. 5 discusses real examples of synthetic decimal numbers applied to tabular data at Statistics Norway. The paper finishes with some concluding remarks.

It is worth pointing out that although the theory outlined in the paper is based on synthetic decimal numbers, we are aware that this method for generating synthetic data is not optimal. We stress that the synthetic data are not the final product but merely used as a means to keep track of the suppression pattern across multiple linked tables and groupings.

2   A Motivating Example

As an example, Table 1 is a two-dimensional frequency table used as a starting point for municipal statistics. The municipalities can be aggregated in two different ways (county and size). A suppressed table of age versus county where frequencies less than four are primarily suppressed is given in Table 2. To make this table, the size categorization was not considered. Suppose that one wants to publish size aggregates afterwards, with the constraint that cells suppressed in Table 2 should not be disclosed. By looking at Table 2 one can derive that the following cells may be published:

- *big:Total*, since this is the sum of already published aggregates
  $(58 + 64 + 37 = 159)$.
- *small:Total* since this can be found by differentiation from the overall total
  $(199 - 159 = 40)$.
- *big:middle*, since this is the sum of *D:middle* and *county-4:middle* already
  published $(12 + 44 = 56)$.
- *small:middle* since this can be found by differentiation from *Total:middle*
  $(72 - 56 = 16)$.

Table 1. Example frequency table, inner cells only.

region			age		
municipality	county	size	young	middle	old
A	county-1	small	1	1	1
B	county-2	small	5	6	8
C	county-3	small	2	9	7
D	county-3	big	35	12	11
E	county-4	big	3	26	35
F	county-4	big	5	18	14

Table 2. Suppressed frequency table. Primary and secondary suppressed cells are denoted by "." and "–" respectively.

region	age			
	young	middle	old	Total
A	.	.	.	.
B	–	–	–	–
C	.	9	–	18
D	–	12	–	58
E	.	–	35	64
F	–	–	14	37
county-1	.	.	.	.
county-2	–	–	–	–
county-3	37	21	18	76
county-4	8	44	49	101
Total	51	72	76	199

Table 3. Synthetic inner cells with aggregates. The synthesis is based on the suppression pattern of Table 2.

region	age			
	young	middle	old	Total
A	2.6845	3.3203	4.4218	10.4266
B	3.3155	3.6797	4.5782	11.5734
C	9.7418	9	−0.7418	18
D	27.2582	12	18.7418	58
E	5.6435	23.3565	35	64
F	2.3565	20.6435	14	37
county-1	2.6845	3.3203	4.4218	10.4266
county-2	3.3155	3.6797	4.5782	11.5734
county-3	37	21	18	76
county-4	8	44	49	101
small	15.7418	16	8.2582	40
big	35.2582	56	67.7418	159
Total	51	72	76	199

Similar reasoning cannot be made for the remainder cells (*small:young*, *big:young*, *small:old*, and *big:old*), and these will be suppressed.

The same information may be obtained in a different way without the use of manual reasoning, but by using the synthetic decimal number approach. Table 3 of synthetic decimal numbers is obtained in two steps:

1. A synthetic version of the 18 frequencies in Table 1, which can be seen as a sub-table of Table 3, is generated. That is, the 14 suppressed frequencies are synthesized. This is done by using the unsuppressed frequencies in Table 2.
2. The result from the first step is used to find the remaining aggregate cells by straightforward summation.

As expected, the unsuppressed cells in Table 2 have become whole numbers in Table 3. As for the size aggregates, which are not included in Table 2, some have become whole numbers, and some have not. All whole numbers in Table 3 match the real numbers. Thus, we may publish values that are whole numbers and suppress the cells containing decimal numbers.

This simple example illustrates how the synthetic data can be used as a stencil tool to determine the suppression pattern of the new aggregates. Note that suppressing new data in this way is based on what can be released according to already published data. This is a quick method that avoids a new round of suppression where everything must be coordinated. In many cases, however, more cells can be published if such a new coordinated suppression is performed. An additional point is that it is advantageous to include all known crossings and groupings in the initial suppression. This includes crossings and groupings that may not be relevant for the initial publication, but that one expect to be relevant in additional table requests from the data.

The synthetic frequencies may also be published directly as a replacement for real frequencies. Note, however, that the underlying linear regression model is not realistic. This is evident in Table 3 where we obtained a negative number.

3   A Theoretical Framework

The microdata underlying tabular data of sum aggregates can be represented by an aggregated data set with one row per unique combination of all the dimensional variables involved. This data set will have one column per aggregated measurement variable (magnitude variable) in addition to a column representing the frequencies of the unique combinations in the microdata. We refer to this type of aggregated data as *the inner table*. Depending on the problem, variable combinations that are not represented in the microdata can be included in the inner table with zeros. Multiple linked tables for publication can be based on a common inner table.

We will now consider a single value variable within the inner table, either a measurement variable or the frequency variable. Here we denote this column in the inner table as the vector y. We denote the vector of all the aggregates based on y for all involved linked tables as z_{all}. Values in z_{all} may also be single values from y.

Now, z_{all} can be computed from y via a dummy matrix X_{all}:

$$z_{all} = X_{all}^T y. \tag{1}$$

We assume that the data have been subjected to suppression. Without loss of generality, we assume that the elements in z_{all} are arranged so that z_{all} and X_{all} can be partitioned so that Eq. (1) can be written as

$$\begin{bmatrix} z \\ z_d \\ z_s \end{bmatrix} = [\, X \ X_d \ X_s \,]^T y. \tag{2}$$

Here z_s consists of the suppressed cells and X_s consists of the corresponding columns of X_{all}. To handle collinearity, the remaining part of X_{all} is divided into two. Matrix X consists of linearly independent columns and X_d depends linearly on X. That is, X_d can be written as

$$X_d = XM \tag{3}$$

where M is a matrix.

Without considering the actual distribution of y, we will use tools from ordinary linear regression to generate a synthetic version of y. By regressing y onto X, we draw synthetic inner cells, y^*, as

$$y^* = \widehat{y} + e^* \tag{4}$$

where

$$\widehat{y} = Hy \tag{5}$$
$$e^* = (I - H)u^* \tag{6}$$
$$u^* \sim \mathcal{N}\left(0, \sigma^2 I\right) \tag{7}$$
$$H = X(X^T X)^{-1} X^T \tag{8}$$

where $\mathcal{N}$ refers to the multivariate normal distribution and where I is the identity matrix. This means that the elements of u^* are independently and identically normally distributed with variance σ^2. Here, H is the hat matrix, which is the projection matrix that maps observed values to fitted values according to a standard regression model ($y = X\beta + \text{error}$). Thus, y^* is generated in a way that preserves the regression fits. The distribution of y^* follows from multivariate normal distribution theory:

$$y^* \sim \mathcal{N}\left(\widehat{y}, \sigma^2(I - H)\right) \tag{9}$$

And further it follows that the univariate distribution of a linear combination is

$$c^T y^* \sim \mathcal{N}\left(c^T \widehat{y}, \sigma^2(c^T c - c^T H c)\right) \tag{10}$$

where c is a vector of linear combination coefficients.

These synthetic data have interesting and useful properties that we will formulate as four propositions.

Proposition 1. *Synthetic inner cells, y^*, depends on y through the unsuppressed published cells only.*

The proposition follows from $Hy = X(X^T X)^{-1} z$.

Proposition 2. *The values of the unsuppressed published cells can be exactly recalculated from y^*.*

To show this proposition one can calculate $X^T y^* = z$ and $X_d^T y^* = z_d$ by insertion from Eqs. (2)–(6) and (8). In particular, $X^T(I - H) = 0$.

Proposition 3. *Assume a fixed $\sigma^2 > 0$. Then, the value of any fixed linear combination of synthetic inner cells, $c^T y^*$, is with probability one a value different from a whole number unless the linear combination can be found as a linear combination of unsuppressed published cells.*

The distribution of $c^T y^*$ is given in (10). It follows from integral theory that for any normal distribution, the probability of a whole number is zero. The only exception is the degenerate case where the variance is zero. Since H is a projection matrix, $c^T H c = (Hc)^T Hc$. It follows that the variance in (10) is zero if and only if $Hc = c$. Thus, c is within the column space of X, which means that c can be written as Xb where $b = (X^T X)^{-1} X^T c$. Then, according to Proposition 2, $c^T y^* = b^T (X^T y^*) = b^T z$.

Proposition 4. *Assume a fixed $\sigma^2 > 0$. Whenever the value of a fixed linear combination of synthetic inner cells is a whole number, it can be determined that this is a combination of published cells.*

This can be said to be a logical consequence of Proposition 3 and we will not introduce more formalism.

Proposition 5. *Assume a fixed $\sigma^2 > 0$. Whenever the value of two independent draws of a fixed linear combination of synthetic inner cells give exactly the same result, it can be determined that this is a combination of published cells.*

The rationale for this proposition is in line with the discussions above. Two draws from a normal distribution will never be identical unless the variance is zero. The zero variance case is also the reason for Proposition 3.

4   Application of the Theory

One possibility is to apply the theory to replace suppressed cells by synthetic values. Then we compute

$$z_{all}^* = X_{all}^T y^*. \tag{11}$$

According to Proposition 2, the non-suppressed cells are preserved. As follows from Proposition 1, the synthetic values of the suppressed cells can safely be published. Synthetic replacements for suppressed cells can be made in several ways. The method proposed here, which is based on ordinary linear regression, may not be the best. In the present paper we focus on another application. The synthetic inner cells can be stored without confidentiality requirements, and they can be aggregated when needed. From these aggregates it can be possible to determine suppression. This is most useful when all inner cells are included in z_{all}. In practice, this mean that some inner cells are suppressed, and some are not.

4.1 Application to Frequency Tables

Application to frequency tables has already been illustrated in Sect. 2. Any group of inner synthetic frequencies can be aggregated. Whenever aggregates are whole numbers, they can safely be published as true unsuppressed frequencies. Other aggregates will be suppressed or alternatively the synthetic values can be published.

4.2 Precision and Implementation

In practice, a precision limit is needed to determine a whole number. With standard double precision numbers, a reasonable choice of the tolerance is 10^{-9}. Then, the probability of obtaining a whole number by chance is relatively low. Usually, the precision error will also be smaller than this tolerance, so that the actual whole numbers are preserved.

For better control and for large data sets, it may be worthwhile to generate multiple (e.g. 3) synthetic values. Thus, the tolerance can be increased (to e.g. 10^{-5}) so that larger precision errors are handled. The probability of incorrect whole numbers is kept low, since identical whole numbers are required for all replicates.

Another possibility is to generate synthetic values that do not represent the true frequencies. This can be done whenever the synthetic values are only used to determine suppression. Then, one can let the y-value be one or zero for all the inner cells. The latter means that e^* (6) is the vector of synthetic values. Looking for whole numbers then boils down to looking for zeros.

When implementing the methodology in practice, X and y should be reduced as much as possible. Rows of unsuppressed inner cells can be removed and the occurring redundant columns of X can also be removed.

4.3 Application to Magnitude Tables

Common magnitude tables consist of sum aggregates of measurement variables. These are not necessarily whole numbers. But even if they are, the decision of suppression based on synthesizing these numbers is not recommended. The reason is that some cell values can be very large, and it can be difficult to control the precision errors. For the same reason, the method that checks equality of two independent draws (Proposition 5) is also not recommended. To overcome these difficulties, one may generate synthetic values that do not represent the real magnitude values in the published table. This was described in detail in Sect. 4.2 above. Such synthetic values should be used only to determine suppression. If the aim of the synthetic values is to replace suppressed values, the main approach based on real inner cell values may be used. The underlying distribution assumptions are, however, questionable.

4.4   Detection of Disclosure Risk

The suppression pattern of the initial data may be re-identified from the synthetic decimal numbers. One must be aware that this may not work if the suppressed data have a residual disclosure risk. Suppressed cells revealable by linear relationships become whole numbers. This can be the case if linked tables are handled by an iteration routine that does not consider possible disclosure by combining information from all the tables. If such a residual disclosure risk is considered acceptable, then synthetic decimal numbers are not recommended. Suppressed cells can also be revealed as whole numbers if there is an error in the program or algorithm underlying the suppression. Therefore, a spin-off usage of synthetic decimal numbers is testing of suppression routines. This is now in use through the R-package easySdcTable [9].

5   Real Applications

5.1   Register-Based Employment Statistics

Statistics Norway's register-based employment statistics are based on employment information about persons between 15 and 74 years old. Confidentiality treatment of the statistics involves suppression. Figures concerning only one or two private enterprises are usually primarily suppressed. The suppression pattern must be consistent across several tables where municipalities and industries are aggregated differently. The latter contains many special groupings based on NACE codes (European classification standard). In addition, commissioned statistics are produced afterwards. Then municipalities and industries can be aggregated in new, previously unknown ways.

Now, all such employment statistics are handled by synthetic decimal numbers. First an inner table with synthetic decimal numbers are produced. Then the suppression pattern for any table, commissioned or not, is found by aggregating this inner table.

To illustrate how large and complex data the synthetic decimal number method can handle, we consider data from year 2019. Here, the dimension of the dummy matrix, X_{all} (1), within this process is 80459×171678. Duplicate columns have then been removed. Explicitly, there were 80459 inner cells and 171678 unique cells considered for publication. Of the 171678 cells, a total of 79373 cells were suppressed, of which 50198 were primarily suppressed.

In this application the frequencies of employees were synthesized to decimal numbers. Due to many and large frequencies, three replicates of the synthetic values were generated. The tolerance was set to 10^{-5}, which ensures that three whole numbers due to chance are very unlikely.

The process of generating and managing the synthetic frequencies was not very computationally intensive compared to the secondary suppression. In this case, the secondary suppression was performed using the relatively fast Gaussian elimination algorithm [10]. This took 127 min on a Linux server, while the generation of the synthetic numbers took an additional 28 min.

5.2   Commissioned Data in Business Statistics

The synthetic decimal number method has also been successfully applied to business statistics involving magnitude tables at Statistics Norway. Business statistics are of wide interest to both the public and decision makers, and after publication of statistics in this field Statistics Norway often receive requests on additional tables. The type of requests are typically unknown at the time of publication, but still an important task for Statistics Norway in order to fulfill their role as information provider.

To ensure that the suppression pattern in the commissioned tables accompany that of the already published tables, we collected a historical set of commissioned tables to investigate which variables where requested, and at what level of detail. We found that the requests often involved tables at more refined levels of geographical regions and/or economic activity (NACE codes). There were also requests on non-standard groupings of both dimensional variables. The vast number of requests involved the four measurement variables turnover, wages and salaries, value added and gross investments. Prior to the main publication of business statistics, we therefore made sure to apply cell suppression and the synthetic decimal number method to these four variables at the most detailed geographical level and on 5-digit NACE level. Two extra levels of geographical granularity were included compared to the tables that Statistics Norway usually publish. There is of course a risk of obtaining much more suppressions in the main publications when allowing a more granular representation. In this application, however, the increase in suppressions as a result of finer dimensional granularity were only between 1% and 2.5% for the four measurement variables.

Since business statistics often contain variables with a large number of digits (e.g. gross investments), we applied the synthetic decimal number method to the frequencies instead of the measurements (c.f. Sect. 4.3). Thus, four sets of synthetic frequencies were made from the suppression patterns of the four measurement variables. Note that these data are of smaller size and complexity compared to the data in Sect. 5.1. Within a Windows environment, the run-time for the suppression algorithm [10] was about 30 min. In addition, the extra time used to generate decimal numbers was about 5 min.

Post publication, we will now be able to accommodate any request on these four variables at various geographical regions and NACE levels. The flexibility of the synthetic decimal number method makes the task of building new and consistently protected tables as simple as aggregating the decimal numbers according to the new table. Many requested tables can be generated very easily, and a demanding new round of coordinating suppression will thus be avoided. However, we are aware that this can be at the expense of the number of suppressed cells.

6 Concluding Remarks

In [8] synthetic decimal numbers were described as a special case of IPSO [2]. This means that e^* (4) is rescaled so that the variance of the residuals is preserved. Such preservation is not necessary in our suppression application. In fact, extra re-scaling afterwards was suggested in [8] to prevent information leakage through the variance. Note that the general multi-response framework of IPSO is also not needed here.

An issue concerning the implementation is how $\hat{y}$ should be calculated. In R, the standard regression method (lm) make use of the QR decomposition [4]. In [8] IPSO was formulated by a generalized QR decomposition. Implementation by QR is not directly generalizable to large sparse matrices. An effective implementation that has now been used avoids QR and corresponds well with the theory described above. That is, the first step is to find the matrix X of linearly independent columns. To do this, Gaussian elimination is applied. The algorithm is implemented in R and is a simplification of the algorithm used for secondary suppression [10, 11]. Thereafter, regression fits can be found efficiently according to [1]. The whole process of generating and using such decimal numbers is made available as extra features in the R package GaussSuppression [10].

All in all, this article has presented synthetic decimal numbers in a simplified way that fits well with effective implementation. The properties are thoroughly documented through Propositions 1–5.

Decimal numbers can be used to determine the suppression pattern of additional tables requested after the publication of pre-planned tables. The aggregation of decimal numbers is all that is needed, and this is an easy alternative to running a new suppression routine that needs to be coordinated with published tables. But it should also be mentioned that extra coordinating suppression can result in fewer suppressed cells. As is often the case, the pros and cons must be considered. The synthetic decimal method is already implemented and in use at Statistics Norway, as described in Sect. 5.

References

1. Bates, D., R Development Core Team: Comparing Least Squares Calculations (2022). https://cran.r-project.org/package=Matrix (r Vignette)
2. Burridge, J.: Information preserving statistical obfuscation. Stat. Comput. **13**(4), 321–327 (2003). https://doi.org/10.1023/A:1025658621216

3. Castro, J., Via, A.: Revisiting interval protection, a.k.a. partial cell suppression, for tabular data. In: Domingo-Ferrer, J., Pejić-Bach, M. (eds.) PSD 2016. LNCS, vol. 9867, pp. 3–14. Springer, Cham (2016). https://doi.org/10.1007/978-3-319-45381-1_1

4. Chan, T.F.: Rank revealing QR factorizations. Linear Alg. Appl. **88**, 67–82 (1987). https://doi.org/10.1016/0024-3795(87)90103-0

5. Fischetti, M., Salazar, J.J.: Solving the cell suppression problem on tabular data with linear constraints. Manag. Sci. **47**(7), 1008–1027 (2001). http://www.jstor.org/stable/822485

6. Giessing, S., de Wolf, P.P., Reiffert, M., Geyer, F.: Considerations to deal with the frozen cell problem in Tau-Argus Modular. In: Joint UNECE/Eurostat Expert Meeting on Statistical Data Confidentiality, 1–3 December 2021, Poznań, Poland (2021)

7. Hundepool, A., et al.: Statistical Disclosure Control. John Wiley & Sons, Ltd., Boca Raton (2012). https://doi.org/10.1002/9781118348239.ch1

8. Langsrud, Ø.: Information preserving regression-based tools for statistical disclosure control. Statist. Comput. **29**(5), 965–976 (2019). https://doi.org/10.1007/s11222-018-9848-9

9. Langsrud, Ø.: easySdcTable: Easy Interface to the Statistical Disclosure Control Package 'sdcTable' Extended with the 'GaussSuppression' Method (2022). https://CRAN.R-project.org/package=easySdcTable (r package version 1.0.3)

10. Langsrud, Ø., Lupp, D.: GaussSuppression: Tabular Data Suppression using Gaussian Elimination (2022). https://CRAN.R-project.org/package=GaussSuppression (r package version 0.4.0)

11. Lupp, D.P., Langsrud, Ø.: Suppression of directly-disclosive cells in frequency tables. In: Joint UNECE/Eurostat Expert Meeting on Statistical Data Confidentiality, 1–3 December 2021, Poznań, Poland (2021)

Disclosure Risk Assessment and Record Linkage

The Risk of Disclosure When Reporting Commonly Used Univariate Statistics

Ben Derrick$^{(\boxtimes)}$ ⓘ, Elizabeth Green ⓘ, Felix Ritchie ⓘ, and Paul White ⓘ

University of the West of England, Bristol, UK
ben.derrick@uwe.ac.uk

Abstract. When basic or descriptive summary statistics are reported, it may be possible that the entire sample of observations is inadvertently disclosed, or that members within a sample will be able to work out responses of others. Three sets of univariate summary statistics that are frequently reported are considered: the mean and standard deviation; the median and lower and upper quartiles; the median and minimum and maximum. The methodology assesses how often the full sample of results can be reverse engineered given the summary statistics. The R package uwedragon is recommended for users to assess this risk for a given data set, prior to reporting the mean and standard deviation. It is shown that the disclosure risk is particularly high for small sample sizes on a highly discrete scale. This risk is reduced when alternatives to the mean and standard deviation are reported. An example is given to invoke discussion on appropriate reporting of summary statistics, also giving attention to the box and whiskers plot which is frequently used to visualise some of the summary statistics. Six variations of the box and whiskers plot are discussed, to illustrate disclosure issues that may arise. It is concluded that the safest summary statistics to report is a three-number summary of median, and lower and upper quartiles, which can be graphically displayed by the literal 'boxplot' with no whiskers.

Keywords: SDC · Statistic · Disclosure · Control · Summary · Quartile · Boxplot

1 Introduction

In statistical analyses there is potential conflict between providing useful results and protecting the confidentiality of individuals within the data [1]. Given commonly reported univariate summary statistics, it may be possible to construct the exact frequencies of values within a sample, which in many contexts would be unwarranted disclosure.

For illustrative purposes, consider a four-point scale for reporting health on a survey (1 = Good health, 2 = Fair health, 3 = Bad health, 4 = Very bad health). In a summary of

© Springer Nature Switzerland AG 2022
J. Domingo-Ferrer and M. Laurent (Eds.): PSD 2022, LNCS 13463, pp. 119–129, 2022.
https://doi.org/10.1007/978-3-031-13945-1_9

the results separated by gender and ethnicity, assume the following means and standard deviations (SD) for males are reported:

White: N = 18 Mean = 2.06 SD = 0.998
Mixed: N = 8   Mean = 2.00 SD = 0.926
Asian: N = 6   Mean = 2.67 SD = 0.816
Black: N = 5   Mean = 2.00 SD = 0.000
Other: N = 1   Mean = 5.00 SD = 0.000

There is debate regarding ascribing numeric values to ordinal data for analyses. This is frequently done in practice, and is might not be unreasonable when pragmatic assumptions of equal distance between groups are stated [2].

In the example, the ethnic groups with standard deviation equal to zero must all have reported the same value (the group mean). It may not be as straightforward to reverse engineer the frequencies for the remaining three groups, but the R package uwedragon will show the plausible frequency distributions for a given sample size, mean and standard deviation [3]. For the 'Asian' and 'Mixed' groups there are only two frequency distributions possible with the stated means and standard deviations. For the 'White' group there are four frequency distributions possible. Table 1 is the table of results as given by Lowthian and Ritchie [4].

Table 1. Health survey responses for males by ethnicity.

	Good	Fair	Bad	Very bad	Total
White	6	7	3	2	18
Mixed	2	2	3	1	8
Asian	1	0	5.	0	6
Black	0	5	0	0	5
Other	0	0	0	1	1
Total	9	14	11	4	38

As Lowthian and Ritchie [4] state, from this table we draw several conclusions:

- The single male who does not identify with any of the ethnic groups has 'Very bad health'. This can cause group attribute disclosure but not necessarily reidentification.
- All of the individuals who identify as Black have 'Fair health'.
- The one Asian who responded that he enjoys 'Good health' knows that his Asian colleagues all report 'Bad health'.

The reporting of mean and standard deviation in addition to, or as an alternative to, the frequency table may also have similar associated disclosure risk. As stated, there are two possible solutions for the Asian category with the given mean and standard deviation. The R package uwedragon shows that sample values are either {1, 3, 3, 3,

3, 3} or {2, 2, 2, 3, 3, 4}. Thus, again if an individual within this grouping reported 1 'Good health', he can work out that his Asian colleagues all reported 3 'Bad health'. If in fact the second solution had been true, then the person reporting a 4 'Very bad health' would know that all of the other respondents in that grouping had reported better health.

The uwedragon package can help identify the level of risk by supplying detail of the possible solutions for a given sample size, mean and standard deviation. Furthermore, the uwedragon package offers suggestions for disguising the mean and standard deviation when the risk level is high but there is still a need to report these figures [3]. The addition of noise in this way, or a similar manner, reduces the risk of reconstruction [5].

It may be that there are other less disclosive summary statistics that could be alternatively reported. An alternative measure of location to the mean could be the median. Likewise, an alternative measure of variability to standard deviation would be interquartile range, taken from lower quartile and upper quartile. These alternatives are based simply on an ordered location point, so will result in a reduced capacity to reconstruct an entire set of values or identify extreme observations. Algorithms for estimating the mean and standard deviation based on actual median, range and sample size can be utilized [6]. Reporting median, range and sample size alongside estimates for the mean and standard deviation would have a disguise effect reducing the risk of reconstruction.

This paper provides methodology and results that raise awareness of the potential disclosure risk when reporting only the mean and standard deviation, particularly for small measurement scales and small sample sizes. This paper further considers the use of alternative summary statistics that may be less disclosive in these situations. The alternative summary statistics considered are: either only the median, lower quartile and upper quartile; or only the median, minimum and maximum.

2 Methodology

We consider a scale restricted to k defined points for a sample of size n. The total sample space is the number of combinations for the values 1 to k in a sample of size n. Univariate summary statistics of each combination within the sample space are calculated and compared to the same summary statistics for each other combination within the sample space. A high proportion of combinations within the sample space that can be uniquely identified by the given summary statistics is a high disclosure risk.

For example, for a $k = 5$-point scale with $n = 5$, the total sample space is 129. The combination of sample values {1, 2, 2, 3, 4}, has mean $\bar{x} = 2.40$ with standard deviation $s = 1.14$. No other combination within the sample space gives this same $\bar{x}$ and s, and this is referred to as a unique identification. In fact, 87 of the 129 possible combinations for k = 5 and n = 5 can be uniquely identified by their mean and standard deviation.

Using the approach by Derrick et. al. [7], we report the total number of possible different sample configurations for sample sizes $n = 3, 4, 5, \ldots 10, 11, 12$. We then report the number of these samples which can be uniquely identified through knowing the mean and standard deviation when reported with full precision, and when reported to two decimal places or one decimal place (divisor of variance used $= n - 1$).

A summary of the results is given for a 7-point scale and a 10-point scale (with additional scales in the appendix). These tables summarise those situations where there is a unique one-to-one correspondence between $(\bar{x}, s, n)$ and a sample configuration leading to $(\bar{x}, s, n)$ uniquely identifying the sample which gives rise to $(\bar{x}, s)$.

This methodology is herein extended to give the number of unique solutions when alternative summary statistics are given. Firstly, if only the median, first quartile (Q1) and third quartile (Q3) are reported. Secondly, if only the median, minimum and maximum are reported.

The minimum and maximum values are the true minimum and true maximum from the sample. The median is calculated as the middle value in the ordered sample (or midpoint of two central values if sample size is an even number). The calculation of quartiles differs in common statistical software. We consider several of these approaches calculated using the quantile function in R [8]. Mathematical definition of the methods is given by Hyndman and Fan [9]. SPSS and Minitab both use 'method 6', the R default is 'method 7', whereas the SAS default is 'method 2'.

3   Results

Table 2 gives the number of unique identifications for the given summary statistics reported when data is from an inherent 7-point scale. Table 3 provides the same information for a 10-point scale.

Table 2. Number of unique solutions, data on 7-point scale.

n	Sample space	Mean and SD reported			Median, Q1 and Q3 reported			Median, Min, Max reported
		Full	2dp	1dp	SPSS	R	SAS	
3	84	76	76	76	84	84	84	84
4	210	143	143	143	210	210	180	85
5	462	206	193	193	80	7	7	7
6	924	246	222	200	440	28	4	13
7	1716	295	253	203	0	24	0	7
8	3003	289	289	201	59	59	16	13
9	5005	405	325	215	0	0	0	7
10	8008	438	361	202	3	59	0	13
11	12376	493	397	198	0	0	0	7
12	18564	533	433	213	3	3	0	13

Reporting the median, Q1 and Q3, theoretically contains a smaller number of disclosive scenarios than reporting the mean and SD, when $n > 6$. By virtue of reporting quantiles to decimals of 0 or 0.5 as per method 2 in [9], the approach to calculating quantiles adopted by SAS is the least disclosive, relative to procedures in Minitab, SPSS and R.

Table 3. Number of unique solutions, data on 10-point scale.

n	Sample space	Mean and SD reported			Median, Q1 and Q3 reported			Median, Min, Max reported
		Full	2dp	1dp	SPSS	R	SAS	
3	220	188	188	188	220	220	220	220
4	715	353	353	343	705	705	468	181
5	2,002	509	422	346	128	10	10	10
6	5,005	564	472	332	1072	52	4	19
7	11,440	747	527	310	0	36	0	10
8	24,310	603	603	344	64	64	16	19
9	48,620	955	676	310	0	0	0	10
10	92,378	944	749	338	0	64	0	19
11	167,960	1134	822	286	0	0	0	10
12	293,930	1143	895	291	0	0	0	19

Reporting the maximum, minimum and median has low disclosive risk in terms of the entire set of sample values being revealed, particularly if the maximum or minimum value within a sample is not unique. However, there may be serious misgivings in practice regarding reporting the minimum and maximum. Paradoxically, revealing these values may protect the rest of the sample from being revealed.

The unique solutions for n > 4 when reporting the median, minimum and maximum represent the cases where all sample values are identical. Due to the standard deviation of zero, such combinations are also identifiable if the mean and standard deviation are instead reported. However, reporting the median, 1st quartile and 3rd quartile in these instances does not necessarily reveal all sample values.

When reporting the mean and SD, if the sample space is large, i.e. $k \geq 10$ and $n \geq 10$, the percentage of times the true underpinning sample is discovered is less than 1%.

Summary statistics assessed above may be reported in different combinations. The methodology could be extended to numerous different statistical reporting combinations. For example, the default descriptive statistics option in SPSS leads users to report all of the univariate summary statistics above and also include statistics for skewness and kurtosis. Note that adding additional summary statistics will increase the disclosure risk. For instance, reporting mean and standard deviation with the median, will result in a higher number of unique combinations being revealed than reporting only the mean and standard deviation.

4 Discussion Example

Consider the following hypothetical set of exam marks {1, 40, 50, 55, 58, 58, 60, 62, 65, 66, 66, 68, 70, 71, 72, 74, 75, 75, 80, 85}. Here there is a duty not to reveal individual exam marks when summarising the results, and particular care will have to be taken regarding the lowest scorer.

In this scenario the mean = 62.55 and SD = 17.90. You may take some comfort that the sample size and possible scale combination is too large for the uwedragon package to identify possible distributions of results. However, the high value of the standard deviation indicates the presence of some extreme values, and work could commence on identifying possible maximum and minimum values [7].

The median is midway between the 10th and 11th observation = 66. How the quartiles differ depending on approach used is shown in Table 4. The different approaches to calculating the quartiles could offer further assistance to protecting the data from being reversed engineered to reveal all values, if the method is chosen at random and not reported to the end user. Reporting Q1 and Q3 gives an idea of the spread of the data without revealing information about any potential extreme observations.

Table 4. Calculation of quartiles.

R function	1st quartile	3rd quartile	Note
Quantile (type = 2)	58	73.0	As per SAS
Quantile (type = 6)	58	73.5	As per Minitab & SPSS
Quantile (type = 7)	58	72.5	Default in R
Summary	58	72.5	
Fivenum	58	73.0	
Boxplot	58	73.0	With true min/max

The true minimum is 1, and the true maximum is 85. Note that some statistical software may present alternative minimum and maximum values with subsequent reporting of 'mild' or 'extreme' outliers. However, values in the extremes may be sensitive information, which may not be appropriate to disclose. Here, reporting the minimum is a risk of revealing that the weakest performer scored 1/100 on the exam. Likewise, for scales where there is no upper limit (e.g. salary) it may be more appropriate to report the upper quartile rather than the maximum.

5 Graphical Representation

Graphically, the summary statistics considered in examples like the above are often displayed in a box and whiskers plot, so natural temptation may be to summarise the data in this way. These depictions are often described as a five-number summary of minimum, lower quartile, median, upper quartile and maximum. However, this description is not always entirely accurate, and in fact can disclose many more than five values when 'outliers' are present.

Six variations of box-and-whisker plots or 'boxplots' are considered. Illustrations of each of the variations are given in Figs. 1 through Fig. 6, for the discussion example data. Included below is a description of each variation with a statement of causes for concern. The graphics, including the applicable quartile calculation, are as per the boxplot function in R [8].

1 **Tukey's schematic plot.** This is the traditional box and whiskers plot with inter-quartile range (IQR) × 1.5 for whiskers [10].

- Extreme observations are explicitly revealed.
- For each 'outlier' that is revealed, in addition to values for the 'minimum' and 'maximum', it slightly increases the opportunity for the sample to be reconstructed by a determined individual.

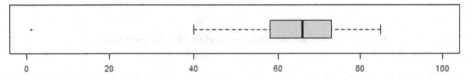

Fig. 1. Turkey schematic plot, traditional box and whiskers plot with IQR*1.5 for whiskers

2 **Box and whiskers plot with mean inserted**

- Reporting both the mean and median may give an indication of the direction and magnitude of extreme observation/s, even if outliers are removed from the plot the position of the mean relative to the median alludes to these extremes.
- The reporting of additional summary statistics increases the opportunity for the sample to be reconstructed by a determined individual.

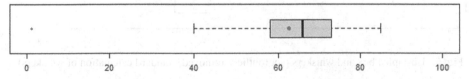

Fig. 2. Box and whiskers plot with mean added

3 Modifications to the traditional calculation of the whiskers within a box and whiskers plot

- Same issues as the traditional box and whiskers plot, but even more observations are explicitly revealed if the multiplier for IQR is <1.5.

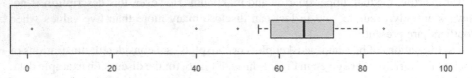

Fig. 3. Modification to traditional calculation IQR*0.5 for whiskers

4 Box and whiskers plot using true minimum and maximum. In this scenario whiskers are not calculated based on IQR, but extend to the full range of the data.

- Maximum and minimum explicitly revealed.
- Distorted impression of distribution if maximum or minimum is an extreme outlier.

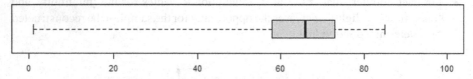

Fig. 4. Box and whiskers plot unmodified for extreme values

5 Unstapled box and whiskers plot. Here whiskers are calculated as per Tukey [10], but outliers are removed. The staples are subsequently removed herein to indicate that there may be extreme values beyond the reach of the whiskers.

- The most extreme observations are not explicitly revealed, but individuals within these missing extremes will be aware that they are 'outliers'.
- Without clear statement of the form of boxplot, incorrect perception of the true maximum and true minimum is possible.

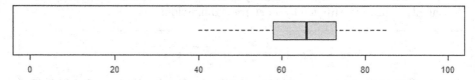

Fig. 5. Unstapled box and whiskers plot (outliers removed, standard calculation of whiskers)

6 **'Boxplot' - literally**. A three-number summary is given: lower quartile; median; and upper quartile. This plot includes no whiskers and no outliers, thus reducing the disclosure risk particularly relating to extreme observations.

- Safest to report, but losing some insight into skewness that may be of interest.

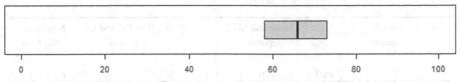

Fig. 6. Literal box plot (whiskers and outliers removed)

6 Conclusion

This paper explores the disclosure risk when reporting univariate summary statistics.

It has been demonstrated that reporting the mean and standard deviation to summarise a sample can result in a disclosure risk. The risk generally decreases with increasing sample size and as the range of possible values on the measurement scale increases. The R package uwedragon can be used to check if reporting the mean and standard deviation for a given sample uniquely identifies the sample values.

To reduce the risk of reconstruction from a sample that uniquely identifies the sample values, noise can be added to summary statistics [5, 7]. In the case of quartiles, the different ways in which these can be calculated, frequently adds what can be described as naturally occurring noise, if the calculation method is not reported.

If concerned about the risk of reporting mean and standard deviation, when $n > 6$ a three-figure summary can instead be reported: median; lower quartile and upper quartile. Although limited to only three values, this can be graphically displayed by the literal 'boxplot' when a basic visualisation of the distribution is desired.

If the sample space is large, and standard deviation is not zero, then the reporting of the mean and standard deviation has a low risk of being fully disclosive of all sample values. However, some indication of extreme values may be apparent for a large standard deviation.

Appendix

See Tables A1, A2, and A3

Table A1. Number of unique solutions, data on 5-point scale.

n	Sample space	Mean and SD reported			Median, Q1 and Q3 reported			Median, Min, Max
		Full	2dp	1dp	SPSS	R	SAS	reported
3	35	33	33	33	35	35	35	35
4	70	56	56	56	70	70	68	41
5	129	87	79	79	48	5	5	5
6	210	105	101	101	151	15	4	9
7	330	131	121	121	0	16	0	5
8	495	141	141	133	39	39	16	9
9	715	177	161	135	0	0	0	5
10	1001	205	181	157	7	39	0	9
11	1365	223	201	130	0	0	0	5
12	1820	243	221	149	7	7	0	9

Table A2. Number of unique solutions, data on 9-point scale.

n	Sample space	Mean and SD reported			Median, Q1 and Q3 reported			Median, Min, Max
		Full	2dp	1dp	SPSS	R	SAS	reported
3	165	145	145	145	165	165	165	165
4	495	271	271	271	493	493	356	145
5	1,287	396	327	286	112	9	9	9
6	3,003	440	364	279	850	44	4	17
7	6,435	527	399	306	0	32	0	9
8	12,870	449	449	284	64	64	16	17
9	24,310	693	499	270	0	0	0	9
10	43,758	701	549	275	0	64	0	17
11	75,582	821	599	246	0	0	0	9
12	125,970	837	649	261	0	0	0	17

Table A3. Number of unique solutions, data on 11-point scale.

n	Sample space	Mean and SD reported			Median, Q1 and Q3 reported			Median, Min, Max reported
		Full	2dp	1dp	SPSS	R	SAS	
3	286	238	238	238	286	286	286	286
4	1,001	443	443	419	971	971	596	221
5	3,003	592	496	386	144	11	11	11
6	8,008	654	530	369	1296	60	4	21
7	19,448	830	580	342	0	40	0	11
8	43,758	652	652	355	64	64	16	21
9	92,378	1080	722	342	0	0	0	11
10	184,756	1044	794	363	0	64	0	21
11	352,716	1263	866	304	0	0	0	11
12	646,646	1232	938	311	0	0	0	21

References

1. Skinner, C.: Statistical disclosure control for survey data. In Handbook of Statistics, vol. 29, pp. 381–396. Elsevier (2009). https://doi.org/10.1016/S0169-7161(08)00015-1
2. Derrick, B., White, P.: Comparing two samples from an individual Likert question. Int. J. Math. Statist. **18**(3) (2017)
3. Derrick, B.: uwedragon: Data Research, Access, Governance Network: Statistical Disclosure Control. R package (2022). https://cran.r-project.org/web/packages/uwedragon/index.html
4. Lowthian, P., Ritchie, F.: Ensuring the confidentiality of statistical outputs from the ADRN. ADRN Technical paper (2017). https://uwe-repository.worktribe.com/output/888435
5. Dinur, I., Nissim, K.: Revealing information while preserving privacy. PODS **2003**, 202–210 (2003)
6. Hozo, S.P., Djulbegovic, B., Hozo, I.: Estimating the mean and variance from the median, range, and the size of a sample. BMC Med. Res. Methodol. **5**(1), 1–10 (2005)
7. Derrick, B., Green, L., Kember, K., Ritchie, F., White, P.: Safety in numbers: Minimum thresholding, Maximum bounds, and Little White Lies: The case of the Mean and Standard Deviation Scottish Economic Society Conference 2022 (2022). www.ses2022.org/sessions/protecting-confidentiality-social-science-research-outputs
8. R Core team: A Language and Environment for Statistical Computing (2021). https://www.R-project.org/
9. Hyndman, R.J., Fan, Y.: Sample quantiles in statistical packages. Am. Stat. **50**(4), 361–365 (1996)
10. Tukey, J.W.: Exploratory Data Analysis, p. 9780201076165. Addison-Wesley, ISBN (1977)

Privacy-Preserving Protocols

Tit-for-Tat Disclosure of a Binding Sequence of User Analyses in Safe Data Access Centers

Josep Domingo-Ferrer[✉]

Department of Computer Engineering and Mathematics,
CYBERCAT-Center for Cybersecurity Research of Catalonia,
UNESCO Chair in Data Privacy, Universitat Rovira i Virgili,
Av. Països Catalans 26, 43007 Tarragona, Catalonia
josep.domingo@urv.cat

Abstract. Safe access centers, a.k.a. trusted research environments, and digital marketplaces allow a user to submit her own statistical analyses on the data stored by controllers. As a responsible organization for the data, the controller may wish to enforce several ethical controls on the user's code. If data are personal or otherwise confidential, the controller needs to make sure the outputs of user analyses do not leak any personal information. Additionally, the data controller may require other ethical properties of the user analyses, like fairness or explainability in case an analysis consists of training a decision model. If a user analysis fails to meet the controller's ethical requirements, the controller stops running the user's code on his data. On the other side, the user would like to achieve fair disclosure of her code, that is, stop disclosing the remainder of her code to the controller in case the controller refuses to return the result of the current analysis. In this concept paper, we present a framework allowing the user: i) to bind herself to a sequence of analyses before receiving any actual output, which precludes disclosure attacks via adaptive analyses; ii) to disclose her next instruction/analysis to the controller only after the controller has returned the output of the current instruction/analysis, which ensures fair disclosure. Our aim is to protect both the controller's and the user's interests.

Keywords: Safe access centers · Digital marketplaces · Trusted research environments · Confidentiality · Output checking · Fairness · Ethics

1 Introduction

There is an increasing demand of data of all sorts for research and decision making. Yet, unrestricted sharing of data is often infeasible, for various reasons. On the one side, personal data fall under the umbrella of data protection legislation, such as the European Union's General Data Protection Regulation [7]. On the other side, there may be data that are confidential for other reasons, such as trade secrets.

© Springer Nature Switzerland AG 2022
J. Domingo-Ferrer and M. Laurent (Eds.): PSD 2022, LNCS 13463, pp. 133–141, 2022.
https://doi.org/10.1007/978-3-031-13945-1_10

Anonymization [5,10] is a possible solution to reconcile the need of data dissemination for secondary uses (different from the primary use that motivated their collection) with the confidentiality constraints. However, anonymization entails some accuracy loss, which may be unaffordable in some secondary analyses. This creates a demand for accurate original data.

Possible workarounds to allow users to compute their analyses on original data that are confidential is for them either to access a remote access center run by the controller (where the users are under surveillance) or to submit the code of the analyses to the data controller who holds the data. This latter case is called remote execution: the controller runs the code on the user's behalf, and returns her the results if these do not leak any of the original data on which they were computed. Examples of such arrangements include:

- National statistical institutes and data archives often feature so-called safe access centers (a.k.a. research data centers or trusted research environments) as an option for researchers who cannot be satisfied with anonymized data. A safe access center may be a physical facility to which the researcher must travel or an on-line service that the researcher can remotely access. For example, Eurostat, the EU statistical office, operates both a physical safe access center in Luxembourg and an on-line safe access center (called KIOSK [6]). Whatever the case, there is an environment in which the researcher runs her analyses using software provided by the controller and is under monitoring by the controller's staff during her entire work session. Furthermore, output checking rules [1,8] are manually enforced by the controller's staff to decide whether the outputs of each statistical calculation by the researcher can be safely returned to her.
- Some decentralized data marketplaces, like Ocean [12], complement their offer of anonymized data (data-as-a-service) with the possibility of running computations on the original data they store (compute-to-data). Yet, they have no solution to thwart data leakages arising from the results of computations.

The approaches to monitoring user analyses sketched above are at best at the level of each single statistical calculation (output checking rules) and aimed only at ensuring the confidentiality of the data held by the controller. In fact, in the case of decentralized data marketplaces, not even the above countermeasures are provided.

Contribution and Plan of This Paper

This is a concept paper where we consider an entire sequence of user analyses rather than a single statistical computation. In our approach:

- The user binds herself to her analysis sequence before seeing any of its outputs. In this way, adaptive analyses aimed at isolating specific confidential records are precluded: all possible strategies to isolate records must be programmed beforehand in the code.

– The user reveals her code on an instruction-by-instruction basis, that is, she discloses her $i+1$-th statistical analysis only after she has received the output of her i-th statistical analysis. In this way, fair disclosure is achieved: if the controller aborts a certain statistical analysis, the user can avoid disclosing the remainder of her planned analyses to the controller.
– Since the user binds herself to a sequence, this sequence can also be certified by some certification authority as being compliant with other ethical properties. For example, if the analyses include training machine learning models, the fairness and explainability of the trained models are relevant ethical properties.

Section 2 gives background on output checking rules. Section 3 presents our scheme for user's binding to a sequence of analyses with tit-for-tat instruction disclosure. Section 4 analyzes how our framework can be leveraged to enforce confidentiality for the controller's data and other ethical properties, while guaranteeing fair disclosure of the user's analyses. Conclusions and future research avenues are gathered in Sect. 5.

2   Background

In [1,8] two slightly different sets of rules of thumb are proposed to decide whether the outputs of various statistical analyses can be safely returned to a user/analyst. "Safely" in this context means without leaking the underlying confidential data.

Each rule can be formalized in terms of the following attributes: *AnalysisType*, *Output*, *Confidential*, *Context* and *Decision*. By way of example, we next quote the rule concerning frequency tables:

AnalysisType: FrequencyTable
Output: Number of units in each cell.
Confidential: YES/NO (YES means the data on which the frequency table is computed are confidential).
Decision: YES/NO

The decision is NO, that is, the output is not returned if data are confidential AND {some cell contains less than 10 units OR a single cell contains more than 90% of the total number of units in a row or column}.

Similar rules are given for other statistical analyses. For example, in [1] rules are specified for magnitude tables, maxima, minima, percentiles, modes, means, indices, ratios, indicators, concentration ratios, variances, skewness, kurtosis, graphs, linear regression coefficients, nonlinear regression coefficients, regression residuals, regression residuals plots, test statistics, factor analyses, correlations and correspondence analyses.

3 Binding to a Sequence of Analyses with Tit-for-Tat Analysis Disclosure

In [3,4], a hash-based coding of programs was proposed to allow a CPU to check the integrity of programs at run-time. The goal of that coding was to detect at run-time any alteration suffered by the code after it was written, *e.g.* due to computer viruses.

We repurpose and adapt that scheme to allow the user, a.k.a. data analyst, to bind herself to a certain sequence of analyses, which she cannot alter as a result of seeing intermediate results. At the same time, the user does not disclose the next analysis in her sequence until she has obtained the results of her previous analysis.

We adapt here the above hash-based coding for a sequence of analyses. The variants in [3,4] for non-sequential programs (with branches) could also be adapted for our purpose in a way analogous to that described in the next sections. Such an adaptation is left for future research.

3.1 Instruction Preparation

Assume the user's sequence of analyses is $i_1, \ldots, i_n$, where each i_j is an instruction describing a certain statistical analysis. Let H be a specified cryptographically strong one-way hash function such that in general $H(X \oplus Y) \neq H(X) \oplus H(Y)$, where $\oplus$ is the XOR operator (addition modulo 2).

We want to transform instructions $i_1, \ldots, i_n$ so that they have a length close but not greater than the output of H. This is necessary for the security properties below.

If all instructions $i_1, \ldots, i_n$ are shorter than the output of H, then fixed-length instructions $I_1, \ldots, I_n$ can be obtained by padding with a known filler or a redundancy pattern.

However, if we want to accommodate complex statistical analyses or even small routines in each instruction i_j, the length of instructions may be greater than the length of the output of H. In this case, the user can compute the hash of i_j, $h_j = H'(i_j)$, using another cryptographically strong secure hash function H' whose output is shorter than the output of H. Then what is padded *with some redundancy* into $I_1, \cdots, I_n$ are, respectively, $h_1 \| p_1, \ldots, h_n \| p_n$, where each key is appended a pointer p_j to a location where instruction i_j is stored. We assume the output of H' concatenated with a pointer is not longer than the output of H.

3.2 Hash-Based Coding and Execution

We encode $I_1, \ldots, I_n$ into a *trace sequence* $T_0, T_1, \ldots, T_n$, where traces are computed in reverse order according to the following equalities:

$$T_0 = H(T_1) \oplus I_1;$$
$$T_1 = H(T_2) \oplus I_2;$$
$$\vdots \tag{1}$$
$$T_{n-1} = H(T_n) \oplus I_n;$$
$$T_n = H(I_n).$$

To "seal" the trace sequence, the first trace T_0 is signed by the user who has written the analysis sequence or by some certification authority. Call t_0 the signed version of T_0.

To start executing the analysis sequence, traces t_0, T_0 and T_1 need to be supplied to the controller/processor, who verifies the signature t_0 on T_0 and retrieves the first instruction as

$$I_1 := H(T_1) \oplus T_0.$$

In general, after instructions $I_1, \ldots, I_{j-1}$ have been retrieved, learning trace T_j allows the controller to retrieve the j-th instruction as

$$I_j := H(T_j) \oplus T_{j-1}. \tag{2}$$

The final step is to run the instruction/analysis:

- If I_j contains an instruction i_j, this instruction/analysis is run by the controller and, if appropriate, its results are returned to the user.
- If I_j contains $h_j \| p_j$, then the controller fetches an analysis instruction i_j at location p_j, checks that $H'(i_j) = h_j$ and, if the check passes, runs i_j and returns its results to the user if appropriate.

The controller can decide on the "appropriateness" of returning a result by (manually or automatically) using an output checking rule such as those recalled in Sect. 2.

3.3 Security Properties

The above coding and execution scheme has two interesting properties.

Proposition 1 (Tit-for-tat instruction disclosure). *The user having constructed the sequence of analyses can disclose her sequence to the controller instruction by instruction, upon receipt of partial results, rather than in one shot.*

Proof. From Expressions (1), it can be seen that each trace T_i encrypts an instruction I_{i+1} by adding to it a pseudorandom value $H(T_{i+1})$ that is as long as I_{i+1}. Thus, the controller cannot decrypt I_{i+1} from T_i until the user reveals T_{i+1} to the controller. $\square$

The second property refers to the binding nature of the sequence.

Proposition 2 (Binding sequence of analyses). *Even if the controller is only revealed the sequence instruction by instruction, he can detect any changes introduced to the sequence since its execution started.*

Proof. We will show this by induction. After the user signs T_0 and the signature t_0 is revealed to the controller, the latter can detect any change in T_0 and abort the sequence execution. For any $j \geq 1$, if $T_0, \dots, T_{j-1}$ are not changed, the controller can detect any instruction substitution, deletion or insertion of T_j at run-time as follows: if T_j^* is supplied to the controller instead of T_j, then Expression (2) will yield a gibberish I_j^* in place of I_j (because H is a strong hash function). Two cases must be distinguished:

- If I_j contained an instruction i_j, then I_j^* will be gibberish as well, rather than a proper instruction, which will cause the execution to abort.
- If I_j contained a hash value plus a pointer, that is $h_j \| p_j$, then I_j^* will be gibberish and will not contain the redundancy that should have been embedded with a valid concatenation of a hash value and a pointer. This will also result in the execution being aborted.

Thus, we have proven that changes in the sequence of traces will be detected.

An alternative for an attacker when I_j contains a hash value h_j plus a pointer p_j is to try to change the instruction i_j at address p_j, say, to i_j^*. If the attacker does this, the controller's check $H'(i_j^*) \stackrel{?}{=} h_j$ will fail and cause the execution to be aborted.

Thus, by just revealing t_0 to the controller, the user is binding herself to the entire (and yet undisclosed) sequence of analyses. $\qquad\square$

4   Using Binding Sequences of User Analyses to Ensure Ethical Compliance

When the user binds herself to a sequence of analyses, she actually binds herself to a certain behavior. As the user discloses one instruction/analysis after the other, it is possible for the controller to monitor that the user's behavior stays compliant with some ethical values. We next discuss compliance with the following values: confidentiality, fairness and transparency-explainability. The latter two values are applicable when an analysis consists of training a machine learning model on the controller's original data.

4.1   Checking Confidentiality

A data controller offering to run statistical analyses on his confidential data on the user's behalf wants to prevent the user from learning specific confidential records. There are two ways in which the user may seek to learn confidential information from statistical queries:

Single statistical query. A single statistical query/analysis may leak individual records if it affects a sufficiently small set of records. For example, computing the average value of an attribute over a query set of size 1 obviously leaks the value of the attribute for the affected record.

Sequence of statistical queries. Tracker attacks [2] have been known for about 40 years: a sequence of queries is constructed such that it pads small query sets with enough extra records to give them a size in the allowable range. Then the effect of padding is subtracted, and statistical outputs over small query sets can be obtained.

In our framework, the controller can defend against leaks from single statistical queries by resorting to output checking, discussed in Sect. 2. The controller uses output checking rules to decide whether the output of the current instruction/analysis can be safely returned to the user. If it cannot, then the controller withholds the output and aborts the execution of the user's analysis sequence.

Regarding leaks caused by a sequence of instructions, they can be prevented by the controller by using query set size controls [11]. This strategy consists in controlling the size of the set of records isolated by the queries processed so far plus the current query. If the query set size is below a threshold set by the controller, he withholds the output of the current query and aborts the execution of the user sequence. Some amount of output perturbation can be used to increase the safety of query set size control, as hinted in [2].

In case of sequence abort, the user does not need to disclose the remaining instructions/analyses of her sequence. In case of non-abort, the user discloses a new trace to allow the controller to recover a new instruction/analysis.

4.2 *Ex ante* Checking of Explainability and Fairness

If the user's sequence of analyses trains a machine learning model on the controller's data, the controller/processor may wish to make sure the resulting model is explainable or fair before returning it to the user or even before running the user's sequence that trains the model.

Whereas confidentiality can be checked as the user discloses one instruction/analysis after the other, checking that a sequence trains an explainable or a fair model may require examining the entire sequence. Yet, this can hardly be done by the controller if the next instruction is disclosed by the user only after the controller returns her the results of the previous instruction.

A way around is for the user to have her sequence certified by a certification authority trusted by the controller/processor and the user:

1. The user binds herself to her sequence of analyses by encoding its instructions into traces and signing the first trace as explained in Sect. 3.
2. Then the user submits her *entire* binding sequence of analyses to the certification authority. The user trust the authority not to disseminate or misuse her sequence of analyses.
3. The authority checks the code for the required ethical value:

- Checking for transparency-explainability can be done by making sure the code corresponds to training an explicitly explainable model (decision rules, decision trees, random forests, etc.).
- Checking for fairness by code analysis can be done by making sure the code includes instructions for pre-processing, in-processing or post-processing anti-discrimination (*e.g.* [9]).

4. If the check passes, the authority appends its signature to the first trace of the binding sequence.
5. Upon finding the signature of the authority when being revealed the first sequence trace, the controller is reassured that the sequence will train an explainable, respectively fair, model.

5   Conclusions and Further Research

We have presented a new concept to make sure the sequence of analyses run by a user in a safe access center or a trusted research environment is ethically compliant. Ethical values to be enforced include respondent privacy (non-disclosure of the underlying original microdata), user privacy (fair disclosure between user and controller), and fairness of trained models if any, *inter alia*. On the one side, the user binds herself to a sequence, which means she cannot modify it depending on the intermediate results obtained. This prevents adaptive attacks and thus thwarts disclosure of the underlying respondent microdata (*respondent privacy*). If the user breaks this bond and tries to change her sequence, the controller stops returning the results of the analyses. On the other side, the user reveals her next analysis only if she obtains the result of her previous analysis. This ensures *user privacy* for the rest of the analysis sequence, and it can be construed as a form of *fairness* between user and controller.

Admittedly, our proposal can only be employed if the user's research has a pre-set purpose, so that it is possible for the user to define in advance the sequence of analyses she wants to perform.

Future work will consist of building a demonstrator in a practical trusted research environment. Also, we plan to extend the approach to non-sequential analysis programs (with loops and subroutine calls).

Acknowledgments and Disclaimer. Thanks go to a reviewer for making comments and raising stimulating questions. Partial support is acknowledged from the European Commission (projects H2020-871042 "SoBigData++" and H2020-101006879 "MobiDataLab"), the Government of Catalonia (ICREA Acadèmia Prize), and UK Research and Innovation (project MC_PC_21033 "GRAIMatter"). The author is with the UNESCO Chair in Data Privacy, but the views in this paper are his own and are not necessarily shared by UNESCO or the above mentioned funding agencies.

References

1. Bond, S., Brandt, M., de Wolf, P.-P.: Guidelines for the Checking of Output Based on Microdata Research. Deliverable D11.8, project FP7-262608 "DwB: Data without Boundaries" (2015). https://ec.europa.eu/eurostat/cros/system/files/dwb_standalone-document_output-checking-guidelines.pdf
2. Denning, D.E., Schlörer, J.: Inference controls for statistical databases. Computer 16(7), 69–82 (1983)
3. Domingo-Ferrer, J.: Software run-time protection: a cryptographic issue. In: Damgård, I.B. (ed.) EUROCRYPT 1990. LNCS, vol. 473, pp. 474–480. Springer, Heidelberg (1991). https://doi.org/10.1007/3-540-46877-3_43
4. Domingo-Ferrer, J.: Algorithm-sequenced access control. Comput. Secur. 10(7), 639–652 (1991)
5. Domingo-Ferrer, J., Sánchez, D., Soria-Comas, J.: Database Anonymization: Privacy Models, Data Utility, and Microaggregation-Based Inter-Model Connections. Morgan & Claypool (2016)
6. Eurostat: Options for decentralised/remote access to European microdata. In: 5th Meeting of the Working Group on Methodology, Luxembourg, 1 April 2020. (videoconference)
7. General Data Protection Regulation. Regulation (EU) 2016/679. https://gdpr-info.eu
8. Griffiths, E., et al.: Handbook on Statistical Disclosure Control for Outputs (version 1.0). Safe Data Access Professionals Working Group (2019). https://ukdataservice.ac.uk/media/622521/thf_datareport_aw_web.pdf
9. Hajian, S., Domingo-Ferrer, J.: A methodology for direct and indirect discrimination prevention in data mining. IEEE Trans. Knowl. Data Eng. 242, 35–48 (2013)
10. Hundepool, A., et al.: Statistical Disclosure Control. Wiley, Hoboken (2012)
11. Nussbaum, E., Segal, M.: Privacy analysis of query-set-size control. In: Domingo-Ferrer, J., Muralidhar, K. (eds.) PSD 2020. LNCS, vol. 12276, pp. 183–194. Springer, Cham (2020). https://doi.org/10.1007/978-3-030-57521-2_13
12. Ocean Protocol. https://oceanprotocol.com. Accessed 14 July 2022

Secure and Non-interactive k-NN Classifier Using Symmetric Fully Homomorphic Encryption

Yulliwas Ameur$^{(\boxtimes)}$, Rezak Aziz , Vincent Audigier ,
and Samia Bouzefrane$^{(\boxtimes)}$

CEDRIC Lab, Cnam, 292 rue Saint Martin, 75141 Paris, France
{yulliwas.ameur,samia.bouzefrane}@lecnam.net

Abstract. "Machine learning as a service" (MLaaS) in the cloud accelerates the adoption of machine learning techniques. Nevertheless, the externalization of data on the cloud raises a serious vulnerability issue because it requires disclosing private data to the cloud provider. This paper deals with this problem and brings a solution for the K-nearest neighbors (k-NN) algorithm with a homomorphic encryption scheme (called TFHE) by operating on end-to-end encrypted data while preserving privacy. The proposed solution addresses all stages of k-NN algorithm with fully encrypted data, including the majority vote for the class-label assignment. Unlike existing techniques, our solution does not require intermediate interactions between the server and the client when executing the classification task. Our algorithm has been assessed with quantitative variables and has demonstrated its efficiency on large and relevant real-world data sets while scaling well across different parameters on simulated data.

Keywords: k-nearest neighbors · Homomorphic encryption · TFHE · Data privacy · IoT · Cloud computing · Privacy-preserving

1 Introduction

Cloud services have become central to data storage and data exploitation. Among these services, a machine-learning service is offered to train different models to predict for decision-making purposes. However, this raises the issue of data security because the data processed by the cloud may be sensitive and confidential while belonging to entities that do not trust the cloud provider.

The most commonly used cryptographic techniques are secret sharing, multiparty computation, and homomorphic encryption (HE) to achieve privacy-preserving for machine learning. Several techniques ensure strong privacy protection, often coming at the expense of reduced in terms of speed and communication.

Homomorphic encryption is an encryption technique that allows computations directly on encrypted data. The results are encrypted and can be revealed/decrypted only by the owner of the secret key. This principle is very useful in many domains where data sharing and privacy preservation are required.

© Springer Nature Switzerland AG 2022
J. Domingo-Ferrer and M. Laurent (Eds.): PSD 2022, LNCS 13463, pp. 142–154, 2022.
https://doi.org/10.1007/978-3-031-13945-1_11

For example, externalizing personal data considered as sensitive such as medical data or banking transactions [10]. This paper considers a scenario where sensitive data are collected by IoT devices and outsourced on a resourceful cloud for Machine Learning (ML) processing. Considering that the cloud provider is not trustworthy, we propose using HE to preserve privacy.

ML over encrypted data has been particularly investigated in the domain of neural networks (NN). However, as state-of-the-art non-linear NN layers (pooling and activation) are too complex to be directly executed in the encrypted world, there is a strong need for approximating them. Much of the subsequent works have been proposed to address the limitation of implementing the non-linear activation function by a polynomial approximation using Taylor series and Chebyshev polynomials, among others [1,3,5,7]. Due to the high cost of the HE systems, those methods do not scale well and are not appropriate for deep neural networks [12], which are time-consuming due to the great depth of the network.

One way to solve the computational cost problem inherent to HE is to investigate less complex supervised methods. The k-nearest neighbors (k-NN) approach presents several advantages. Indeed, for a predefined number of neighbors k, the model does not require any training step, the value of the response variable for a given individual is obtained directly from the values observed on the neighbors without needing to estimate new parameters. In addition, the method can handle continuous, categorical, and mixed data. Furthermore, as a non-parametric method, k-NN can be relevant for many data structures as long as the number of observations is sufficiently large.

The prediction for a new observation is obtained by:

- Identifying the k nearest neighbors (according to a given distance)
- Computing the majority class among them (for a classification problem) or by averaging values (for a regression problem).

HE has been recently investigated by various authors for k-NN [9,13,14,16].

[9] suggested a Homomorphic additive encryption scheme [11]. They investigated the privacy preservation in an outsourced k-NN system with various data owners. The untrusted entity securely computes the computations of distances by using HE. However, the comparison and classification phases require interactions. Given that the computational and communication difficulties scale linearly, they admit that the method may not be practical for massive data volumes. The cost of communications between the entities is also a limitation in the deployment of this work [13].

[14], used an "asymmetric scalar-product-preserving encryption" (ASPE). However, the client has the ciphertext, and the server can decrypt it. The proposed solution is vulnerable to Chosen Plaintext Attacks as stated by [15].

Recently, [16] proposed a secure k-NN algorithm in quadratic complexity concerning the size of the database completely non-interactively by using a fully homomorphic encryption [6]. However, they assume that the majority vote is done on a clear-text domain, which is a significant security flaw that we will address here. Doing a majority vote on a clear-text domain imposes interaction between entities, which causes information leakage.

Unlike other existing works, in this paper, we propose a new methodology to apply k-NN on encrypted data by using fully homomorphic encryption avoiding interaction between entities.

We consider a client/service provider architecture that supports scenarios described here. The service provider is a company that owns a labeled training dataset D composed of sensitive data allowing predict for novel observation by k-NN. This model is offered as a service.

We assume a context that is concerned by some privacy issues as in the following:

- Because the training data are sensitive, they cannot be shared with a third party such as a client.
- The model is an intellectual property of the service provider. Hence, the service provider does not want to share the used model with his clients.
- A client who needs to perform classification on the service-provider platform does not trust the service provider.

This work aims to do a classification using k-NN algorithm on sensitive data using HE. Our solution assumes that the service provider, which is the dataset owner, has all the necessary resources to perform the data classification and storage. This assumption ensures that encrypting the training dataset is not necessary since these data are kept with the data owner. Only the client will need to encrypt his query that includes his data by using a private key and by sending it to the data owner for classification. The goal is to protect the training dataset, the query, and the model parameter k. Our solution meets the following privacy requirement as in the following:

- The contents of D are known only by the data owner since they are not sent to other parties.
- The client's query is not revealed to the data owner.
- The client knows only the predicted class.
- The index of the k nearest neighbors is unknown from the data owner or the client.

Our solution has a greater added value than the existing literature solutions. First, it guarantees that no information leakage occurs during the process: the only things known by the data owner are the dataset and the model used. The only things that the client knows are the query and the class. All intermediate results are encrypted. In addition, our solution is fully non-interactive since prediction is performed by the data owner and do not need any decryption during the process. Finally, it supports multi-label classification.

The rest of this paper is organized as follows: Sect. 2 presents the background of HE before highlighting the principle of Fast Fully homomorphic encryption over the Torus (TFHE). Our proposed solution is then described in Sect. 3. A simulation study is presented in Sect. 4 to assess our methodology based on real datasets. Finally, Sect. 5 concludes the paper.

2   Background

2.1   Homomorphic Encryption

HE allows a third party (a service provider in our scenario) to compute functions on ciphertexts to preserve the confidentiality of the data. An encryption scheme is called homomorphic over an operation $*$ if it supports the following property:

$$E(m_1) * E(m_2) = E(m_1 * m_2)$$

where E is the encryption algorithm and m_1, m_2 belong to M the set of all possible messages. An additional algorithm is needed for homomorphic encryption schemes, called the *Eval* algorithm. This algorithm is defined as follows:

$$Eval(f, C_1, C_2) = f(m_1, m_2)$$

where $Dec(C_1) = m_1$ and $Dec(C_2) = m_2$ and f is a function that determines the type of the HE scheme. In case f supports the evaluation of arbitrary functions for an unlimited number of times, the HE is called *Fully* homomorphic encryption (FHE).

This paper uses "TFHE: Fast Fully homomorphic encryption over the Torus" as an RLWE-based scheme in his fully homomorphic setting, especially in gate bootstrapping mode. The bootstrapping procedure is the homomorphic evaluation of a decryption circuit on the encryption of a secret key.

2.2   Functional Bootstrap in TFHE

TFHE defines three types of ciphertexts, TLWE Sample, TRLWE Sample and TRGSW Sample.

There are two bootstraps algorithms in TFHE. Gate Bootstrap was introduced to implement logic gates, and the Circuit Bootstrap, which converts TLWE samples to TRGSW samples. In our work, we use Functional Bootstrap. By "Functional" we mean that the bootstrap can evaluate functions using a BlindRotate algorithm to perform a lookup table (Lut) evaluation. LUTs are a simple, efficient way of evaluating discretized functions. For instance the sign function was used in [2] and [8].

3   Our Contribution

3.1   The System Model

Our system uses the client-server architecture (see Fig. 1). The client is the querier, and the server is the data owner.

1. The data owner: owns the data and can do heavy calculations. For example, it receives the query in an encrypted way, performs an encrypted k-NN algorithm then sends the result to the querier for decryption.
2. The querier: generates the keys, encrypts the query that contains its data and sends it to the data owner for computations before decrypting the result. The querier can be an ordinary computer or any IoT device that collects data.

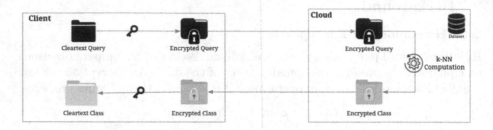

Fig. 1. The system model: the client is the querier, and the server is the data owner: the data owner receives the query in an encrypted way, performs an encrypted k-NN algorithm then sends the result to the querier for decryption.

3.2   Encrypted k-NN Challenges

In order to propose an encrypted version of k-NN, we should substitute the challenging operations used in the standard k-NN with equivalent operations in encrypted domains. As seen before, k-NN is composed of three parts: the distance calculation, the distance sorting, the selection of the k nearest neighbors, and the majority vote.

This subsection will introduce the equivalent operations as integrated with our solution.

Distance Calculation. The euclidean distance calculation between the dataset entries x_i as well as the query q are necessary in order to find the k nearest neighbors to the query. We can use the standard formula of the distance as in (1).

$$d^2(x_i, q) = \sum_{j=0}^{p} x_{ij}^2 + \sum_{j=0}^{p} q_j^2 - 2 * \sum_{j=0}^{p} x_{ij} q_j \tag{1}$$

What is relevant in our case is the difference between two distances to compare them. So, we get the formula (2) as follows:

$$d^2(x_i, q) - d^2(x_{i'}, q) = \sum_{j=0}^{p} (x_{ij}^2 - x_{i'j}^2) - 2 * \sum_{j=0}^{p} (x_{i'j} - x_{ij}) q_j \tag{2}$$

Since the dataset is a clear text, we can easily calculate formula (2) using the TFHE scheme. However, we need to adapt it. Using TFHE, the difference between the distances should be in the range of $[-\frac{1}{2}, \frac{1}{2}]$. Another constraint is that the multiplication is done between a clear-text integer and a ciphertext. Two rescaling values are required to resolve these constraints. Let v be the first one. It is used to have values of the differences between $[-\frac{1}{2}, \frac{1}{2}]$. Let p be the second one. It indicates the precision of the differences. Each attribute of the dataset as well as the query are rescaled using v. p is used when calculating the product $(x_{i'j} - x_{ij}) q_j$.

Sorting. Sorting computed distances is a crucial step in k-NN. The standard algorithm for sorting, like the bubble sort, can be used while considering encrypted data. However, these algorithms are time-consuming in an encrypted world because the worst case is computed every time. The authors [4] propose two methods to sort an array of values. The method of the direct sort is used in [16]. It is based on a matrix of comparison called delta matrix:

$$\begin{pmatrix} m_{1,1} & m_{1,2} & \cdots & m_{1,n} \\ m_{2,1} & m_{2,2} & \cdots & m_{2,n} \\ \vdots & \vdots & \ddots & \vdots \\ m_{n,1} & m_{n,2} & \cdots & m_{n,n} \end{pmatrix}$$

with

$$m_{i,j} = \bar{sign}(X_i - X_j) = \begin{cases} 1 \text{ if } X_i < X_j \\ 0 \text{ else.} \end{cases}$$

When summing columns of this matrix, we will have a sorting index of the distances.

Majority Vote. The majority vote can cause a problem because the operation requires comparison to detect the class. To determine the predicted class, we need to know k nearest neighbors' classes. This step is challenging in two ways: first, we need to avoid information leakage, unlike the solutions in the literature. Second, the majority vote requires comparison in order to predict the class.

To the best of our knowledge, no solution in the literature studied this point in an encrypted way without information leakage. Therefore, in the following subsection, we will demonstrate a solution to process k-NN with the majority vote in an encrypted way while supporting a multi-label classification.

3.3   Our Proposed k-NN Algorithm

Our proposed algorithm, called "HE-kNN-V", is composed of three steps: the construction of the delta matrix, the selection of k-nearest neighbors, and the majority vote. The two first steps are similar to the solution of [16] even if we adapt the existing formulas in order to eliminate unnecessary calculations. [16] use polynomials to define the formulas, while what interests us is just one term of those polynomials to eliminate un-necessary calculations. The majority vote is our added value and is specific to our solution. We will discuss in this subsection the design of our solution, including each building block.

Building the Delta Matrix. To build the delta matrix, we need to know the sign of the differences between the distances to sort. Since we defined a method to calculate the differences in the last subsection, the sign can easily be achieved using the standard bootstrapping sign function in TFHE. However, the standard bootstrapping function returns $+1$ if the phase is greater than 0 and -1 if the phase is lower than 0. Therefore, since we need to have 0 or 1 in the matrix, we need to adapt the bootstrapping operation to return $\frac{1}{2}$ and $-\frac{1}{2}$ then by adding $\frac{1}{2}$ to the result we will have 0 or 1.

Even if building this matrix is time-consuming, it is highly parallelizable.

Selecting the k-Nearest Neighbors. To select the k–nearest neighbors, we use the scoring operation proposed by Zuber [16]. By using the delta matrix, the principle is as follows:

1. Sum m values in each column with m the number of possible operations without bootstrapping.
2. If there are still values to sum: do a bootstrapping operation using the modified sign bootstrapping function (see Algorithm 1 in [16]) and go to Step 1.
3. Otherwise, execute the modified sign bootstrapping and return the last sign returned by this operation.

Finally, we obtain an encrypted vector where the position i equals the cipher of 1 if the individual with index i is among the k-nearest neighbors, the cipher of 0 otherwise. We call this vector the "mask" (see Fig. 2 for more clarity).

Majority Vote. The majority vote is the most important added value in our work. We propose to do the majority vote without any leakage of information, unlike existing works like that of [16] in which the majority value is done in clear text or by using other alternative solutions proposed in the literature.

First, we illustrate the issue with the method of [16]. We consider the scenario where the querier does the calculations. The majority vote is done in clear text, but we need to decrypt the vector of indexes of the nearest neighbor. The data owner does the decryption. Significant information leakage occurs if the data owner knows the vector of indexes. Then, he will know the classification of the query, and by doing some triangulation, he can approximate the query. In addition, the solution will be interactive. If we consider the scenario where the data owner does the calculation, the decryption of the vector is done by the querier. However, to do the classification, the querier should know the labels of the dataset, which is also a critical leakage of information. In addition, the querier will know the size of the dataset and the k parameter of nearest neighbors considered. This information is considered as internal information of the model used, and it should be protected.

In our solution, the majority vote is done by the data owner in an encrypted way. First, the data owner encodes the labels using one hot encoding. Having the mask and the matrix of labels in one hot form, it is easy to do an AND operation between the mask and each column of the labels, as in Fig. 2. We get a matrix A (for affectation) with A_{ij} equal to 1 if the individual i is among the k-nearest neighbors and its class is j. Using this matrix, it is possible to sum the columns and obtain the probability of each class. We can now return only the class and guarantee no information leakage and no interactivity.

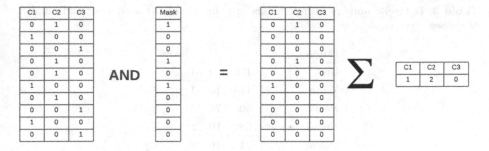

Fig. 2. Majority vote illustration by using the mask

4 Performance Evaluation

In this section, we discuss the experiments of our solution. First, we describe the technical and the setup of the environment. Then, we will evaluate the performances of our solution according to different criteria: execution time, accuracy, bandwidth consumption.

4.1 Test Environment

Setup. Our solution is implemented using the TFHE scheme in C/C++ and Python for training k-NN in clear text and for tests. To test the effect of parallelism, we used OpenMP to do some parallelization. The source code is available in the following github "https://github.com/Yulliwas/HE-kNN-V". Our solution is tested on Linux Ubuntu 64-bit machine with i7-8700 CPU 3.20 GHz.

Table 1 shows the parameters used to setup TFHE scheme (Table 2).

Table 1. TFHE parameters: λ for the overall security, N for the size of the polynomials, σ for the Gaussian noise parameter.

λ	N	σ
110	1024	10^{-9}

Table 2. HE-kNN parameters: the number of operations m without needing a bootstrapping, the bootstrapping base b, and the rescaling factors v and p

m	v	p	b
64	4	1000	4 * m−4

Datasets. To test our solution, we choose to use 6 datasets: Iris, Breast Cancer, Wine, Heart, Glass and MNIST as in Table 3. The goal is to test the performances of our algorithm in different distributions of data, so that to confirm that our solution works with any dataset and that has performances that are equivalent to those of clear-text domains.

Table 3. Datasets: number of individuals (n), the size of the model (d) and number of classes

Dataset	n	d	Classes
Iris	150	4	3
Wine	178	13	3
Heart	303	75	5
Breast cancer	699	10	2
Glass	214	10	2
MNIST	1797	10	3

Simulation Procedure. First, we preprocess the data by rescaling each attributes to a value between 0 and 1. Our dataset and the query should be rescaled by a factor of v as seen above. We must also multiply the dataset vectors by the precision factor τ and then rounded. In the other hand, the query vector is divided by this same factor. To obtain the classification rate, first we need to divide our dataset to a training set and a test set. We choose to use 20% of our dataset as a test set and the rest as a training set. Among the training set, we select a certain number of points that represent as well as possible our dataset. The process for choosing the best points that represent our training set is as follows:

1. choose n individuals randomly;
2. calculate the classification rate;
3. Repeat the previous Step 1 and Step 2 a certain amount of time and keep the best accuracy and the best individuals.

To select the k parameter, we use the same procedure as in the clear domain. In our case, we tested different values of k and we keep the best k value that gives the best results.

4.2   Performance Results

To position our approach according to existing works, and especially regarding the voting step that is performed without information leakage, we compare in Table 4 our solution with Zuber's solution and with a clear-text version based on the Iris dataset and a fixed $k = 3$. The comparison is done in terms of complexity (C), Information Leakage (L), accuracy (A), interactivity (I) and execution time (T). The accuracy and the prediction time are indicated only when it is possible.

Empirical Study

Classification Rate. To evaluate the classification rate, we have chosen the accuracy instead of other metrics like: recall or F1-score. We studied the accuracy according to two parameters: the number of data sampled from the dataset and the number k of neighbors. The goal is to choose the best points that represent the datasets and the best k parameters for each dataset.

Table 4. Comparison between solutions for Iris dataset: complexity (C), information leakage (L), accuracy (A), interactivity (I) and execution time (T).

Work	C	L	I	A	T
HE-kNN-V	$O(n^2)$	N	N	0.97	1.72 s
HE-kNN-VP	$O(n^2)$	N	N	0.97	0.46 s
Zuber	$O(n^2)$	Y	Y	0.98	1.74 s
Clear k-NN	$O(n)$	Y	N	0.95	1.8 ms

We chose real-world datasets in order to see the evolution of the accuracy and compared it to clear-text accuracy.

In one hand, we know that the accuracy depends on the k parameter and we can confirm it easily in the graphs. On the other hand, the assumption that the accuracy depends on the number of data used is not complete. For the dataset where the data is well separated (like Iris), having a lot of data is not necessary, the best accuracy can be achieved using only few data. But, in the case where data is not well separated (like in Heart dataset), the accuracy seems to depend on the number of data.

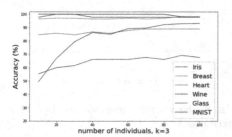

Fig. 3. Encrypted accuracy vs number of individuals

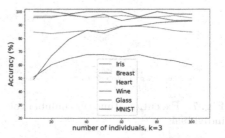

Fig. 4. Clear-text accuracy vs number of attributes

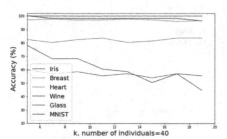

Fig. 5. Encrypted accuracy vs k-parameter

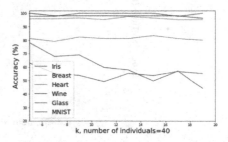

Fig. 6. Clear-text accuracy vs k-parameter

According to our different simulations illustrated in Fig. 3 and Fig. 4, we do not lose accuracy when we apply our HE-kNN-V method on the encrypted data compared to the application of the kNN on the plain data. This is possible by varying the number of individuals and by fixing k to 3.

We also notice that by setting the number of individuals to 40 and varying k, (see Fig. 5 and Fig. 6) the accuracy behaves in the same way between the application of the kNN on the plain data and the application of our method HE-kNN-V on the encrypted data.

Execution Time. In our solution, the execution time is independent of the content of the dataset, it does not depend on the values, but does depend on the content, since it depends on the number of tuples. We can use either simulated dataset or real world dataset. To visualize the evolution of the execution time according to k, n and d, we choose to use the Breast Cancer dataset instead of simulating a new dataset. We change n, k, d and we see the evolution of the execution time.

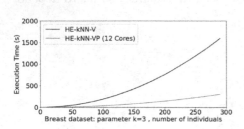

Fig. 7. Execution time vs number of individuals

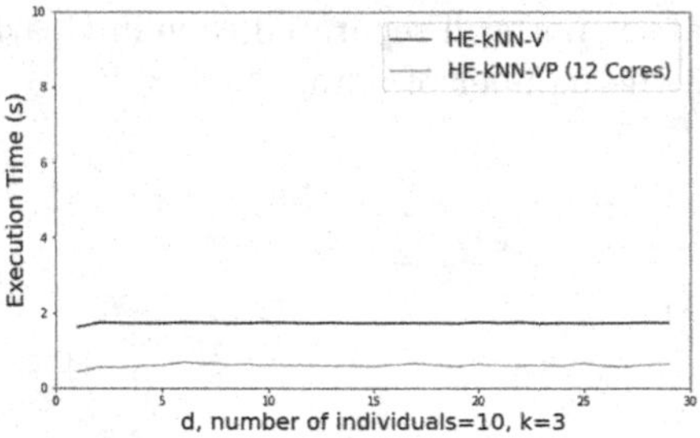

Fig. 8. Execution time vs number of attributes

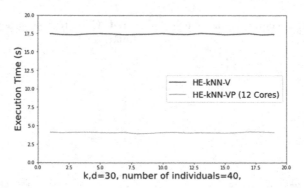

Fig. 9. Execution time vs k-parameter

Our simulations, as depicted in Fig. 7, illustrate that HE-kNN-V is parallelizable, and also that the number of individuals strongly impacts the execution time
unlike the two simulations of Fig. 8 and Fig. 9 where the variation of respectively
d the number of attributes and k does not impact the execution time.

Bandwidth. In our solution, the only thing that is communicated is the query
in the ciphertext and the response in the ciphertext. The size of the query is
proportional to the number of attributes d. Each attribute is a TLWE Sample
with the size of 4 KB and the size of the response (number of classes) * 4 KB.
The bandwidth according to each dataset is illustrated in Table 5.

Table 5. Bandwidth

Dataset	Bandwidth (KB)
Iris	28
Wine	64
Heart	64
Breast cancer	128
Glass	60
MNIST	296

Discussion. According to our experiments, we can say that the accuracy in
our case depends on three factors: the number of individuals, the representativ-
ity of these individuals and the k parameter. To have a better model that fits
our dataset, we must select the individuals that are more representative of our
dataset and the best k parameter. We also should take care of the number of
individuals because most of the execution time depends on that number.

5   Conclusion

We proposed HE-kNN-V a method for performing k-NN on encrypted data
that includes a majority vote for class-label assignment. The proposed solution
addresses all stages of k-NN algorithm with fully encrypted data. It guarantees
that no information leakage occurs during the process. Unlike other techniques,
our solution eliminates the need for intermediate interactions between the server
and the client when performing classification tasks. Our algorithm has been evaluated using quantitative variables and demonstrated its efficiency on large and
relevant real-world data sets. As a perspective, it would be interesting to see
how a hardware acceleration of the TFHE scheme could improve the computation time of our proposed solution HE-kNN-V.

References

1. Al Badawi, A., et al.: Towards the AlexNet moment for homomorphic encryption: HCNN, the first homomorphic CNN on encrypted data with GPUs. IEEE Trans. Emerg. Top. Comput. **9**(3), 1330–1343 (2021)
2. Bourse, F., Minelli, M., Minihold, M., Paillier, P.: Fast homomorphic evaluation of deep discretized neural networks. In: Shacham, H., Boldyreva, A. (eds.) CRYPTO 2018. LNCS, vol. 10993, pp. 483–512. Springer, Cham (2018). https://doi.org/10.1007/978-3-319-96878-0_17
3. Brutzkus, A., Elisha, O., Gilad-Bachrach, R.: Low latency privacy preserving inference. ArXiv, abs/1812.10659 (2019)
4. Çetin, G.S., Doröz, Y., Sunar, B., Savaş, E.: Depth optimized efficient homomorphic sorting. In: Lauter, K., Rodríguez-Henríquez, F. (eds.) LATINCRYPT 2015. LNCS, vol. 9230, pp. 61–80. Springer, Cham (2015). https://doi.org/10.1007/978-3-319-22174-8_4
5. Chabanne, H., de Wargny, A., Milgram, J., Morel, C., Prouff, E.: Privacy-preserving classification on deep neural network. IACR Cryptology ePrint Archive 2017/35 (2017)
6. Chillotti, I., Gama, N., Georgieva, M., Izabachène, M.: TFHE: fast fully homomorphic encryption library, August 2016. https://tfhe.github.io/tfhe/
7. Hesamifard, E., Takabi, H., Ghasemi, M.: CryptoDL: deep neural networks over encrypted data. ArXiv, abs/1711.05189 (2017)
8. Izabachène, M., Sirdey, R., Zuber, M.: Practical fully homomorphic encryption for fully masked neural networks. In: Mu, Y., Deng, R.H., Huang, X. (eds.) CANS 2019. LNCS, vol. 11829, pp. 24–36. Springer, Cham (2019). https://doi.org/10.1007/978-3-030-31578-8_2
9. Li, F., Shin, R., Paxson, V.: Exploring privacy preservation in outsourced k-nearest neighbors with multiple data owners. In: Proceedings of the 2015 ACM Workshop on Cloud Computing Security Workshop, CCSW 2015, pp. 53–64. Association for Computing Machinery, New York (2015)
10. Masters, O., Hunt, H., Steffinlongo, E., Crawford, J., Bergamaschi, F.: Towards a homomorphic machine learning big data pipeline for the financial services sector. IACR Cryptology ePrint Archive 2019/1113 (2019)
11. Paillier, P.: Public-key cryptosystems based on composite degree residuosity classes. In: Stern, J. (ed.) EUROCRYPT 1999. LNCS, vol. 1592, pp. 223–238. Springer, Heidelberg (1999). https://doi.org/10.1007/3-540-48910-X_16
12. Pulido-Gaytan, B., et al.: Privacy-preserving neural networks with homomorphic encryption: challenges and opportunities. Peer-to-Peer Netw. Appl. **14**(3), 1666–1691 (2021). https://doi.org/10.1007/s12083-021-01076-8
13. Samanthula, B.K., Elmehdwi, Y., Jiang, W.: k-nearest neighbor classification over semantically secure encrypted relational data. IEEE Trans. Knowl. Data Eng. **27**(5), 1261–1273 (2015)
14. Wong, W.K., Cheung, D.W., Kao, B., Mamoulis, N.: Secure kNN computation on encrypted databases. In: Proceedings of the 2009 ACM SIGMOD International Conference on Management of Data, SIGMOD 2009, pp. 139–152. Association for Computing Machinery, New York (2009)
15. Xiao, X., Li, F., Yao, B.: Secure nearest neighbor revisited. In: Proceedings of the 2013 IEEE International Conference on Data Engineering (ICDE 2013), USA, pp. 733–744. IEEE Computer Society (2013)
16. Zuber, M., Sirdey, R.: Efficient homomorphic evaluation of k-NN classifiers. In: Proceedings on Privacy Enhancing Technologies 2021, pp. 111–129 (2021)

Unstructured and Mobility Data

Automatic Evaluation of Disclosure Risks of Text Anonymization Methods

Benet Manzanares-Salor[1], David Sánchez[1]([⊠]), and Pierre Lison[2]

[1] Department of Computer Engineering and Mathematics, UNESCO Chair in Data Privacy,
CYBERCAT, Universitat Rovira i Virgili, Tarragona, Spain
{benet.manzanares,david.sanchez}@urv.cat
[2] Norwegian Computing Center, Oslo, Norway
plison@nr.no

Abstract. The standard approach to evaluate text anonymization methods consists of comparing their outcomes with the anonymization performed by human experts. The degree of privacy protection attained is then measured with the IR-based recall metric, which expresses the proportion of re-identifying terms that were correctly detected by the anonymization method. However, the use of recall to estimate the degree of privacy protection suffers from several limitations. The first is that it assigns a uniform weight to each re-identifying term, thereby ignoring the fact that some missed re-identifying terms may have a larger influence on the disclosure risk than others. Furthermore, IR-based metrics assume the existence of a single gold standard annotation. This assumption does not hold for text anonymization, where several maskings (each one encompassing a different combination of terms) could be equally valid to prevent disclosure. Finally, those metrics rely on manually anonymized datasets, which are inherently subjective and may be prone to various errors, omissions and inconsistencies. To tackle these issues, we propose an automatic re-identification attack for (anonymized) texts that provides a realistic assessment of disclosure risks. Our method follows a similar premise as the well-known record linkage methods employed to evaluate anonymized structured data, and leverages state-of-the-art deep learning language models to exploit the background knowledge available to potential attackers. We also report empirical evaluations of several well-known methods and tools for text anonymization. Results show significant re-identification risks for all methods, including also manual anonymization efforts.

Keywords: Text anonymization · Re-identification risk · Language models · BERT

1 Introduction

The availability of textual data is crucial for many research tasks and business analytics. However, due to its human origin, textual data often includes personal private information. In such case, appropriate measures should be undertaken prior distributing the data to third parties or releasing them to the public in order to comply with the General Data

© Springer Nature Switzerland AG 2022
J. Domingo-Ferrer and M. Laurent (Eds.): PSD 2022, LNCS 13463, pp. 157–171, 2022.
https://doi.org/10.1007/978-3-031-13945-1_12

Protection Regulation (GDPR) [1]. These measures involve either obtaining explicit consent of the individuals the data refer to (which may be infeasible in many cases), or applying an anonymization process by which the data can no longer be attributed to specific individuals. The latter renders data no longer personal and, therefore, outside the scope of the GDPR.

Data anonymization has been widely employed to protect structured databases, in which the individuals' data consist of records of attributes. In this context, a variety of well-stablished anonymization methods and privacy models have been proposed, such as k-anonymity and its extensions [2–4], or ε-differential privacy [5]. However, plain (unstructured) text anonymization is significantly more challenging [6, 7]. The challenges derive from the fact that the re-identifying personal attributes mentioned in the text are unbounded and, quite often, not clearly linked to the individual they refer to.

Most approaches to text anonymization rely on natural language processing (NLP) techniques –named entity recognition (NER)– [8–20] to detect and mask words of potentially sensitive categories, such as names or addresses. Since these methods limit masking to (a typically reduced set of) pre-established categories, they usually offer weak protection against re-identification, the latter being caused by a large variety of entity types. Alternately, methods proposed in the area of privacy preserving data publishing (PPDP) [21–27] consider *any* information that jeopardizes individual's anonymity. However, the damage they cause to the data and several scalability issues make them unpractical in many scenarios [6].

Moreover, because most text anonymization methods do not offer formal privacy guarantees, the degree of protection they offer should be empirically evaluated, as done in the statistical disclosure control (SDC) literature [28]. The standard way to evaluate text anonymization methods consists of comparing their outcomes with manually anonymized versions of the documents to be protected [8, 10–15, 18, 20, 21]. The performance of anonymization methods is then measured through IR-based metrics, specifically precision and recall. Whereas precision accounts for unnecessarily masked terms (which would negatively affect the utility and readability of the anonymized outcomes), recall, which accounts for the amount of undetected re-identifying terms, is roughly equaled as the inverse of disclosure risk. However, recall is severely limited because i) not all (missed) re-identifying terms contribute equally to disclosure, ii) several maskings (each one encompassing a different combination of terms) could be equally valid to prevent disclosure, and iii) it relies on manual anonymization, which may be prone to errors and omissions [6, 29].

In contrast, in the SDC field, the disclosure risk of anonymized databases is empirically measured by subjecting the anonymized data to re-identification attacks, more specifically, *record linkage attacks* [30–33]. Record linkage matches records in the protected database and a background database containing publicly available identified information of the protected individuals. Because successful matchings between both databases results in re-identification, the percentage of correct record linkages provides a realistic an objective measure of the disclosure risk, and an accurate simulation of what an external attacker may learn from the anonymized outcomes.

Because assessing disclosure risks by measuring the performance of automatic re-identification attacks is more convenient and realistic than relying on (limited and human-dependent) IR-based metrics, in this paper we propose a re-identification attack for text anonymization methods grounded on the same formal principles as the record linkage attack employed in structured databases. On that basis, we also provide an intuitive disclosure risk metric based on the re-identification accuracy, which overcomes the limitations of the commonly employed recall-based risk assessment.

To maximize re-identifiability, our attack leverages state-of-the-art machine learning techniques for NLP [34]. These techniques have proved to obtain human or above-human level in several language-related tasks, thereby making our method a realistic representation of an ideal human attacker. We also show the application of our attack to evaluate the level of protection offered by a variety of widely used and state-of-the-art text anonymization methods and tools, in addition to a sample of human-based anonymization employed in a previous work as evaluation ground truth [27].

The remainder of this paper is organized as follows. Section 2 provides background on privacy evaluation for anonymized text. Section 3 presents our attack and metric for assessing the re-identification risk of anonymized texts. Section 4 reports and discusses the empirical evaluation of a variety of automated and manual anonymization approaches. The final section gathers the conclusions and depicts lines of future research.

2 Background

In the context of document anonymization, recall is used as standard to evaluate the level of privacy protection attained by automatic anonymization methods [8, 10–15, 18, 20, 21]. Recall is an IR-based completeness metric, which is defined as the fraction of relevant instances that were properly identified by the method to be evaluated:

$$Recall = \frac{\#TruePositives}{\#TruePositives + \#FalseNegatives} \tag{1}$$

where *#TruePositives* is the number of relevant instances identified and *#FalseNegatives* represents the missed ones. In text anonymization, the relevant instances correspond to words or n-grams that should be masked. These are identified via manual annotation, which is considered the ground truth.

Because IR-based metrics (precision and recall) are the standard way to evaluate many NLP tasks (and NER in particular), and NER techniques are the most common way to tackle text anonymization, perhaps by inertia, the vast majority of methods employ recall to assess the level of attained privacy protection. Nevertheless, this suffers from a variety of issues [29, 35]. First, recall does not measure the actual residual disclosure risk of anonymized documents, but just compares the outputs with manual annotations. Manual anonymization is by definition, subjective and non-unique, and may be prone to errors, bias and omissions [6, 29]. On top of that, manual annotation is costly and time consuming, and usually involves several human experts, whose annotations should be integrated through a non-trivial process. Another limitation of recall-based evaluation is that it assumes that all identified/missed entities contribute equally to mitigate/increase the risk, which is certainly inaccurate [29]. Obviously, failing to mask identifying information (such as a proper name) is much more disclosive on the individual to be protected than just missing her job or her whereabouts.

On the other hand, in the area of SDC, the level of privacy protection attained by anonymization methods on a structured database is measured according to the success of a re-identification attack (*record linkage* [33]) that a hypothetical attacker could perform on the anonymized outcomes. Record linkage tries re-identify anonymized records by linking the masked quasi-identifiers present in those records with those available on publicly available identified sources. Then, the re-identification risk is measured as the percentage of correct linkages:

$$Re\text{-}identification\ risk \approx Linkage\ accuracy = \frac{\#CorrectLinkedRecords}{\#Records} \qquad (2)$$

Compared to recall, the record linkage accuracy offers an automatic and objective means to evaluate privacy that does not rely on manual annotations.

3 A Re-identification Attack for Evaluating Anonymized Text

In this section, we present a re-identification attack for (anonymized) text based on state-of-the-art NLP machine learning techniques. Our attack aims to provide a practical, realistic and objective mean to evaluate the privacy protection offered by anonymization methods for textual data.

In broad terms, the attack aims to re-identify the individuals referred in a set of anonymized documents by leveraging a classifier trained on a collection of identified and publicly available documents encompassing a population of subjects in which the individuals referred in the anonymized documents are contained. For example, one may use publicly available social media publications from a city's inhabitants to re-identify anonymized medical reports from that city's hospital. By construction, the publicly available data should be a superset of the anonymized set. The protected documents would contain confidential attributes (e.g., diagnoses) and masked quasi-identifiers (e.g., age intervals) from unidentified individuals, whereas the publicly available documents would contain identifiers (e.g., a complete name) and clear quasi-identifiers (e.g., a specific age) from known individuals. Consequently, unequivocal matchings of the (quasi-) identifiers of both types of documents (due to a weak anonymization), would allow re-identifying the protected documents and, therefore, disclose the confidential attributes of the corresponding individuals.

Our method can be seen as an adaptation of the standard record linkage attack from structured databases to textual data, where documents correspond to records, words (or n-grams) roughly correspond to attribute values and the classifier provides the criterion to find the best match/linkage between the anonymized and public documents.

The attack is designed with the aim of recreating as realistically as possible what a real attacker would do to re-identify the protected individuals. This also accounts for the amount of resources (computation and background data) that a real attacker may reasonably devote and have available to execute the attack. This is in line with the GDPR (Recital 26), which specifies that, to assess the risk of re-identification, one should take into account the reasonable means that can be employed to perform such re-identification. This makes our attack and the derived risk metric more realistic.

Formally, let A_D be the set of anonymized (non-identified) documents and B_D the set of identified publicly available documents (i.e., background documents). Each document describes or refers to a specific individual, thereby defining the sets of individuals A_I and B_I, and the mapping bijective functions $F_A: A_D \rightarrow A_I$ and $F_B: B_D \rightarrow B_I$. Assuming $A_I \subseteq B_I$ (as in the original record linkage attack), $F_C: A_D \rightarrow B_I$ is the re-identification function that matches protected documents with the corresponding known individuals. On this basis, from the point of view of an attacker, A_D, B_D, B_I and F_B are known, and A_I, F_A and F_C are unknown. Therefore, the purpose of the attack is obtaining F_C' (an approximation of F_C) by exploiting the similarities between A_D and B_D sets.

In Algorithm 1 we formalize our proposal, which returns the number of correct re-identifications achieved by the attack on an input collection of anonymized documents. First, a machine learning classifier is built and trained to predict F_C (line 1, more details in Sect. 3.1). Using the formal notation above, the classifier would implement F_C' by learning which individuals from B_I correspond to the documents in B_D according to the knowledge available to the attacker. Subsequently, the same classifier is evaluated with the set of anonymized documents A_D (line 4). A correct re-identification would happen if the prediction (i.e., F_C') matches F_C (lines 5–6). Finally, the number of re-identifications are returned (line 9).

Algorithm 1. Re-identification risk assessment for anonymized text documents

---

```
Input: Aᴅ  // set of anonymized documents
    Bᴅ  // set of background documents
    Bᵢ  // set of individuals from background documents
    Fʙ  // mapping function from Bᴅ to Bᵢ
    Fᴄ  // groundtruth mapping function from Aᴅ to Bᵢ
Output: numReIds // number of correct re-identifications

1  classf = build_classifier(Bᴅ, Bᵢ, Fʙ, Aᴅ);
2  numReIds = 0; // Number of correct re-identifications
3  for each d in Aᴅ do // Evaluation loop for all documents
4      pred_ind = classf.predict(d); // Predicted Bᵢ individual for d
5      if (pred_ind == Fᴄ(d)) then // If correct re-identification
6          numReIds++;
7      end if
8  end for
9  return numReIds;
```

Similarly to the record linkage method (Eq. 2), we assess the re-identification risk of A_D according to the accuracy of the re-identification attack:

$$Re\text{-}identification\ risk \approx Re\text{-}identification\ accuracy = \frac{numReIds}{|A_I|} \quad (3)$$

3.1 Building the Classifier

We next detail the internals of the *build_classifier* method (line 1 of Algorithm 1). Its goal is to reproduce as faithfully as possible the techniques that a potential attacker may employ to conduct the re-identification attack. This includes considering state-of-the-art NLP classification models and taking advantage of the data available to the attacker.

To select the model, we consider state-of-the-art *word embedding* and *transformer*-based models, which have recently revolutionized the area of NLP. Word embeddings [36] map words (tokens) to real-valued vector representations that capture their meaning, so that words closer in the vector space are expected to be semantically related. The initial approaches to word embeddings produced a fixed vector for each token. Nevertheless, in many cases, words' meaning is affected by the context (especially for polysemic words) and, therefore, they cannot be properly defined through unique embeddings. This led to the creation of contextual word embeddings [37], where the embedding depends on the context of the word instance. Since our classifier requires non-ambiguous words representations, which allow to determine if a word is related with a particular individual, using contextual word embeddings is the best strategy.

Word embedding models require from large training corpora in order to build general and robust word representations. This has led to the popularization of *pre-trained* models [34, 38], which are trained once with an enormous corpus and then are used in multiple NLP tasks. Even though the results obtained from these pre-trained models are good enough for a variety of problems, better performance can be achieved through *fine-tuning*, a procedure in which word embeddings are further trained with the task's specific corpus. We expect the attacker to follow this paradigm, which provides high quality results while significantly reducing cost of training models from scratch.

Another technology that took a step forward in NLP is the *transformer* architecture [39]. The strengths of this approach are the capability of handling long-range dependencies with ease and a reduced processing time based on parallelism. One of the most popular and well-established transformer-based model for NLP is BERT (Bidirectional Encoder Representations from Transformers) [34], which is pre-trained with a huge corpora (Wikipedia and the BookCorpus), and is capable of learning high quality contextual word embeddings. After simple modifications and fine-tuning, BERT is capable of obtaining human-level or even better performance in multiple language-related tasks, including document classification. On this basis, we consider BERT (or its variations) a well-suited model for our attack, since it can obtain outstanding results with neither a huge cost nor unfeasible knowledge assumptions from the attacker.

In addition to build her own classifier, we also expect the attacker to define a development set to have an intuition of the classifier's performance. In this way, it would be also possible to tune the classifier's hyperparameters to maximize the re-identification accuracy. This configures training as a best model search, in which multiple hyperparameters are evaluated according to the accuracy obtained on the development set.

Going back to our algorithm, the classifier returned by the *build_classifier* method is such that, after the further pre-training and fine-tuning steps, obtains the best accuracy on the development set. To this end, multiple trainings with different hyperparameters are performed, searching the best combination. A fixed number of epochs is defined for further pre-training, and fine-tuning is run until a pre-defined maximum number of epochs is achieved or development accuracy does not improve (early stopping).

Regarding the data that can be employed to build the classifier and the development set, recall that the attacker knowledge is limited to B_D, B_I, F_B and A_D. On the one hand, documents in B_D provide knowledge of the individuals' specific vocabulary, which improves understanding of domain-specific words. Additionally, B_D can be labeled on B_I by using F_B, thereby providing useful information about the relationship between the publicly available background data and the individuals' identity. This can lead to the detection of (quasi-)identifying attributes (e.g., the person's name or her demographic attributes), which are the base of the re-identification attack. On the other hand, unlabeled documents in A_D convey knowledge on the anonymized vocabulary. This includes information such as the co-occurrence of words left in clear with those subjected to masking, which may allow inferring the latter from the former.

On this basis, a straightforward approach would be to use all documents in B_D and A_D for further pre-training, and documents in B_D labeled on B_I for fine-tuning. This produces a model with domain-specific knowledge capable of mapping documents to B_I, as it is required for the attack. Nonetheless, it is important to note that the goal of the model is to correctly classify documents in A_D, which come from a different data distribution than the documents in B_D. Concretely, B_D are clear texts (such as identified posts in social media) whereas A_D are anonymized texts (such as non-identified medical reports with some words masked via suppression or generalization). Because machine learning algorithms are sensitive to differences between training and test data distributions, this could hamper the accuracy. For example, during the fine-tuning step, the classifier may learn to focus on identifying words or structures that are not present in the anonymized documents, which would be useless for the attack. To tackle this problem, we propose creating an anonymized version of B_D called B_D' by using any off-the-shelf text anonymization method available to the attacker. Ideally the same method used for A_D should be employed but, because such method would be usually unknown, a standard NER-based method (being NER the most common approach for practical text anonymization), can be used instead. As a result, documents in B_D' would provide an approximation of how data are anonymized, by employing documents more similar to those in A_D. This offers useful information on how known documents (B_D) are anonymized, thereby facilitating disclosure of masked words based on their context. In addition, B_D' can be labeled on B_I (since $B_D' \rightarrow B_D$ is known), therefore facilitating the discovery of the identities underlying the masked documents; for instance, by discovering identifying words neglected by the anonymization method (e.g., a particular street name) that are also present in documents from A_D. Taking this into consideration, we propose using B_D, B_D' and A_D documents for further pre-training and the union of B_D and B_D' labeled on B_I for fine-tuning, thereby obtaining a classifier model better adapted to the content of the anonymized documents.

For the development set, we propose to extract a random subset of configurable size from the documents in B_D, which we call C_D, and transform it to match, as much as possible, the data distribution of A_D. An intuitive approach would be to anonymize C_D; however, this would result into identical documents to those in B_D', which are already present in training data. Thereupon, a previous step is required, aiming to differentiate

C_D texts from the B_D ones and, if possible, to assimilate them to those in A_D prior anonymization. To this end, we propose to perform a summarization-like process on documents from C_D, obtaining $\hat{C}_D$. On this basis, abstractive or hybrid summarization methods are preferred rather than extractive ones [40], so that they produce summarizations that do not include sentences present in documents from B_D. After that, the summarized documents in $\hat{C}_D$ are anonymized (obtaining $\hat{C}_D{}'$) in the same way as done for $B_D{}'$. Finally, the documents in $\hat{C}_D{}'$ are used as the development set of the attack.

4   Empirical Experiments

This section reports empirical results on the application of our re-identification attack to a variety of text anonymization methods, both NLP-oriented and PPDP-grounded. We also test the risk resulting from a manual anonymization effort.

As introduced above, NLP methods [8–20] tackle anonymization as a NER task, in which allegedly private information categories (names, locations, dates, etc.) are detected and masked. Detection is based on rules and models trained to identify the specific categories, and masking consists of replacing the detected entities by their corresponding categories. We considered the following systems and tools that have been employed for NER-based text anonymization [6]:

- *Stanford NER* [41]: provides three pre-trained NER models: *NER3*, which detects ORGANIZATION, LOCATION and PERSON types; *NER4*, which adds the MISC (miscellaneous) type; and *NER7*, which detects ORGANIZATION, DATE, MONEY, PERSON, PERCENT and TIME types.
- *Microsoft Presidio*[1]: a NER-based tool specifically oriented towards anonymization. Among the variety of types supported by Presidio, we enabled those corresponding to quasi-identifying information: NRP -person's nationality, religious or political group-, LOCATION, PERSON and DATE_TIME types.
- *spaCy NER*[2]: we used the *en_core_web_lg,* model, which is capable of detecting named entities of CARDINAL, DATE, EVENT, FAC (e.g., buildings, airports, etc.), GPE (e.g., countries, cities, etc.), LANGUAGE, LAW (named documents made into laws), LOC (non-GPE locations such as mountain ranges), MONEY, NORP (nationalities or religious political group), ORDINAL, ORG, PERCENT, PERSON, PRODUCT, QUANTITY, TIME and WORK_OF_ART types.

Regarding PPDP text anonymization methods, most of them are on the theoretical side [23, 25, 26], suffer from severe scalability issues [21, 42, 43] or seriously damage data utility [22, 24], making them hardly applicable. The only practical method we found is [27], which is based on word embedding models. Due to the lack of a name, this method will be referred to as Word2Vec, this being the backbone neural model employed by this work.

In addition to automatic methods, we also considered the manual anonymization conducted by the authors of [27], which allows us to assess the robustness of manual

---

[1] https://github.com/microsoft/presidio.

[2] https://spacy.io/api/entityrecognizer.

effort against our re-identification attack. Finally, we also report re-identification results on the unprotected versions of the documents in A_D. This constitutes the baseline risk that anonymization methods should (significantly) reduce.

As evaluation data, we employed the corpus described in [27], which consists of 19,000 Wikipedia articles under the "20th century actors" category. To simulate the scenario described in Sect. 3, we considered the article abstracts as the private documents to be anonymized, whereas the article bodies (whose content overlap with the abstracts, even though presented in a different, more detailed way) were assumed to be the identified publicly available information. From this corpus, 50 article abstracts corresponding to popular, contemporary and English speaking actors were extracted in [27] as the set to be subjected to both automatic and manual anonymization. In terms of our attack, the 50 actors in the extracted set constitute A_I, the 50 abstracts anonymized with a method m define $A_D{}^m$, and the article bodies in the corpus constitute B_D (with a population of B_I actors that should encompass A_I).

The amount of background documents B_D used to perform the attack, and their overlap with A_I, have a critical role in the success of the attack. To test this aspect, we defined several attack scenarios by setting increasingly larger B_Ds:

- *50_eval*: a worst case scenario for privacy, in which B_I exactly matches A_I, thereby constituting the easiest re-identification setting. In this case B_D comprises the 50 article bodies of the 50 anonymized abstracts.
- *500_random*: a synthetic scenario consisting of 500 random article bodies taken from the total of 19,000 in the corpus plus those corresponding to the 50 actors in A_I that were not included in the initial random selection. This ensures that $A_I \subseteq B_I$.
- *500_filtered*: a set of 581 article bodies obtained by systematically filtering the initial 19,000 according to several features related to the actors in A_D. In particular, we discarded non-native English speakers, non-actors (e.g., directors), dead individuals, those born before 1950 or after 1995 (latter included) and those whose article included less than 100 links and was present in less than 40 languages (the latter two being related to the 'popularity' of the actor). These criteria aim to maximize the number of individuals in A_I present in B_I, even without knowing A_I, as it would happen in practice. As a result, 40 out of the 50 actors in A_I appeared in B_I. This limits the re-identification accuracy to 80%.
- *2000_filtered*: a set of 1,952 article bodies obtained by using the same criteria as in the prior set but omitting the filter on the number of languages. This results in 41 actors from A_I appearing in B_I, which limits the re-identification accuracy to 82%.

Once B_D is set for a particular scenario, the corresponding B_D', C_D, $\hat{C}_D$ and $\hat{C}_D$' sets required to define the training and development sets should be created as detailed in Sect. 3.1. To create B_D', we anonymized the documents in B_D by using spaCy NER. On the other hand, $\hat{C}_D$ comprised a subset of the abstracts corresponding to the bodies in B_D. Being the abstracts summaries of the article bodies, this procedure follows the summarization-based approach proposed in Sect. 3.1, thus not requiring explicitly building C_D. The size of $\hat{C}_D$ was set to 10% for the *2000_filtered*, *500_filtered* and *500_random* scenarios, and 30% for *50_eval*. Finally, the documents in $\hat{C}_D$ were anonymized by following the same method employed for B_D', thus obtaining the $\hat{C}_D'$ set that constitutes the development set.

To realistically simulate the implementation of our method by a potential attacker, we considered the resources that such attacker would reasonably devote. On this basis, we employed Google Colaboratory, which offers the most powerful free platform for building and running machine learning models. Resources at Google Colaboratory may vary depending on the actual demand. In our tests, the running environment consisted of an Nvidia Tesla K80 GPU with 16GB of VRAM, an Intel Xeon CPU and 12GB of RAM. Google Colaboratory's free tier limits the maximum duration of a run to 12 h. Trainings with a longer duration require from saving the current model and manually restoring the process, resulting in a new environment with a potentially different hardware allocation. In order to ensure that all the computation is made on the same hardware (and also to avoid the tedious manual restoring of the test), we didn't consider scenarios with training runtimes longer than 12 h. This discarded a potential scenario using the whole 19,000 articles as B_D, whose fine-tuning runtime is estimated at about 21 h for 10 epochs. The other scenarios had runtimes of 31, 99, 297 and 301 min, respectively. Note that *500_filtered* took 2.5 times longer to train than *500_random* because the length of the documents in the former was 3 times larger, since the popularity filters applied resulted in longer articles.

Out of the wide variety of pre-trained models based on BERT[3], we have considered those that stand out for their accuracy and/or efficiency, and that can be fine-tuned with the limitations of our execution environment (e.g., GPU memory). Under this premise, we selected DistilBERT (*distilbert-base-uncased*), a distilled version of the original BERT which reduces 40% the model's size but keeps a 97% of its performance in multiple tasks; this provides a great trade-off between accuracy and cost.

As discussed in Sect. 3.1, the model training included performing a best model search based on model's hyperparameters. Considering the number of tests to be conducted, their runtime and their similarities, we applied it to the *50_eval* scenario and used the obtained parameters in the remaining scenarios. Specifically, the hyperparameters that provided the best accuracy for the development set were: *learning rate* 5e-5, *batch size* 16, *sliding window length/overlap* 512/128 and *sliding window length/overlap for classification* 100/25. Additionally, the Hugging Face's AdamW optimizer was used with default parameters except for the learning rate (*betas* 0.9 and 0.999, *eps* 1e-8 and *weight decay* 0).

Pre-training was performed during 3 epochs and fine-tuning during a maximum of 20 epochs. Using the accuracy at the development set for early stopping criteria with a patience of 5 epochs, fine-tuning was run for ~20 epochs for the *50_eval*, *500_random* and *500_filtered* scenarios and during ~10 epochs for the *2000_filtered* scenario. Additionally, it is important to note that the pre-training only used B_D and B_D' without performing the optimal fine-tuning using each one of the A_{DS}. Doing so would increase the number of tests by a factor of 8 (the number of methods/configurations tested), and we observed no noticeable benefits in the worst-case scenario *50_eval*.

---

[3] https://huggingface.co/docs/transformers/index.

4.1 Results

Figure 1 depicts the re-identification risk of each combination of background knowledge and anonymization approach.

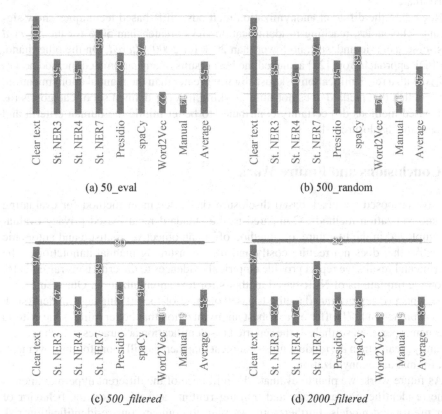

(a) 50_eval

(b) 500_random

(c) *500_filtered*

(d) *2000_filtered*

Fig. 1. Re-identification risk percentages of several anonymization approaches with different sets of background documents. In (c) and (d) the maximum possible re-identification accuracy is depicted as a horizontal line.

First, we notice that the re-identification risk of $A_D{}^{Clear\ text}$ (that is, non-anonymized documents) is close to the maximum, which is, 100% for *50_eval* and *500_random*, and 80% and 82% for *500_filtered* and *2000_filtered*, respectively. This proves the effectiveness of the tuned DistilBERT model as classifier. For the case of anonymized documents, we observe that the attack is capable of re-identifying individuals even from $A_D{}^{Manual}$, with accuracies well-above the random guess, which is 2% for *50_eval*, 0.2% for *500_random*, 0.17% for *500_filtered* and 0.05% for *2000_filtered*. This illustrates that manual anonymization efforts are prone to errors and omissions, and are limited when used as evaluation ground truth.

On the other hand, the average re-identification risk illustrate how B_D influences the results. In particular, the *500_random* scenario provides just slightly less re-identification risk than *50_eval*, because the common features of the 50 protected individuals make them easily differentiable within the random set. In contrast, the risk of the filtered B_{DS} is significantly lower because i) not all the protected individuals are present in B_D and ii) those present are more similar to the other individuals in B_D, thereby being harder to discriminate.

Regarding the different anonymization methods, NER-based techniques show significant deficiencies, reaching re-identification risks greater than 50% for the *50_eval* worst-case scenario and, still, no lower than 20% for *2000_filtered*. On the other hand, the PPDP approach from [27] achieved the best results of any automated method across all B_{DS}, with a re-identification risk just slightly greater than the manual anonymization. That fact that this method does not limit masking to a pre-defined set of categories (as NER-based methods do) certainly contributes to better mimic the human criteria and decrease the disclosure risk.

5   Conclusions and Future Work

We have proposed an attack-based disclosure risk assessment method for evaluating text anonymization methods. Compared to the standard recall-based privacy evaluation employed in the literature, our method offers an objective, realistic and automatic alternative that does not require costly and time consuming manual annotations. The experimental results we report provide empirical evidences to the criticisms raised in [6, 27] on the limitations of NER-based methods for text anonymization. Our results also suggest that privacy-grounded methods based on state-of-the-art language models (such as the approach in [27]) offer more robust anonymization that better mimics the criteria of human experts. Nevertheless, the reported re-identification accuracies, which are significantly greater than the random guess, suggest that there is still room for improvement, even for manual anonymization.

As future work, we plan to evaluate the influence of the different hypermarameters in the re-identification accuracy and training runtime and, also, test the behavior of other pre-trained models. Furthermore, we plan to compare our re-identification risk assessment to the standard recall metric.

Acknowledgements. Partial support to this work has been received from the Norwegian Research Council (CLEANUP project, grant nr. 308904), the European Commission (projects H2020-871042 "SoBigData++" and H2020-101006879 "MobiDataLab") and the Government of Catalonia (ICREA Acadèmia Prize to D. Sánchez). The opinions in this paper are the authors' own and do not commit UNESCO or any of the funders.

References

1. Regulation (EU) 2016/679 of the European Parliament and of the Council of 27 April on the Protection of Natural Persons with Regard to the Processing of Personal Data and on the Free Movement of Such Data and Repealing Directive 95/46/EC. In: Commission, E. (ed.) (2016)

2. Li, N., Li, T., Venkatasubramanian, S.: t-closeness: privacy beyond k-anonymity and l-diversity. In: 2007 IEEE 23rd International Conference on Data Engineering, pp. 106–115. IEEE (2007)
3. Machanavajjhala, A., Kifer, D., Gehrke, J., Venkitasubramaniam, M.: l-diversity: privacy beyond k-anonymity. ACM Trans. Knowl. Disc. Data (TKDD) **1**, 3-es (2007)
4. Sweeney, L.: k-anonymity: A model for protecting privacy. Int. J. Uncertain. Fuzz. Knowl. Based Syst. **10**, 557–570 (2002)
5. Dwork, C.: Differential privacy. In: International Colloquium on Automata, Languages, and Programming, pp. 1–12. Springer (2006)
6. Lison, P., Pilán, I., Sánchez, D., Batet, M., Øvrelid, L.: Anonymisation models for text data: state of the art, challenges and future directions. In: Proceedings of the 59th Annual Meeting of the Association for Computational Linguistics and the 11th International Joint Conference on Natural Language Processing, vol. 1, Long Papers, pp. 4188–4203 (2021)
7. Csányi, G.M., Nagy, D., Vági, R., Vadász, J.P., Orosz, T.: Challenges and open problems of legal document anonymization. Symmetry **13**, 1490 (2021)
8. Aberdeen, J., et al.: The MITRE identification scrubber toolkit: design, training, and assessment. Int. J. Med. Informatics **79**, 849–859 (2010)
9. Chen, A., Jonnagaddala, J., Nekkantti, C., Liaw, S.-T.: Generation of surrogates for de-identification of electronic health records. In: MEDINFO 2019: Health and Wellbeing e-Networks for All, pp. 70–73. IOS Press (2019)
10. Dernoncourt, F., Lee, J.Y., Uzuner, O., Szolovits, P.: De-identification of patient notes with recurrent neural networks. J. Am. Med. Inform. Assoc. **24**, 596–606 (2017)
11. Johnson, A.E., Bulgarelli, L., Pollard, T.J.: Deidentification of free-text medical records using pre-trained bidirectional transformers. In: Proceedings of the ACM Conference on Health, Inference, and Learning, pp. 214–221 (2020)
12. Liu, Z., Tang, B., Wang, X., Chen, Q.: De-identification of clinical notes via recurrent neural network and conditional random field. J. Biomed. Inform. **75**, S34–S42 (2017)
13. Mamede, N., Baptista, J., Dias, F.: Automated anonymization of text documents. In: 2016 IEEE Congress on Evolutionary Computation (CEC), pp. 1287–1294. IEEE (2016)
14. Meystre, S.M., Friedlin, F.J., South, B.R., Shen, S., Samore, M.H.: Automatic de-identification of textual documents in the electronic health record: a review of recent research. BMC Med. Res. Methodol. **10**, 1–16 (2010)
15. Neamatullah, I., et al.: Automated de-identification of free-text medical records. BMC Med. Inform. Decis. Mak. **8**, 1–17 (2008)
16. Reddy, S., Knight, K.: Obfuscating gender in social media writing. In: Proceedings of the First Workshop on NLP and Computational Social Science, pp. 17–26 (2016)
17. Sweeney, L.: Replacing personally-identifying information in medical records, the Scrub system. In: Proceedings of the AMIA Annual Fall Symposium, p. 333. American Medical Informatics Association (1996)
18. Szarvas, G., Farkas, R., Busa-Fekete, R.: State-of-the-art anonymization of medical records using an iterative machine learning framework. J. Am. Med. Inform. Assoc. **14**, 574–580 (2007)
19. Xu, Q., Qu, L., Xu, C., Cui, R.: Privacy-aware text rewriting. In: Proceedings of the 12th International Conference on Natural Language Generation, pp. 247–257 (2019)
20. Yang, H., Garibaldi, J.M.: Automatic detection of protected health information from clinic narratives. J. Biomed. Inform. **58**, S30–S38 (2015)
21. Sánchez, D., Batet, M.: C-sanitized: a privacy model for document redaction and sanitization. J. Am. Soc. Inf. Sci. **67**, 148–163 (2016)

22. Mosallanezhad, A., Beigi, G., Liu, H.: Deep reinforcement learning-based text anonymization against private-attribute inference. In: Proceedings of the 2019 Conference on Empirical Methods in Natural Language Processing and the 9th International Joint Conference on Natural Language Processing (EMNLP-IJCNLP), pp. 2360–2369 (2019)
23. Chakaravarthy, V.T., Gupta, H., Roy, P., Mohania, M.K.: Efficient techniques for document sanitization. In: Proceedings of the 17th ACM Conference on Information and Knowledge Management, pp. 843–852 (2008)
24. Fernandes, N., Dras, M., McIver, A.: Generalised differential privacy for text document processing. In: Nielson, F., Sands, D. (eds.) Principles of Security and Trust. LNCS, vol. 11426, pp. 123–148. Springer, Cham (2019). https://doi.org/10.1007/978-3-030-17138-4_6
25. Cumby, C., Ghani, R.: A machine learning based system for semi-automatically redacting documents. In: Proceedings of the AAAI Conference on Artificial Intelligence, pp. 1628–1635 (2011)
26. Anandan, B., Clifton, C., Jiang, W., Murugesan, M., Pastrana-Camacho, P., Si, L.: t-Plausibility: generalizing words to desensitize text. Trans. Data Priv. **5**, 505–534 (2012)
27. Hassan, F., Sanchez, D., Domingo-Ferrer, J.: Utility-preserving privacy protection of textual documents via word embeddings. IEEE Trans. Knowl. Data Eng. 1 (2021)
28. Hundepool, A., et al.: Statistical Disclosure Control. Wiley, New York (2012)
29. Pilán, I., Lison, P., Øvrelid, L., Papadopoulou, A., Sánchez, D., Batet, M.: The Text Anonymization Benchmark (TAB): A Dedicated Corpus and Evaluation Framework for Text Anonymization. arXiv preprint arXiv:2202.00443 (2022)
30. Domingo-Ferrer, J., Torra, V.J.S.: Computing: disclosure risk assessment in statistical microdata protection via advanced record linkage. Statist. Comput. **13**, 343–354 (2003)
31. Nin Guerrero, J., Herranz Sotoca, J., Torra i Reventós, V.: On method-specific record linkage for risk assessment. In: Proceedings of the Joint UNECE/Eurostat Work Session on Statistical Data Confidentiality, pp. 1–12 (2007)
32. Torra, V., Abowd, J.M., Domingo-Ferrer, J.: Using Mahalanobis distance-based record linkage for disclosure risk assessment. In: DomingoFerrer, J., Franconi, L. (eds.) Privacy in Statistical Databases. LNCS, vol. 4302, pp. 233–242. Springer, Heidelberg (2006). https://doi.org/10.1007/11930242_20
33. Torra, V., Stokes, K.J.I.J.o.U., Fuzziness, Systems, K.-B.: A formalization of record linkage and its application to data protection. Int. J. Uncert. Fuzz. Knowl. Based Syst. **20**, 907–919 (2012)
34. Devlin, J., Chang, M.-W., Lee, K., Toutanova, K.: Bert: Pre-training of deep bidirectional transformers for language understanding. arXiv preprint arXiv:1810.04805 (2018)
35. Mozes, M., Kleinberg, B.J.: No Intruder, no Validity: Evaluation Criteria for Privacy-Preserving Text Anonymization (2021)
36. Bengio, Y., Ducharme, R., Vincent, P., Jauvin, C.: A neural probabilistic language model. J. Mach. Learn. Res. **3**, 1137–1155 (2003)
37. Liu, Y., Liu, Z., Chua, T.-S., Sun, M.: Topical word embeddings. In: Twenty-Ninth AAAI Conference on Artificial Intelligence (2015)
38. Mikolov, T., Chen, K., Corrado, G., Dean, J.: Efficient estimation of word representations in vector space. arXiv preprint arXiv:1301.3781 (2013)
39. Vaswani, A., et al.: Attention is all you need. In: Advances in Neural Information Processing Systems, pp. 5998–6008 (2017)
40. El-Kassas, W.S., Salama, C.R., Rafea, A.A., Mohamed, H.K.: Automatic text summarization: a comprehensive survey. Expert Syst. Appl. **165**, 113679 (2021)
41. Manning, C.D., Surdeanu, M., Bauer, J., Finkel, J.R., Bethard, S., McClosky, D.: The Stanford CoreNLP natural language processing toolkit. In: Proceedings of 52nd Annual Meeting of the Association for Computational Linguistics: System Demonstrations, pp. 55–60 (2014)

42. Sánchez, D., Batet, M.: Toward sensitive document release with privacy guarantees. Eng. Appl. Artif. Intell. **59**, 23–34 (2017)
43. Staddon, J., Golle, P., Zimny, B.: Web-based inference detection. In: USENIX Security Symposium (2007)

Generation of Synthetic Trajectory Microdata from Language Models

Alberto Blanco-Justicia(✉)🆔, Najeeb Moharram Jebreel🆔, Jesús A. Manjón🆔,
and Josep Domingo-Ferrer🆔

Department of Computer Engineering and Mathematics,
Universitat Rovira i Virgili, Av. Paisos Catalans 26,
43007 Tarragona, Catalonia
{alberto.blanco,najeeb.jebreel,jesus.manjon,josep.domingo}@urv.cat

Abstract. Releasing and sharing mobility data, and specifically trajectories, is necessary for many applications, from infrastructure planning to epidemiology. Yet, trajectories are highly sensitive data, because the points visited by an individual can be identifying and also confidential. Hence, trajectories must be anonymized before releasing or sharing them. While most contributions to the trajectory anonymization literature take statistical approaches, deep learning is increasingly being used. We observe that natural language sentences and trajectories share a sequential nature that can be exploited in similar ways. In this paper, we present preliminary work on generating synthetic trajectories using machine learning models typically used for natural language processing. Our empirical results attest to the quality of the generated synthetic trajectories. Furthermore, our methods allow discovering natural neighborhoods based on trajectories.

Keywords: Privacy · Synthetic data generation · Mobility data

1 Introduction

Personal mobility data in their simplest form are data about individuals that include their locations at specific times. Sources of real-time raw individual location data include, but are not limited to, cell towers, Wi-Fi access points, RFID tag readers, location-based services, or credit card payments. Historical location data, in the form of data sets in which each of the records corresponds to an individual and includes her location data for some time periods, are referred to as trajectory microdata sets. Such trajectory microdata sets are often interesting to transport authorities, operators, and other stakeholders to evaluate and improve their services, the state of the traffic, etc. Recently, due to the COVID-19 pandemic, the health authorities have also become interested in mobility data to predict the spread of infectious diseases.

The above landscape motivates the need to share or even publicly release mobility data. Sharing is occasionally done at an aggregate level (*e.g.*, heat maps),

© Springer Nature Switzerland AG 2022
J. Domingo-Ferrer and M. Laurent (Eds.): PSD 2022, LNCS 13463, pp. 172–187, 2022.
https://doi.org/10.1007/978-3-031-13945-1_13

rather than at an individual level. Whichever the specific type of mobility data, sharing them entails a potential privacy risk. Mobility data are highly unique and regular. *Unicity* refers to the data of different individuals being easily differentiable, particularly at some specific locations. The starting and ending locations of an individual's trajectories are often their home and work locations which, again, are highly unique and can lead to reidentification. In [4] it is shown that individual full trajectories can be uniquely recovered with the knowledge of only 2 locations, and knowledge of 4 locations can lead to full reidentification of 95% of individuals in data set containing trajectories for 1.5 million individuals. The *regularity* of trajectories implies that each individual's data follows periodic patterns. Namely, individuals tend to follow the same trajectories during workdays–home to work and back to home.

These features may allow attackers with publicly available data or background knowledge about an individual (such as place of work) to infer sensitive information about that individual, including health status, religious beliefs, social relationships, sexual preferences, etc.

Our interest in this paper is trajectory microdata. A trajectory is a list of spatio-temporal points visited by a mobile object. A trajectory microdata set contains a set of trajectories, where each trajectory normally corresponds to a different individual. The points in each trajectory are both quasi-identifiers and confidential information: indeed, some locations can be very identifying (*e.g.* the trajectory origin can be the individual's home) and other locations can be very confidential (*e.g.* if the individual visited a hospital, a church or a brothel). Thus, anonymizing trajectories is not easy. Several anonymization mechanisms have been proposed, but most of them do not provide solid privacy guarantees or distort the data too much [9]. An alternative to releasing anonymized trajectories is to generate synthetic trajectories that do not correspond to any specific real trajectory [18].

Contribution and Plan of this Paper

We propose to leverage deep learning models used in natural language processing, and in particular for next-word prediction, to generate synthetic trajectory data. A key idea in the proposal is that road networks impose a context to people's movements, and so there is semantics connected to the transition from one trajectory point to the next. Capturing this semantics is something that modern language models based on deep learning have shown to excel at.

Section 2 reviews related work on trajectory data protection and synthetic generation, including methods based on deep learning. Section 3 describes our mechanism for synthetic trajectory data generation. Section 4 shows the results of our experimental evaluation. Finally, Sect. 5 concludes the paper and highlights ideas for future work.

2   Related Work

2.1   Sequential Models for Trajectory Prediction

Since trajectories have the same sequential nature as natural language sentences, sequence models used for next-word prediction have also been extended to next location prediction task. [2,21] use recurrent neural networks (RNNs) [19] and their variations in the next location prediction task. RNNs have shown superior performance due to their ability to capture trajectory data's latent spatial and temporal features. [25] utilise the bidirectional long-short-term memory (BiL-STM) model and the similarity-based Markov model (SMM) to predict the individuals' next locations while maintaining the semantic and spatial patterns of the individuals' trajectories.

2.2   Privacy-Preserving Tajectory Data Publishing

Existing methods for privacy-preserving trajectory publishing can be divided into statistical methods and deep learning (DL)-based methods.

Statistical methods rely on one of the following principles [9,13]: (i) suppression by removing points of trajectories that can identify individuals; (ii) generalization by making the trajectories indistinguishable via grouping them into larger ranges; (iii) distortion by using differential privacy (DP) to ensure that the presence of a record in a data set leaks a controlled amount of information; and (iv) perturbation by using techniques like location merging, clustering, or generating virtual trajectories.

Most of the proposed works in the literature adopt one or more of the techniques above to release privacy-preserving trajectory data. For instance, NWA [1] anonymizes trajectories following a two-step procedure: 1) building clusters of at least k similar trajectories, and 2) anonymizing trajectories in each cluster to produce k-anonymous trajectory data. GLOVE [12] adopts a different procedure with two steps as well: 1) computing trajectory-wise merge costs and 2) iteratively building clusters by merging two trajectories with the smallest cost until satisfying k-anonymity. [6] use microaggregation clustering to group trajectories according to their similarity and then replace them with group representatives. [7] group similar trajectories and remove some of them to ensure k-anonymity. KTL [22] adapts both $l-$diversity and $t-$closeness to trajectory data to counter attacks facilitated by k-anonymity (e.g., attribute linkage). [3,14] adopt DP-based methods to release distorted trajectories. [14] merge coexistent points from different trajectories using a partitioning procedure based on the exponential DP mechanism, whereas [3] propose a mechanism for perturbing semantic trajectories that satisfies ϵ-local DP.

However, statistical methods generally do not provide a proper trade-off between the utility and privacy of their published trajectory data [24]. Non-DP methods, to some extent, maintain the utility of the published data, but they are vulnerable to several privacy attacks (e.g., attribute linkage and background knowledge attacks). Although l-diversity and t-closeness methods offer better protection against privacy attacks, they can have a negative impact on the utility [16]. On the other hand, DP-based methods attempt to make the presence or absence of any single record unnoticeable from the protected output, which makes such methods ill-suited to protect microdata (records corresponding to individual subjects) without (i) severely reducing the data utility or (ii) significantly degrading the privacy guarantee being offered [5].

DL-based methods aim to generate synthetic trajectories that can realistically reproduce the patterns of individuals' mobility [18]. The intuition is that the generated synthetic data come from the same distribution of real trajectories (thereby preserving utility). At the same time, they do not correspond to real trajectories (thereby preserving privacy).

Existing DL methods leverage sequence natural language processing (NLP) models, such as RNNs [19], or generative models, such as generative adversarial networks (GANs) [11], to approximate the distribution of the real trajectory data and then sample synthetic trajectories from that distribution.

[10,17] exploit the ability of RNNs to model problems over sequential data having long-term temporal dependencies. Like training a next-word prediction model, they train a next location prediction model using the real trajectory data as training data. Then, they construct a synthetic trajectory by starting at some arbitrary location and iteratively feeding the current output trajectory sequence as input to the next step in the trained model.

GANs [11] set up a game between two neural networks: the generator G and the discriminator D. G's goal is to generate "synthetic" data classified as "real" by D, whereas D's goal is to correctly distinguish between real and synthetic data and provide feedback to G to improve the realism of the generated data. trajGAN [24] consists of a generator G which generates a dense representation of synthetic trajectories from a random input vector z and a discriminator D, which classifies input trajectory samples as "real" or "fake". To capture contextual and hidden mobility patterns and generate more realistic trajectories, trajGAN [24] uses RNNs to create dense representations of trajectories. SVAE [15] builds its generator G based on an LSTM and a Variational Autoencoder (VAE) to combine the ability of LSTMs to process sequential data with the ability of VAEs to construct a latent space that captures key features of the training data. MoveSim [8] uses a self-attention-based sequential model as a generator to capture the temporal transitions in human mobility. In addition, the discriminator uses a mobility regularity-aware loss to distinguish real from synthetic trajectories. [23] propose a two-stage GAN method (TSG) to generate fine-grained and plausible trajectories. In the first stage, trajectories are transformed into a discrete grid representation and passed as input for a generative model to learn the general pattern. In the second stage, inside each grid, an encoder-decoder generator is used to extract road information from the map image and then embed it into two parallel LSTMs to generate trajectory sequences.

Although DL-based methods have shown promising performance in generating high-utility synthetic trajectories, privacy issues are likely to arise due to overfitting the trained models on the original training data [18]. Consequently, a synthetic trajectory may resemble a real one and give an attacker the chance to use this information for re-identification. In our proposed work, we adopt a DL-based method to generate plausible synthetic trajectories and also mitigate the above-mentioned privacy risk by integrating a randomization mechanism during the synthetic trajectory generation phase.

3   Synthetic Trajectory Generation Method

This section presents our proposed mechanism for synthetic trajectory data generation. We first explain our approach to preprocess the original data, to convert them from lists of spatio-temporal points into sequences of labels. Then, we describe the BiLSTM neural network architecture, which we use to train a next-point prediction model. Finally, we present the data generation process, in which we use the randomization of next-point predictions to limit or prevent the release of trajectories or subtrajectories present in the original data.

3.1   Data Preprocessing

The first step is to preprocess the trajectory data so that they are amenable to be used as training data for a natural language processing model. Trajectory microdata contain spatio-temporal points (id, x, y, t), where id is a trajectory identifier, (x, y) is a latitude-longitude location and t is the time at which the location was visited by the moving object. Analogously, NLP models take sequences of tokens representing words (or parts of words) and punctuation. In our preprocessing, we first define a bounding box around the area of interest and discard outlying points. For example, when dealing with trajectory data in a given city, trajectories that depart from the city to a far away area are not of great interest.

Then, we build a grid of an arbitrary resolution within the bounding box. The more resolution, the better accuracy we can obtain from further analysis of the generated data, but also the more resources we will need in order to train the generator model. The grid resolution also has an effect on the privacy properties of the generated data. Continuing with the NLP example, the more resolution, the bigger the dictionary of words that our model has to deal with. Next, we assign each of the points in the data set to cells in the grid and label cells using an invertible encoding function (such as alphabetic labeling or a number computed as row × number of columns + column). Once each of the points is encoded as a label, we discard all grid labels that do not appear in the data set, so as to reduce the dictionary size, and recode the labels to the range $[0 \ldots \#labels - 1]$. At this point, each of the trajectories is a sequence of labels, similar to what sentences are.

In addition, we compute and store the distribution of trajectory lengths and discard trajectories with outlying lengths, again to save training resources. This length distribution will later be used during the trajectory generation process.

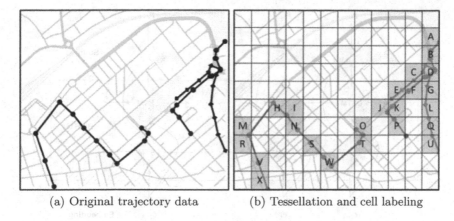

(a) Original trajectory data   (b) Tessellation and cell labeling

Fig. 1. From trajectories to sequences of labels

Finally, we obtain the training (and validation) data by extracting n-grams from the trajectories, using a growing window, and taking the last point in each n-gram as the label for next-word prediction.

3.2 Next-Point Prediction Model

After preprocessing the training data, we use the bidirectional long short-term memory (BiLSTM) model to solve the trajectory next-point prediction task. BiLSTM's main advantage over the other sequence models is that its input flows from the past to the future and vice versa, making BiLSTMs a powerful tool for modeling the sequential dependencies between trajectory points. Since the presence of an individual at a specific location is usually influenced by the previous and next locations the individual visits, BiLSTM is expected to capture those local contextual patterns. Moreover, training BiLSTM on the trajectory data of many individuals' visited points within a limited geographic area is expected to capture the global pattern of the individuals' mobility in that area. Therefore, BiLSTM is expected to predict the individuals' next points more accurately than other models.

Figure 2 shows the architecture of the proposed BiLSTM-based model for the next-point prediction.

The first layer of the model takes a processed trajectory as input, which is represented as sequence of points $(p_{t-1}, p_t, \ldots, p_{t+k})$, where p_t is the point an individual visited as time t. The embedding layer then maps each point in the processed trajectory to a multidimensional vector with a predefined length so the BiLSTM units can process it. Then, the BiLSTM units run the obtained embeddings in two ways, one from past to future and one from future to past, to preserve information from both past and future. Finally, the softmax layer uses the output vectors of the BiLSTM units to produce a vector of probabilities $\in \mathcal{R}^L$ of the next point, where L is the total number of labels.

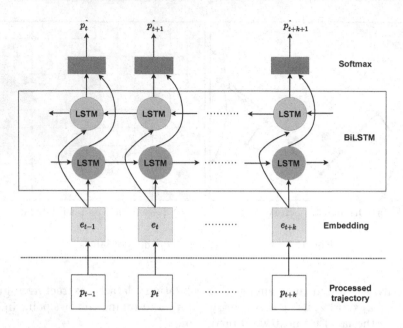

Fig. 2. Proposed BiLSTM-based model for next-point prediction

3.3 Synthetic Data Generation

The generation process starts by choosing the trajectory lengths (number of points per trajectory) from the original length distribution obtained during the preprocessing of the training data. Then, we generate batches of synthetic trajectories of the same length to leverage on the parallelization capabilities of DL models.

To generate each of the synthetic trajectories, we start by drawing a random location label from the dictionary and feeding it to the trained model to predict the next point. Then, we append the predicted point to the trajectory and feed it again to the model. We repeat this process until we obtain trajectories of the desired length.

One potential issue with this approach is that the model learns a 1-to-1 representation or mapping of the training data. This is especially (but not only) possible when ML models overfit the training data. In these cases, the synthesized trajectories are likely to mimic trajectories or sections of trajectories from the training data (which would be akin to sampling them from the training data), and thus not be truly synthetic. The solution we propose in this case is to collect the top-k predictions of the next point and choose one of them uniformly at random. In this case, even if the model is a 1-to-1 mapping of the original data, the probability of a trajectory or sub-trajectory being equal to one in the training data falls to $(1/k)^{length}$. Even so, if the data were big and diverse enough to contain any single possible trajectory in a grided area, all synthetic data generated from them would necessarily be a sample of the original data.

The above described top-k fix can still have issues if the distribution of probabilities for the top k next-point candidates is close to uniform, or if there is a clear peak for the first candidate and very similar probabilities for the rest (which is often a symptom of overfitting). In such cases, the quality of the generated data can decrease, introducing artifacts in the trajectories, such as very long transitions from one point to the following one. The k parameter has to be adequately tuned to avoid these issues. In our experiments, we took $k = 3$.

4   Experimental Analysis

We run experiments that generate synthetic trajectory data out of two public mobility data sets, namely the Cabspotting data set and the GeoLife data set.

4.1   Data Sets and Preprocessing

The Cabspotting data set [20] contains trajectories of occupied taxi cabs in San Francisco and the surrounding areas in California. The data set contains trajectories of nearly 500 taxis collected over 30 days. The trajectories consist of points containing each a GPS location, a timestamp and an identifier. In our preprocessing, we just keep points with longitudes between $-122.6°$ and $-122.0°$ and latitudes between $37.0°$ and $37.4°$. These points define an area of $2,938.38$ km^2 around the San Francisco Bay area. Next, we define two grids with different resolutions to generate two different data sets:

San Francisco 128: This is a grid of 128×128 cells, which results in $16,384$ cells, of which $4,944$ are visited at least once. These will become the labels in our location dictionary for the first data set. They are shown in Fig. 3a.

San Francisco 256: This is a grid of 256×256 cells, or a total $65,536$ cells, of which $12,393$ are visited at least once. These are the labels for the second data set. Figure 3b shows the unique locations in the dictionary.

In both cases, we keep only trajectories with a length between 3 and 45 locations, which result in $437,335$ distinct trajectories. Figure 4 shows the distribution of the lengths of trajectories in the San Francisco 128 and 256 data sets.

The GeoLife data set [26–28] is a trajectory microdata set collected by Microsoft Research Asia during the Geolife project. The data collection process was carried out by 182 users during a period of over 3 years (2007–2012). Each of the trajectories consists of a sequence of timestamped locations, collected with varying sampling rates of one point every 1–5 s or every 5–10 m. The data set collects different mobility modes and activities. Again, we keep only points within longitudes $115.9904°$ and $116.8558°$ and latitudes $39.67329°$ and $40.22726°$, which yields an area of $4,548.78$ km^2 around the city of Beijing. We generate two data sets by defining two different grids of spatial resolutions 128×128 and 256×256:

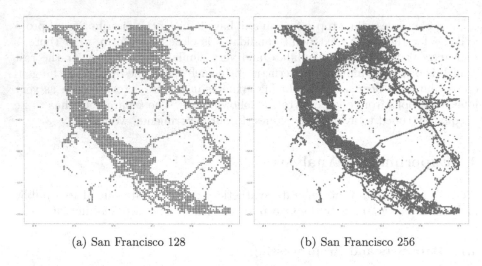

(a) San Francisco 128   (b) San Francisco 256

Fig. 3. Unique locations in the San Francisco data set for resolutions 128 and 256

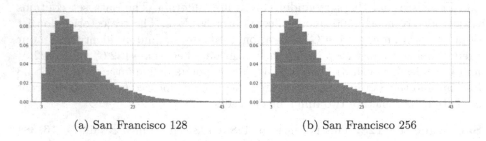

(a) San Francisco 128   (b) San Francisco 256

Fig. 4. Distribution of the lengths of trajectories in the San Francisco 128 and San Francisco 256 data sets

Beijing 128: The first data set is generates by tessellating the given area using a grid of 128×128 cells, which results in $16,384$ cells, of which $7,079$ are visited at least once. Thus, the location dictionary of the Beijing 128 data set consists of those $7,079$ unique locations. Figure 5a shows these unique locations in the dictionary. We keep trajectories with a number of points between 3 and 70, which results in a total $36,827$ distinct trajectories.

Beijing 256: The second grid consists of 256×256 cells, or a total $65,536$ cells, of which $15,831$ are visited at least once. These are the labels for the Beijing 256 data set. In this case, we keep $43,741$ trajectories, which have lengths between 3 and 100 points. Figure 5b shows the unique locations in the dictionary.

Figure 6 shows the distribution of the lengths of trajectories in the Beijing 128 and 256 data sets.

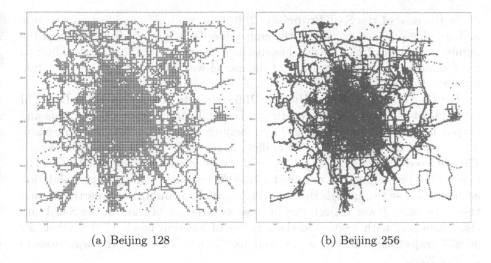

(a) Beijing 128   (b) Beijing 256

Fig. 5. Unique locations in the Beijing data set for resolutions 128 and 256

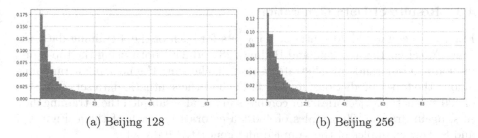

(a) Beijing 128   (b) Beijing 256

Fig. 6. Distribution of the lengths of trajectories in the Beijing 128 and Beijing 256 data sets

In all four data sets, we completed the preprocessing phase by generating the n-grams from the trajectories and keeping the last point in each of the n-grams as the classification label for the next-point prediction models.

4.2   Model Training

For the **San Francisco 128** data set, we trained a BiLSTM consisting of an embedding layer of input dimension 4,944 and output dimension 64, followed by two BiLSTM layers of 256 and 128 units, respectively, and a dense layer of 128 units. The output layer consisted of 4,944 units with the softmax activation function. Dropout layers were included before and after the second BiLSTM layer, with a dropout rate of 20%. The model was trained for 250 epochs with a batch size of 250 and a validation split of 20%, using the Adam optimizer on the sparse categorical crossentropy loss function. The model obtained an accuracy of 48.53% and a validation accuracy of 48.7%.

In the case of the **San Francisco 256** data, we used a similar architecture and training process, except for the embedding layer's input dimension and the number of units of the output layer, which were set to $12,393$, according to the dictionary size. The accuracy of the model was 33.86%, while its validation accuracy was 34.3%.

For the **Beijing 128** and **Beijing 256** data sets, we increased the BiLSTM layer sizes to 512 and 256, respectively. The embedding layer's dimensions and output layer sizes were set to $7,079$ for Beijing 128 and $15,831$ for Beijing 256, according to the sizes of their label dictionaries. The model trained on Beijing 128 obtained an accuracy of 79.3% and a validation accuracy of 62.29%. The one trained on Beijing 256 obtained a 75.40% accuracy and a 47.81% validation accuracy. These two models show overfitting to the training data, partly because there are many fewer trajectories in these two data sets than in the San Francisco data sets with respect to the number of location labels ($7,079$ labels for $36,827$ trajectories in Beijing 128, and $15,831$ labels for $43,741$ trajectories in Beijing 256).

4.3   Results of Data Generation

Finally, we generated 20 synthetic data sets for each of the 4 training data sets (San Francisco 128 and 256, and Beijing 128 and 256) both using the top-1 prediction and our randomized strategy, that is, choosing the next point randomly among the top-k predictions, for $k = 3$. Each of the synthetic data sets consisted of 400 synthetic trajectories. For comparison, we also sampled the 4 training data sets, again drawing 20 samples of 400 trajectories for each of them. Figures 7 and 8 show examples of the sampled and generated data sets.

As mentioned before, both Beijing synthetic data sets show some artifacts in the shape of long jumps across points. This is partly the effect of being smaller data sets and also the effect of overfitting.

In order to assess the quality of the generated data, we compared the original samples with the synthetic data sets, according to the following metrics:

- d_{SL}, computed as the mean of the sum of the distances (in km) between all consecutive locations over all trajectories.
- Δr, defined as the average distance (in km) between any two consecutive points over all trajectories.
- d_{max}, computed as the mean maximum distance (in km) between any two consecutive locations over all trajectories.
- $\#locs$, obtained as the mean number of distinct locations visited in each trajectory.
- The mean number of visits per unique location $\#V/loc$ for all trajectories.

Table 1 shows the results for each of the data sets, averaged over the 20 different samples drawn or generated for each of them.

In the case of the San Francisco synthetic data sets, the results show smaller values for d_{SL}, Δr, and d_{max} under the top-1 and top-3 predictions than in

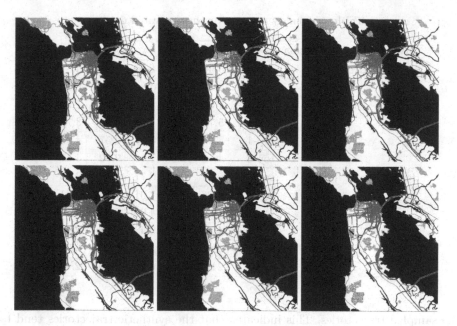

Fig. 7. Examples of San Francisco trajectories. Top figures use a 128×128 grid and bottom figures use a 256×256 grid. Figures in the left are samples from the original data set, figures in the middle show the results of using top-1 predictions, while figures in the right use random top-3 predictions.

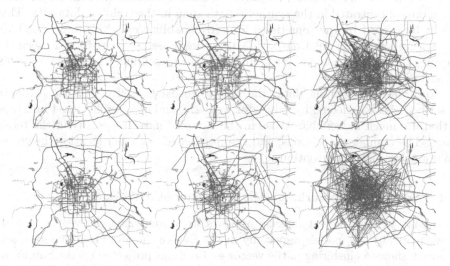

Fig. 8. Examples of Beijing trajectories. Top figures use a 128×128 grid and bottom figures use a 256×256 grid. Figures in the left are samples from the original data set, figures in the middle show the results of using top-1 predictions, while figures in the right use random top-3 predictions.

Table 1. Quality metrics for the generated data

City	Resolution	Method	d_{SL}	Δr	d_{max}	#locs	#V/loc
San Francisco	128	sample	5.94	0.78	1.16	8.40	6.87
		top-1	4.50	0.60	0.84	8.30	8.84
		top-3	4.96	0.67	1.09	8.13	7.20
	256	sample	5.73	0.64	1.11	9.83	3.78
		top-1	3.94	0.45	0.81	9.57	4.81
		top-3	4.70	0.54	1.16	9.52	4.11
Beijing	128	sample	5.91	0.56	0.71	10.37	3.43
		top-1	6.26	0.62	1.31	7.83	4.01
		top-3	14.5	1.45	4.90	8.38	3.41
	256	sample	4.79	0.35	0.48	13.4	2.47
		top-1	5.81	0.45	1.82	9.47	3.00
		top-3	21.47	1.69	6.71	10.93	2.47

the sampled trajectories. This indicates that the synthetic trajectories tend to be slightly shorter than the original ones, especially under the top-1 prediction. The #locs metric shows similar results among the sampled and synthetic data sets, while the #V/loc metric show slightly higher values for the synthetic data. This, together with the shorter trajectories, seem to indicate that the trajectories are more concentrated in the synthetic data than in the training data sets. This might be caused by the sometimes limited capability of RNNs and BiLSTMs to capture long-range relationships. ML models based on transformers seem to capture these relationships better, and we plan to conduct experiments in this regard in the future.

Regarding the Beijing data, the results reveal the effects of the artifacts described above, reflected by values for d_{SL}, Δr, and d_{max} under top-1 and top-3 that are much higher than those in the sampled data (especially in the top-3 case). The number of locations, and the number of visits per location, however, do not show such big deviations.

4.4   Additional Remarks on the Experimental Work

One point of independent interest appears when analyzing the vector embeddings learnt by the models, especially in the case of the San Francisco data set. Figure 9 shows a clustering of the vector embeddings projected on the San Francisco map together with a neighborhood division of the same city[1]. While not exactly aligned, the clustering shows three big areas in the west and three in the east, surrounding a central area with several divisions, similar to that in the neighborhood map.

---

[1] A SF local's guide to the neighborhoods of San Francisco. https://sfgal.com/sf-locals-guide-to-neighborhoods-of-san-francisco/.

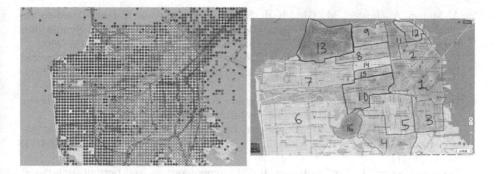

Fig. 9. Clustering of embeddings compared to a map of neighborhoods

Such results suggest that analyzing the vector embeddings resulting from trajectory data sets might be of use to other applications based on city areas, possibly together with additional information. One possible example of application would be the prediction of house prices in different areas of a city.

5 Conclusions and Future Work

In this paper, we have shown preliminary work on the generation of synthetic trajectory microdata using machine learning models typically used for natural language processing and time series. We have shown the potential of such approaches and proposed a strategy to limit the possible data leakages from the data generation. An independent result arising from the study of the vector embeddings learnt during the training process might be of interest in other related areas.

As future work, we plan to replicate the experiments using modern architectures for NLP based on transformers, which have shown a higher performance in NLP tasks than BiLSTMs, especially regarding long-term relationships between concepts. Additionally, we plan to study the effects of differential privacy in the synthetic data, using different approaches such as DP-SGD and PATE. Other potential research directions involve including the time dimension in the generated data and replicating the process in distributed or federated learning scenarios.

Acknowledgements. This research was funded by the European Commission (projects H2020-871042 "SoBigData++" and H2020-101006879 "MobiDataLab"), and the Government of Catalonia (ICREA Acadèmia Prize to J. Domingo-Ferrer, FI grant to N. Jebreel). The authors are with the UNESCO Chair in Data Privacy, but the views in this paper are their own and are not necessarily shared by UNESCO.

References

1. Abul, O., Bonchi, F., Nanni, M.: Never walk alone: Uncertainty for anonymity in moving objects databases. In: 2008 IEEE 24th International Conference on Data Engineering, pp. 376–385. IEEE, 7 April 2008

2. Al-Molegi, A,. Jabreel, M., Ghaleb, B.: STF-RNN: space time features-based recurrent neural network for predicting people next location. In: 2016 IEEE Symposium Series on Computational Intelligence (SSCI), pp. 1–7. IEEE, 6 December 2006
3. Cunningham, T., Cormode, G., Ferhatosmanoglu, H., Srivastava, D.: Real-world trajectory sharing with local differential privacy. arXiv preprint arXiv:2108.02084. 4 August 2021
4. De Montjoye, Y.-A., Hidalgo, C.A., Verleysen, M., Blondel, V.D.: Unique in the crowd: the privacy bounds of human mobility. Sci. Rep. **3**(1), 1–5 (2013)
5. Domingo-Ferrer, J., Sáinches, D., Blanco-Justicia, A. The limits of differential privacy (and its misuse in data release and machine learning). Commun. ACM **64**(7), 33–35 (2021)
6. Domingo-Ferrer, J., Trujillo-Rasua, R.: Microaggregation- and permutation-based anonymization of movement data. Inf. Sci. **15**(208), 55–80 (2012)
7. Dong, Y., Pi, D.: Novel privacy-preserving algorithm based on frequent path for trajectory data publishing. Knowl. Based Syst. **15**(148), 55–65 (2018)
8. Feng, J., Yang, Z., Xu, F., Yu, H., Wang, M,. Li, Y.: Learning to simulate human mobility. In: Proceedings of the 26th ACM SIGKDD International Conference on Knowledge Discovery & Data Mining, pp. 3426–3433. 25 August 2020
9. Fiore, W., et al.: Privacy in trajectory micro-data publishing: a survey. Trans. Data Privacy **13**, 91–149 (2020)
10. Gao, Q., Zhou, F., Zhang, K., Trajcevski, G., Luo, X., Zhang, F.: Identifying human mobility via trajectory embeddings. In: IJCAI, vol. 17, pp. 1689–1695, 19 August 2017
11. Goodfellow, I., et al.: Generative adversarial nets. Adv. Neural Inf. Process. Syst. **27** (2014)
12. Gramaglia, M., Fiore, M.: Hiding mobile traffic fingerprints with glove. In: Proceedings of the 11th ACM Conference on Emerging Networking Experiments and Technologies, pp. 1–13, 1 December 2015
13. Guerra-Balboa, P., Pascual, A.M., Parra-Arnau, J,. Forné, J.: Strufe. Anonymizing trajectory data: limitations and opportunities (2022)
14. Hua, J., Gao, Y., Zhong, S.: Differentially private publication of general time-serial trajectory data. In: 2015 IEEE Conference on Computer Communications (INFOCOM), pp. 549–557, IEEE, 26 April 2015
15. Huang, D., et al.: A variational autoencoder based generative model of urban human mobility. In: 2019 IEEE Conference on Multimedia Information Processing and Retrieval (MIPR), pp. 425–430. IEEE, 28 March 2019
16. Jin, F., Hua, W., Francia, M., Chao, P., Orlowska, M., Zhou, X.: A survey and experimental study on privacy-preserving trajectory data publishing. TechRxiv (2021)
17. Kulkarni, V., Garbinato, B.: Generating synthetic mobility traffic using RNNs. In: Proceedings of the 1st Workshop on Artificial Intelligence and Deep Learning for Geographic Knowledge Discovery, pp. 1–4, 7 November 2017
18. Luca, M., Barlacchi, G., Lepri, B., Pappalardo, L.: A survey on deep learning for human mobility. ACM Comput. Surv. (CSUR) **55**(1), 1–44 (2021)
19. Mikolov, T., Karafiát, M., Burget, L., Cernocký, J., Khudanpur, S.: Recurrent neural network based language model. In Interspeech **2**(3), 1045–1048 (2010)
20. Piorkowski, M, Sarafijanovic-Djukic, N., Grossglauser, M.: CRAWDAD data set EPFL/mobility (v. 2009–02–24). Traceset: cab, downloaded from February 2009
21. Rossi, A., Barlacchi, G., Bianchini, M., Lepri, B.: Modelling taxi drivers' behaviour for the next destination prediction. IEEE Trans. Intell. Transp. Syst. **21**(7), 2980–2989 (2019)

22. Tu, Z., Zhao, K., Xu, F., Li, Y., Su, L., Jin, D.: Protecting trajectory from semantic attack considering k-anonymity, l-diversity, and t-loseness. IEEE Trans. Netw. Serv. Manag. **16**(1), 264–78 (2018)
23. Wang, X., Liu, X., Lu, Z., Yang, H.: Large scale GPS trajectory generation using map based on two stage GAN. J. Data Sci. **19**(1), 126–41 (2021)
24. Xi, L., Hanzhou, C., Clio, A.: trajGANs: using generative adversarial networks for geo-privacy protection of trajectory data. Vision paper (2018)
25. Xu, M., Han, J.: Next location recommendation based on semantic-behavior prediction. In: Proceedings of the 2020 5th International Conference on Big Data and Computing, pp. 65–73, 28 May 2020
26. Zheng, Y., Li, Q., Chen, Y., Xie, X., Ma, W.-Y.: Understanding mobility based on GPS data. In: Proceedings of ACM Conference on Ubiquitous Computing (UbiComp 2008), Seoul, Korea, pp. 312–321. ACM Press (2008)
27. Zheng, Y., Xie, X., Ma, W.-Y.: GeoLife: a collaborative social networking service among User, location and trajectory. IEEE Data Eng. Bull. **33**(2), 32–40 (2010)
28. Zheng, Y., Zhang, L., Xie, X., Ma, W-.Y.: Mining interesting locations and travel sequences from GPS trajectories. In: Proceedings of International conference on World Wild Web (WWW 2009), Madrid, Spain, pp. 791–800. ACM Press (2009)

Synthetic Data

Synthetic Individual Income Tax Data: Methodology, Utility, and Privacy Implications

Claire McKay Bowen[1]([envelope]), Victoria Bryant[2], Leonard Burman[1], John Czajka[3], Surachai Khitatrakun[1], Graham MacDonald[1], Robert McClelland[1], Livia Mucciolo[1], Madeline Pickens[1], Kyle Ueyama[4], Aaron R. Williams[1], Doug Wissoker[1], and Noah Zwiefel[5]

[1] Urban Institute, Washington, D.C. 20024, USA
cbowen@urban.org
[2] Internal Revenue Services, Washington, D.C. 20002, USA
[3] Bethesda, MD 20816, USA
[4] Coiled, New York, NY 10018, USA
[5] University College London, London, UK

Abstract. The United States Internal Revenue Service Statistics of Income (SOI) Division possesses invaluable administrative tax data from individual income tax returns that could vastly expand our understanding of how tax policies affect behavior and how those policies could be made more effective. However, only a small number of government analysts and researchers can access the raw data. The public use file (PUF) that SOI has produced for more than 60 years has become increasingly difficult to protect using traditional statistical disclosure control methods. The vast amount of personal information available in public and private databases combined with enormous computational power create unprecedented disclosure risks. SOI and researchers at the Urban Institute are developing synthetic data that represent the statistical properties of the administrative data without revealing any individual taxpayer information. This paper presents quality estimates of the first fully synthetic PUF and shows how it performs in tax model microsimulations as compared with the PUF and the confidential administrative data.

Keywords: Disclosure control · Synthetic data · Utility · Classification and regression trees

1 Introduction

The United States Internal Revenue Service (IRS) possesses invaluable administrative tax data from individual income tax returns that could vastly expand our understanding of how tax policies affect behavior and how those policies could be made more effective. For decades, the IRS Statistics of Income (SOI) Division has released an annual public use file (PUF), a privacy-protected database of

© Springer Nature Switzerland AG 2022
J. Domingo-Ferrer and M. Laurent (Eds.): PSD 2022, LNCS 13463, pp. 191–204, 2022.
https://doi.org/10.1007/978-3-031-13945-1_14

sampled individual income tax returns. Several organizations, including the American Enterprise Institute and the Urban-Brookings Tax Policy Center use the PUF as the basis of microsimulation models that help the public understand the potential impacts of policy proposals. However, awareness of the growing threats to public use microdata and a general concern for protecting participants' privacy have led the IRS to increasingly restrict and distort information in the PUF. This makes the PUF less useful for policy analysis and academic research.

To address the threats to privacy, we generate a fully synthetic public use file or SynPUF, consisting of pseudo records of individual income tax returns that are statistically representative of the original data (Little 1993; Rubin 1993). Our methodology is an extension of Bowen et al. (2020), which synthesized nonfiler data from tax year 2012, but not all lessons directly apply to the PUF, because individual taxpayer data are much more complex and diverse than the nonfiler data.

Our most important contribution is our methods for addressing these additional challenges to synthesize data, such as synthesizing the survey weights, selecting the order of variables for synthesis, and applying variable constraints. Our ultimate goal is to generate a SynPUF that maintains strong data confidentiality while providing better data utility than the PUF that SOI traditionally releases. For this purpose, we rigorously evaluate the SynPUF against both the administrative tax data and the PUF on various disclosure risk metrics and utility measures.

We organize this paper as follows. Section 2 describes how we generate the SynPUF data source from the confidential administrative tax data and details the data synthesis methodology for the SynPUF, covering our measures for disclosure risk. Section 3 evaluates and compares the SynPUF to the original PUF and confidential administrative tax data. Conclusions and discussions about future work are in Sect. 4.

2   Data Synthesis Methodology

In this section, we describe how the SOI creates the administrative tax data and the PUF. We then outline the subsampling procedure used to generate the most recent PUF. We also provide an overview of our synthetic data generation process and how we address several of the challenges in synthesizing complex tax data.

2.1   Administrative Tax Data

The IRS processes all federal individual income tax returns – Form 1040, its schedules and supplement forms, and relevant informational return – and stores them in the IRS Master File, a massive tax database of more than 100 million unedited tax returns (145 million records in 2012). However, the Master File has limited use for tax policy analysis due to potential data inconsistencies (e.g., original tax forms filed by taxpayers) and its size. To produce a dataset more suitable for analytical purposes, SOI annually produces the INSOLE (Individual and Sole Proprietor), a stratified sample of all individual income tax returns drawn from the Master File and cleaned by the SOI (Burman et al. 2019). In 2012, for example, the INSOLE contained 338, 350 records. The selected returns are edited to add items not captured in the Master File and to be internally consistent.

To produce the PUF, SOI draws a sample from the INSOLE and applies traditional statistical disclosure control (SDC) techniques (Bryant et al. 2014; Bryant 2017). To protect privacy, many INSOLE records are dropped to reduce the sampling rate, making any individual return unlikely to be in the PUF, and records with an extremely large value for a key variable are aggregated into one of four aggregate records (Burman et al. 2019). The 2012 PUF has 172,415 records.

2.2   Synthetic Data Generation

We describe our synthetic generation in three steps: a synthesis preparation step, a synthetic data generation step, and a post-processing step. We classify any data preparation as our synthesis preparation step. Our synthetic data generation step encompasses the synthetic data generation process and any additional noise or modifications we introduced. The post-processing step includes any changes we made to the synthesized data, such as ensuring consistency among variables.

Synthesis Preparation: We draw from the INSOLE to create a new file called the modINSOLE. This approach allows us to keep more of the original records from the INSOLE than those used in the PUF for developing our synthesis model. To generate the 2012 modINSOLE, we use tax returns for tax year 2012 that were filed in calendar year 2013. This excludes a small number of tax-year 2012 returns filed after 2013, because some people filed their taxes late.

Since the INSOLE is a weighted sample of the IRS Master File, we sample the new modINSOLE records within each INSOLE stratum to preserve the pattern of survey weights. To increase data privacy, we combine the 98 INSOLE strata into 25 modINSOLE strata, grouped by weight and the greater of gross positive or the absolute value of gross negative income. Because the combined strata have different weights, we sample with replacement using sampling weights larger than the stratum's smallest sampling weight. We then recalculate the weights until every record in the new modINSOLE stratum has an identical sampling weight. This process allows us to vary the synthesis order and other synthesis strategies by stratum, and implement a wider range of machine learning algorithms, particularly ones that cannot accommodate a sample of records with different weights.

For strata that include records sampled at a rate greater than 0.2 (i.e., 20% of the population), we randomly remove observations and then increase the weights of the remaining observations to maintain the correct population count until the sampling rate of those strata equals 0.2.

Overall, the sampling procedure discards fewer records from the INSOLE than the PUF, which increases the amount of information we can use to train our model. This presents little increase in disclosure risk because, unlike the PUF, no actual tax returns are included in the SynPUF. The final modINSOLE file contains 265, 239 records, which we will refer to as the confidential data.

Synthetic Data Generation Step: For our synthetic data step, we implement a sequence of Classification and Regression Tree (CART) models within each modINSOLE stratum (Breiman et al. 1984; Reiter 2005).

Our synthesis procedure begins with 28 categorical variables (e.g., tax filing status), integer counts (e.g., number of dependents), or integer numeric variables (e.g., age in years). These variables create at least two key challenges. First, the sequential synthesis must avoid impossible combinations of variables (e.g., a spouse age for a record that has a single filing status). Second, our method for adding additional noise to variables (outlined later in the sections) only works for plausibly continuous variables.

To deal with these challenges, we first sort categorical variables into essential, non-essential, and modeled (i.e., variables calculated from the essential variables). We then draw new records from observed combinations of essential variables to guarantee that the synthesized combinations match observed combinations. This preserves the distribution of categorical variables and prevents the creation of impossible combinations. We also check that the joint frequencies of these categorical variables are sufficiently dense (i.e., counts are not ones or twos). Based on the set of values from each draw of essential variables, we generate the modeled variables before synthesizing sequentially the non-essential variables.

To synthesize continuous variables, we apply a sequence of CART models. These models use previously synthesized variables as their explanatory variables. We select the order of variables to be synthesized within each stratum. The synthesis order is important because variables synthesized later in the sequence tend to produce noisier values (Bonnéry et al. 2019). After testing, we find that the order produced by the greatest to least weighted sum of absolute values yields the best synthesis based on our utility metrics.

The application of CART in data synthesis dates to Reiter (2005), who proposed to use a collection of nonparametric models to generate partially synthetic data. CART sorts data through a sequence of binary splits that end in nodes that are intended to be homogenous. When the target variable is categorical, CART predicts the outcome using classification trees. This method builds the tree through an iterative process of splitting the data into binary partitions. For continuous variables, CART uses regression trees to choose the value that splits the continuous values into the partitions. A regression tree creates nodes with the smallest sum of squared errors (calculated as squared deviations from the mean). Because this data-driven method is more flexible than parametric approaches, such as regression-based models, it can account for unusual variable distributions and nonlinear relationships that can be hard to identify and model explicitly. Recent research has demonstrated that it tends to outperform regression-based parametric methods and it is computationally feasible (Bonnéry et al. 2019; Drechsler and Hu 2021).

We estimate CART models for each variable with all previously synthesized outcome variables as potential predictors. Our synthetic-data generation method is based on the insight that a joint multivariate probability distribution can be represented as the product of a sequence of conditional probability distributions. We define this mathematically as

$$f(\boldsymbol{X}|\boldsymbol{\theta}) = f(X_1, X_2, ..., X_k|\theta_1, \theta_2, ..., \theta_k)$$
$$= f_1(X_1|\theta_1) \cdot f_2(X_2|X_1, \theta_2)...f_k(X_k|X_1, ..., X_{k-1}, \theta_k) \tag{1}$$

where X_i for all $i = 1, ..., k$ are the variables to be synthesized, θ are vectors of model parameters, such as regression coefficients and standard errors, and k is the total number of variables. As mentioned above, we use the weighted sum of absolute values to determine our synthesis order.

Here we navigate the tree splits until a unique final node is identified to predict each observation. Traditionally, predictions are made with a conditional mean, but that method may result in incorrect variances and covariances, and too few values in the tails of the marginal distributions (Little and Rubin 2019). Similar to other implementations, we sample from the final node to predict observations, which remedies this issue.

Since the ith CART model would need to consider $i - 1$ variables, and i can be as large as 150, we drop predictors with very low variance (Kuhn and Johnson 2019). These are variables in which the most common value occurs in about 95% of observations, the second most common value occurs in about five percent of the observations and unique values make up no more than one percent of the observations. In addition, we pre-select predictors to be considered based on subject matter expertise, as is common for parametric models like linear regression. Bonnéry et al. (2019) used that approach in their synthesis of education data from the Maryland Longitudinal Data System.

Noise Addition to Protect Against Disclosure: Sampling from final nodes of a fitted CART model reproduces the observed values of the confidential data, creating disclosure risk. Reiter (2005) used a kernel smoother, replacing each value with a local average. We apply the method similar to Bowen et al. (2020), where the noise is added to each value instead, completely obscuring it. A brief summary of the process is the following:

1. Split the outcome variable from the confidential data into equal sized and ordered groups. If a group has an identical minimum and maximum (i.e., the group has no variation), then combine with adjacent groups.
2. Estimate a Gaussian kernel density on each group.
3. Map the predicted value from the CART algorithm to the corresponding group from step 1.
4. Draw a value from a normal distribution with mean equal to the value predicted by the CART algorithm and variance equal to the optimal variance for the kernel density estimator calculated in step 2. The derived value becomes part of the synthetic observation.

This process results in a smooth and unbounded distribution of synthetic data. Furthermore, the procedure adds more noise in sparse parts of the distribution where individual values are distinct and less noise in parts of the distribution where values are common and unidentifiable. No noise is added to very common values, such as zeros.

Mid-synthesis Constraints: Without any constraints, the synthesis may generate values that fall outside the bounds of the data. We outline three types of constraints to address this issue (Drechsler 2011):

1. Unconditional bracketed constraints are univariate constraints on the minima and maxima of continuous variables. For example, net capital losses may only take values in the range of $0 to −$3,000.
2. Conditional bracketed constraints are multivariate constraints on the minima and maxima of continuous variables. For example, educator expenses have a maximum of $500 for married taxpayers filing jointly and $250 for other taxpayers. A variable may need to simultaneously satisfy many conditional bracketed constraints.
3. Some variables must be greater than or equal to another variable or variables. For example, total dividends must be greater than or equal to qualified dividends, total IRA distributions must be greater than or equal to taxable IRA distributions, and total pensions and annuities must be greater than or equal to taxable pensions and annuities. To impose these linear constraints, we calculate some variables during post-processing (e.g., child tax credit), synthesize component parts (e.g., taxable and nontaxable IRA distributions), or model some variables as proportions of other variables. For example, we model the wage split between primary and secondary taxpayers on married filing jointly return as a function of total reported wage and salaries and other variables.

Synthesized Components: If a variable is a sum of component variables and it does not have a bracket constraint for the maximum, the component variables should be synthesized and summed to generate the overall variable. If a constraint cannot be applied with post-processing or synthesizing components, we apply two different approaches: hard bounding and z-bounding (Drechsler 2011). When a value falls outside of a lower or upper bound, we can set that value to the closest bound, which is called hard bounding. While this approach is easy to implement, it causes biases because values may cluster at those bounds. However, hard bounding can work well for certain variables that already have values clustered at the bounds. For instance, net capital loss is capped at $3,000 and the underlying values naturally cluster at that bound.

We also implement the Z-bounding technique that resamples the problematic value up to z times. If none of the sampled values satisfies the constraint the value is hard bounded (Drechsler 2011). This approach can potentially have a high computational cost if many values fail to satisfy the constraint on the first synthesis, because a smaller but still significant proportion may fail on subsequent re-syntheses.

Post-processing Step: To avoid internal inconsistencies, the variables synthesized thus far do not include calculated variables. In this step, we compute these variables based on previously synthesized values. In addition, several variables are capped to protect data privacy or to preserve relationships between variables in the INSOLE.

Calculated Tax Variables: We calculate some variables during the post-processing step instead of synthesizing the variable. For example, the Child Tax Credit is a function of synthesized variables, including the number of qualifying children, AGI, and filing status.

Capping Variables: We restrict the values of some variables to protect data privacy, to preserve relationships between variables in the INSOLE, and to resemble the PUF. For example, we cap age at 85.

Data Modifications: We limit or recode some variable values because these values are not needed to calculate tax liability, are restricted due to SOI policy, or pose a disclosure risk due to too few observations having those values. For example, we restrict the number of dependents to four based on SOI policy. This minimally affects the data utility because there are very few households that have over four dependents. Additionally, all tax variables are rounded according to SOI rounding rules applied to the current PUF.

Reweighting the Strata: Our final post-processing step is ensuring the SynPUF is a representative sample of the confidential data. We plan to reweight synthetic tax return records to ensure means and counts of selected variables approximately match SOI published totals by income group. Guaranteeing that key aspects of the synthetic data, like the distribution of capital gains income, match the US population facilitates the use of SynPUF as a data source for building microsimulation models. Note that this reweighting would not incur any additional privacy loss given that the targets are already published.

2.3   Disclosure Risk Measures

Replacing actual data with fully synthetic data protects against identity disclosure or a data intruder associating an individual with a specific record in the released data. This is because no real observations are released. Similarly, fully synthetic data helps prevent attribute disclosure because no actual values are released while limiting a data intruder's confidence in any given value of a sensitive variable (Reiter 2002). But synthetic data may still risk disclosing information if they are not carefully constructed for various reasons, such as perfectly replicating the observations (Raab et al. 2017). We, therefore, evaluate our processes with two risk measures based on the modINSOLE and before the noise addition step to verify that the risk of disclosure is extremely small.

Frequency: We count the number of records from the original data that are replicated completely in the synthetic data. But because it is very difficult to exactly replicate observations with 175 variables, these counts should also be performed on a meaningful subset of variables, such as the variables on the front and back of Form 1040. We do this in three different ways. First, we count the number of observations from the modINSOLE replicated in the SynPUF. However, the modINSOLE may contain many identical records. If these records were replicated in the SynPUF, there would not be any disclosure risk because the records represent many tax returns and thus could not uniquely identify any individual tax unit. Many of these records consist almost entirely of zero entries, which also reduces the disclosure threat. Our second measure adjusts for this by counting the number of unique observations from the modINSOLE replicated in the SynPUF. In principle, an attacker can also gain some information about the modINSOLE from the SynPUF because a unique record on the SynPUF might reflect a unique record on the modINSOLE. Consequently, our third measure counts the number of observations that are identical and unique in both the confidential data and the synthetic data before smoothing (so-called unique-uniques). Adding noise changes these values, so testing the unique-uniques beforehand (as we do) is a conservative measure.

Sample Heterogeneity: The second risk metric concerns heterogeneity or a check for attribute disclosure risk. In other words, when a data intruder can associate sensitive data characteristics to a particular record or group of records without identifying any exact records. We check for heterogeneity in the SynPUF using a measure known as l-diversity (Machanavajjhala et al. 2007). For each node in the CART models, this measure counts the number of distinct non-zero values in the modINSOLE. For each variable, we count the number of records that contain a value derived from a node with fewer than 3 values (the so-called "rule of 3" used as a rough screen for disclosure risk by government agencies).

We calculate the l-diversity of the final node for each synthetic value for each synthesized variable because more diversity provides more privacy protection. The measures are applied to data before the noise addition in the synthetic data generation step and before IRS rounding rules are applied. These measures are therefore unrealistically pessimistic and the actual privacy protection in the Syn-PUF is greater than they indicate. We count the number of observations in the SynPUF generated from nodes with fewer than three unique values, excluding all-zero nodes.

3   Evaluation

For disclosure risk, the synthesizer does a good job of creating records that are plausible but do not match records in the confidential data too closely. Table 1 in the Appendix shows that no observations in the synthetic dataset exactly match observations in the confidential data under different matching conditions. When examining just variables from the Form 1040, four observations are recreated out of the possible 265,239 records, and one unique observation is recreated. As noted, adding noise and rounding will eliminate these few matches.

Table 2 in the Appendix demonstrates that the overwhelming majority of synthetic values come from heterogenous nodes. More than 75% of values in all variables come from nodes with three or more unique values and 99% of values in 59% of the variables come from nodes with fewer than three unique values.

Table 3 (also in the Appendix) measures node heterogeneity within observations. For example, all the 170 numeric tax variables (out of the 175 variables) come from nodes with at least three unique values in 9,181 observations (which represents 11% of all synthesized observations). Similarly, for 50,106 observations, or 19% of all synthesized observations, all but three variables come from nodes with at least three unique values. This demonstrates that the overwhelming majority of values in each observation come from nodes with heterogeneity. In particular, 89.5% of all observations have at most eight variables whose values come from nodes with fewer than three unique values, and all observations have at most 20 variables whose values come from such nodes.

Next, we show that utility of the SynPUF is generally high. Figure 1a in the Appendix shows that the weighted mean for each variable in the SynPUF closely matches the weighted mean in the confidential data. In contrast, the standard deviations of some variables are markedly smaller in the SynPUF than in the confidential data (Fig. 1b in the Appendix). We suspect this is due to some of our post-processing procedures (e.g., capping variables). We plan to investigate this divergence further in several ways, such as examining the error from the CART models, noise addition, mid-synthetic restrictions, and post-processing step.

Figure 2 in the Appendix shows the density of pairwise differences in correlation coefficients. A value of zero means the correlation between two variables in the synthetic data exactly matches the correlation in the confidential data. While there are a few outliers, most of the correlations are extremely close. 96.1% of differences are less than 0.01 and 55.7% of differences are less than 0.001.

Finally, we compare tax microsimulation results from the PUF, SynPUF, and confidential data. To do this, we apply a tax calculator designed by the Open Source Policy Center and maintained by the Policy Simulation Library[1] to the modINSOLE, the PUF, and the SynPUF. For each dataset, we calculate forms of income, such as adjusted gross income, taxable income, and income from capital gains, taxes on ordinary income and capital gains. We then simulate a uniform increase in tax rates and compare how well tax calculations on the PUF and the SynPUF match the same calculations on the modINSOLE. Figure 3a shows that the synthetic data closely matches the confidential data's distribution of Adjusted Gross Income (AGI) except for the top one percent. Figure 3b illustrates that the synthesizer closely reproduces the number of filers with taxable income. For more microsimulation results, see Bowen et al. (Forthcoming).

4   Conclusions and Future Work

This paper demonstrates the feasibility of producing a fully synthetic public use file of individual income tax return information. The SynPUF matches key characteristics of the confidential data, such as univariate means and most correlations between variables, fairly well. The data also adhere to hundreds of logical constraints that reflect the complexity of the tax code complexity. Based on preliminary tests, the SynPUF appears promising for microsimulation

---

[1] See https://pslmodels.org/.

modeling, but certain aspects, such as the correlation between certain variables, need improvement.

The SynPUF have several advantages over the PUF produced using traditional SDC methods. First, the synthesis is designed to provide a robust protection against disclosure of individual data for every variable and record. In constrast, traditional SDC methods require identifying privacy risks associated with particular records or variables and designing targeted approaches to mitigate those risks.

Second, the process of manual risk assessment and mitigation is labor-intensive and time consuming. In principle, the process of creating a synthetic PUF can be largly automated, especially in years when the tax law does not change, making the production of subsequent PUFs faster. SOI will still need to carefully assess the utility and privacy of the resultant file and probably submit it to trusted users for testing before release. But, the synthesis process should allow fast creation of synthetic PUFs. More importantly, given the resource constraints at the IRS, this process requires much less staff time than the current methods.

Third, a synthetic data file can safely include all the variables in the INSOLE including variables that are not currently released in the PUF. These variables allow analysts at the Joint Committee on Taxation and Office of Tax Analysis to model the effects of a wide range of policies. Including those variables on the SynPUF would allow analysts outside of government to analyze the same range of policies as the official scorekeepers. That independent vetting of the effects of current and proposed policies better informs the public and can strengthen confidence in the information released from official sources. Further, SynPUF can safely include observations with extreme (albeit synthesized) values which facilitate microsimulation analyses that encompass this subpopulation of taxpayers. Currently, records with extreme values have been aggregated into a handful of aggregate records in the PUF.

Finally, because the synthetic PUF does not include any actual tax records, SOI can safely distribute it to a larger audience than the select few institutions who currently have access.

Acknowledgments. The projects outlined in this paper relied on the analytical capability that was made possible in part by a grant from Arnold Ventures. The findings and conclusions are those of the authors and do not necessarily reflect positions or policies of Internal Revenue Service, the Urban Institute, or its funders.

Appendix

Table 1. Number of duplicate records out of the possible 265, 239 records.

Duplicate	All variables count	1040 variables count
Records recreated	0	4
Unique records recreated	0	1
$n < 3$ records recreated	0	1
Unique records recreated as a unique record	0	1

Table 2. l-diversity results within each variable.

	More than 75%	More than 95%	More than 99%
Percent of variables	100%	80%	59.4%

Table 3. l-diversity results across observations.

Variables with l-diversity < 3	Observation count	Percent of total	Cumulative percentage
0	9,181	11.0%	11.0%
1	11,495	4.3%	15.3%
2	14,894	5.6%	21.0%
3	50,106	18.9%	39.8%
4	32,025	12.1%	51.9%
5	28,230	10.6%	62.6%
6	22,620	8.5%	71.1%
7	28,732	10.8%	81.9%
8	20,088	7.6%	89.5%
9	9,322	3.5%	93.0%
10	4,616	1.7%	94.7%
11	4,582	1.7%	96.5%
12	6,244	2.4%	98.8%
13	1,293	0.5%	99.3%
14	1,504	0.6%	99.9%
15	208	0.1%	100.0%
16	67	0.0%	100.0%
17	20	0.0%	100.0%
18–20	12	0.0%	100.0%

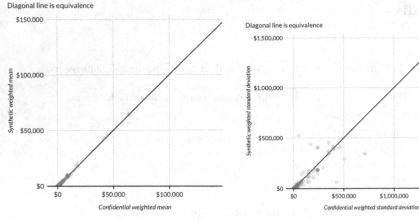

(a) Variable weighted means in the Syn-PUF versus modINSOLE.

(b) Standard deviations in the SynPUF versus modINSOLE.

Fig. 1. Each dot represents a variable, such as wages and salaries or interest income. The diagonal line represents equivalence, and dots off of it indicate that the SynPUF and modINSOLE have different variable weighted means and standard deviations.

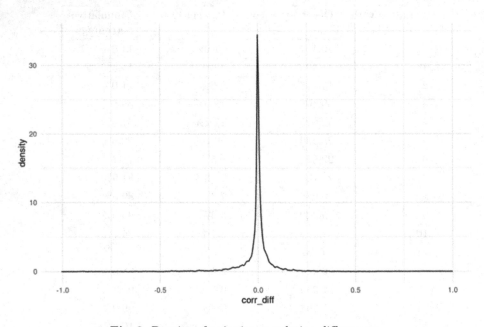

Fig. 2. Density of pairwise correlation differences.

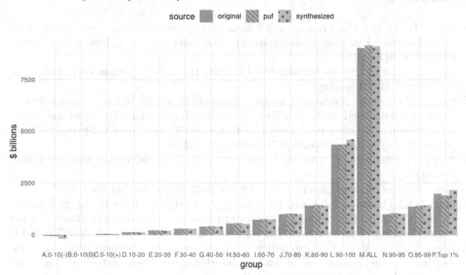

(a) Distribution of Adjusted Gross Income (AGI).

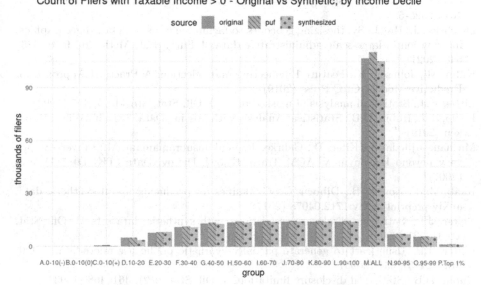

(b) Distribution of filers with taxable income.

Fig. 3. Tax microsimulation results from the confidential data, PUF, and SynPUF, which are grouped in that order for these plots.

References

Bonnéry, D., et al.: The promise and limitations of synthetic data as a strategy to expand access to state-level multi-agency longitudinal data. J. Res. Educ. Effect. **12**(4), 616–647 (2019)

Bowen, C.M., et al.: Synthetic individual income tax data: promises and challenges. Natl. Tax J. (Forthcoming)

Bowen, C.M.K., et al.: A synthetic supplemental public use file of low-income information return data: methodology, utility, and privacy implications. In: Domingo-Ferrer, J., Muralidhar, K. (eds.) PSD 2020. LNCS, vol. 12276, pp. 257–270. Springer, Cham (2020). https://doi.org/10.1007/978-3-030-57521-2_18

Breiman, L., Friedman, J., Olshen, R., Stone, C.: Cart. Classification and Regression Trees (1984)

Bryant, V.: General description booklet for the 2012 public use tax file (2017)

Bryant, V.L., Czajka, J.L., Ivsin, G., Nunns, J.: Design changes to the SOI public use file (PUF). In: Proceedings. Annual Conference on Taxation and Minutes of the Annual Meeting of the National Tax Association, vol. 107, pp. 1–19. JSTOR (2014)

Burman, L.E., et al.: Safely expanding research access to administrative tax data: creating a synthetic public use file and a validation server. Technical report, Technical report US, Internal Revenue Service (2019)

Drechsler, J.: Synthetic Datasets for Statistical Disclosure Control: Theory and Implementation, vol. 201. Springer, New York (2011). https://doi.org/10.1007/978-1-4614-0326-5

Drechsler, J., Hu, J.: Synthesizing geocodes to facilitate access to detailed geographical information in large-scale administrative data. J. Surv. Stat. Methodol. **9**(3), 523–548 (2021)

Kuhn, M., Johnson, K.: Feature Engineering and Selection: A Practical Approach for Predictive Models. CRC Press (2019)

Little, R.J.: Statistical analysis of masked data. J. Off. Stat. Stockh. **9**, 407 (1993)

Little, R.J., Rubin, D.B.: Statistical Analysis with Missing Data, 3rd edn. Wiley, Hoboken (2019)

Machanavajjhala, A., Kifer, D., Gehrke, J., Venkitasubramaniam, M.: l-diversity: privacy beyond k-anonymity. ACM Trans. Knowl. Discov. Data (TKDD) **1**(1), 3-es (2007)

Raab, G.M., Nowok, B., Dibben, C.: Guidelines for producing useful synthetic data. arXiv preprint arXiv:1712.04078 (2017)

Reiter, J.P.: Satisfying disclosure restrictions with synthetic data sets. J. Off. Stat. **18**(4), 531 (2002)

Reiter, J.P.: Using cart to generate partially synthetic public use microdata. J. Off. Stat. **21**(3), 441 (2005)

Rubin, D.B.: Statistical disclosure limitation. J. Off. Stat. **9**(2), 461–468 (1993)

On Integrating the Number of Synthetic Data Sets m into the *a priori* Synthesis Approach

James Jackson[1]([ID]), Robin Mitra[2]([ID]), Brian Francis[1]([ID]), and Iain Dove[3]([ID])

[1] Lancaster University, Lancaster, UK
j.jackson3@lancaster.ac.uk
[2] Cardiff University, Cardiff, UK
[3] Office for National Statistics, Titchfield, UK

Abstract. The synthesis mechanism given in [4] uses saturated models, along with overdispersed count distributions, to generate synthetic categorical data. The mechanism is controlled by tuning parameters, which can be tuned according to a specific risk or utility metric. Thus expected properties of synthetic data sets can be determined analytically *a priori*, that is, before they are generated. While [4] considered the case of generating $m = 1$ data set, this paper considers generating $m > 1$ data sets. In effect, m becomes a tuning parameter and the role of m in relation to the risk-utility trade-off can be shown analytically. The paper introduces a pair of risk metrics, $\tau_3(k, d)$ and $\tau_4(k, d)$, that are suited to $m > 1$ data sets; and also considers the more general issue of how best to analyse $m > 1$ categorical data sets: average the data sets pre-analysis or average results post-analysis. Finally, the methods are demonstrated empirically with the synthesis of a constructed data set which is used to represent the English School Census.

Keywords: Synthetic data · Privacy · Categorical data · Risk metrics · Contingency tables

1 Introduction

When disseminating data relating to individuals, there are always two conflicting targets: maximising utility and minimising disclosure risk. To minimise risk, statistical disclosure control (SDC) methods, which typically involve either suppressing or perturbing certain values, are applied to a data set prior to its release. One such method is the generation of synthetic data sets [6,14], which involves simulating from a model fit to the original data. These methods, while reducing risk, adversely impact the data's utility resulting in a clear trade-off between risk and utility.

This paper focuses on the role of multiple data sets when synthesizing categorical data (that is, data consisting of only categorical variables) at the aggregated level using saturated count models [4]. Saturated synthesis models allow the synthesizer to generate synthetic data with certain pre-specified properties,

© Springer Nature Switzerland AG 2022
J. Domingo-Ferrer and M. Laurent (Eds.): PSD 2022, LNCS 13463, pp. 205–219, 2022.
https://doi.org/10.1007/978-3-031-13945-1_15

thus allowing them to easily tailor the synthesis to suit the data environment [3]. For example, if the intention is to release open data, relatively more noise can be applied to the data than if the data are released in a secure environment. While the Poisson model is often used to model categorical data, for synthesis this is not necessarily an optimal choice, because the synthesizer - that is, the person(s) responsible for synthesizing the data - has no control over the variance and has, therefore, no way to add additional noise to at-risk records in the data. For this reason, the negative binomial (NBI), a two-parameter count distribution, is much more effective for synthesis. As the NBI distribution's variance is not completely determined by the mean - though the variance is always greater than the mean - the variance can be increased accordingly. Nevertheless, there are still restrictions and these are discussed later on.

Specifically, this paper explores how flexibility can be incorporated into the mechanism through the use of multiple synthetic data sets. In some cases (as explained in Sect. 3), $m > 1$ synthetic data sets must be generated; while in other cases, though it may be sufficient to generate just $m = 1$ synthetic data set, the optimal m can still be considered in relation to the risk-utility trade-off: does the improvement in utility sufficiently outweigh the cost in terms of greater risk? This is because, since it reduces simulation error, increasing m leads to greater utility but also, inevitably, greater risk [11,12]. More generally, considering $m > 1$ introduces another tuning parameter for the synthesizer to set, thereby providing further flexibility.

This paper is structured as follows: Sect. 2 summaries the (σ, α)-synthesis mechanism, on which the results in this paper are based; Sect. 3 extends the mechanism to incorporating $m > 1$; Sect. 4 introduces the $\tau_3(k, d)$ and $\tau_4(k, d)$ metrics, developed to assess risk in multiple categorical synthetic data sets; Sect. 5 presents an illustrative example; and lastly Sect. 6 ends the paper with a discussion and areas of future research.

2   Review of the Use of Saturated Models for Synthesis

The discrete nature of categorical data allow it to be expressed as a multi-dimensional contingency table (multi-way table). As a multi-way table, the data consist of a structured set of cell counts $f_1, \ldots, f_K$, which give the frequencies with which each combination of categories is observed.

Synthetic data sets can then be generated by replacing these observed counts (known henceforth as "original counts") with synthetic counts. There are two distinct modelling methods for contingency tables: multinomial models and count models. The multinomial approach ensures that the total number of individuals in the original data n is equal to the total number of individuals in the synthetic data n_{syn}. The syn.catall function in the R package synthpop [7] can be used to generate synthetic data via a saturated multinomial model.

The (σ, α)-synthesis mechanism [4] uses saturated count models for synthesis; specifically, either a saturated negative binomial (NBI) model or a saturated Poisson-inverse Gaussian (PIG) [13] model. In this paper, for brevity, only the

NBI has been considered. Besides, the NBI and PIG distributions are broadly similar, as they share the same mean-variance relationship.

The (σ, α)-synthesis mechanism has two parameters which are set by the synthesizer. The first, $\sigma > 0$, is the scale parameter from a two-parameter count distribution (such as the NBI). The parameter σ can be tuned by the synthesizer to adjust the variability in the synthetic counts, thus increasing or decreasing their expected divergence from the original counts. More noise is required for sensitive cells - usually small cell counts, which correspond to individuals who have a unique (or near-unique) set of observations - to generate sufficient uncertainty to mask the original counts' true values.

The mechanism's second parameter, denoted by $\alpha \geq 0$, relates to the size of the pseudocount - in practice, this is not actually a count but a small positive number such as 0.01 - which is added to zero cell counts (zero cells) in the original data. This assigns a non-zero probability that a zero cell is synthesized to a non-zero. The pseudocount α is only applied to non-structural zero cells (known as random or sampling zeros), which are zero cells for which a non-zero count *could* have been observed. Throughout this paper it has been assumed, for brevity, that $\alpha = 0$.

Given an original count $f_i = N_i \; i = 1, \ldots, K$, the corresponding synthetic count f_i^{syn} is drawn from the following model:

$$f_i^{\mathrm{syn}} \mid f_i = N_i, \sigma \sim \mathrm{NBI}(N_i, \sigma), \quad \text{and therefore,}$$

$$p(f_i^{\mathrm{syn}} = N_2 \mid f_i = N_1, \sigma) = \frac{\Gamma(N_2 + 1/\sigma)}{\Gamma(N_2 + 1) \cdot \Gamma(1/\sigma)} \cdot \left(\frac{\sigma N_1}{1 + \sigma N_1} \right)^{N_2} \cdot \left(\frac{1}{1 + \sigma N_1} \right)^{1/\sigma}.$$

Using a saturated count model has certain advantages in data synthesis. Firstly, it guarantees the preservation of relationships between variables, as no assumptions are made as to which interactions exist. Secondly, the method scales equally well to large data sets, as no model fitting is required - the model's fitted counts are just the observed counts. Finally, as the fitted counts are just equal to the observed counts, it allows expected properties of the synthetic data to be determined *a priori* (that is, prior to synthesis). The (unwelcome) uncertainty around model choice is, in effect, minimised, and instead uncertainty is injected where it is most needed: to add noise to sensitive cells in the original data.

2.1   The τ Metrics

The following τ metrics [4], give a basic quantification of risk (and utility) in tabular data:

$$\tau_1(k) = p(f^{\mathrm{syn}} = k) \qquad\qquad \tau_3(k) = p(f^{\mathrm{syn}} = k \mid f = k)$$
$$\tau_2(k) = p(f = k) \qquad\qquad \tau_4(k) = p(f = k \mid f^{\mathrm{syn}} = k),$$

where f and f^{syn} are arbitrary original and synthetic counts, respectively. The metric $\tau_2(k)$ is the empirical proportion of *original* counts with a count of k, and $\tau_1(k)$ is the proportion of *synthetic* counts of size k. The metric $\tau_3(k)$ is the probability that an original count of size k is synthesized to k; and $\tau_4(k)$ is the probability that a synthetic count of size k originated from a count of size k. The metrics $\tau_3(1)$ and $\tau_4(1)$, in particular, are the most associated with risk, as these relate to uniques and can be viewed as outliers in the data. When, for example, $\tau_4(1)$ is close to 1, it is possible to identify, with near certainty, uniques in the original data from the synthetic data.

When saturated models are used, the expected values of these τ metrics can be found analytically as functions of the tuning parameters (σ, α and, as later described, m). Hence the synthesizer knows, *a priori*, the noise required to achieve a given $\tau_3(1)$ or $\tau_4(1)$ value.

3   The Role of m as a Tuning Parameter

The original inferential frameworks for fully and partially synthetic data sets [9,10] relied on the generation of $m > 1$ synthetic data sets, because they required the computation of the between-synthesis variance b_m (see below). However, when the original data constitute a simple random sample, and the data are completely synthesized, valid inferences can be obtained from $m = 1$ synthetic data set [8]. In this instance, while $m > 1$ data sets are not intrinsic to obtaining *valid* inferences, the *quality* of inferences - for example, the width of confidence intervals - can, nevertheless, be improved upon by increasing m - but at the expense of higher risk. It is less a question, therefore, of which m allows valid inferences to be obtained, but rather a question of which value of m is optimal with respect to the risk-utility trade-off?

Thus m can be viewed as a tuning parameter, and, as with the other tuning parameters σ and α, expected risk and utility profiles can be derived analytically, *a priori*. When saturated models are used for synthesis, ignoring the small bias arising from $\alpha > 0$, simulation error is the only source of uncertainty - and increasing m reduces simulation error. The notion is that $m > 1$ may allow a more favourable position in relation to the risk-utility trade-off than when $m = 1$; in short, it increases the number of options available to the synthesizer.

The use of parallel processing can substantially reduce the central processing unit (CPU) time when generating multiple data sets. Besides, the CPU time taken is typically negligible anyway; the synthesis presented in Sect. 5 took 0.3 s for the NBI with $m = 1$ on a typical laptop running R.

3.1   Obtaining Inferences from $m > 1$ Data Sets

Analysing the $m > 1$ Data Sets Before Averaging the Results. When analysing multiple synthetic data sets, traditionally the analyst considers each data set separately before later combining inferences. While point estimates are simply averaged, the way in which variance estimates are combined depends on the type of synthesis carried out: such as whether fully or partially synthetic data sets are generated and also whether synthetic counts are generated by simulating from the Bayesian posterior predictive distribution or by simulating directly from the fitted model. The combining rules also depend on whether an analyst is using the synthetic data to estimate a population parameter Q, or an observed data estimate $\hat{Q}$: the former needs to account for the sampling uncertainty in the original data whereas the latter does not.

Suppose, then, that an analyst wishes to estimate a univariate population parameter Q from $m > 1$ synthetic data sets. A point estimate $q^{(l)}$, and its variance estimate $v^{(l)}$, is obtained from each synthetic data set, $l = 1, \ldots, m$. Before these estimates are substituted into a set of combining rules, it is common, as an intermediary step, to first calculate the following three quantities [2]:

$$\bar{q}_m = \frac{1}{m} \sum_{l=1}^{m} q^{(l)}, \qquad b_m = \frac{1}{(m-1)} \sum_{l=1}^{m} (q^{(l)} - \bar{q}_m)^2, \qquad \bar{v}_m = \frac{1}{m} \sum_{l=1}^{m} v^{(l)},$$

where $\bar{q}_m$ is the mean estimate, b_m is the 'between-synthesis variance', that is, the sample variance of the $m > 1$ estimates, and $\bar{v}_m$ is the mean 'within-synthesis variance', the mean of the estimates' variance estimates.

The quantity $\bar{q}_m$ is an unbiased estimator for $\hat{Q}$, and is so regardless of whether fully or partially synthetic data sets are generated. When using the synthesis method described in Sect. 2, partially - rather than fully - synthetic data sets are generated, because a synthetic population is not constructed and sampled from, as stipulated in [9]. Hence, the following estimator T_p [10], is valid when estimating $\mathrm{Var}(\hat{Q})$,

$$T_p = \frac{b_m}{m} + \bar{v}_m.$$

The sampling distribution (if frequentist) or posterior distribution (if Bayesian) of $\hat{Q}$ is a t-distribution with $\nu_p = (m - 1)\left(1 + m\bar{v}_m/b_m\right)^2$ degrees of freedom. Often, ν_p is large enough for the t-distribution to be approximated by a normal distribution. However, when the between-synthesis variability is much larger than the within-synthesis variability, that is, when b_m is much larger than $\bar{v}_m$ - as may happen when large amounts of noise are applied to protect sensitive records - then ν_p is crucial to obtaining valid inferences.

As the data sets are completely synthesized in the sense of [8] - that is, no original values remain - the following estimator T_s is valid, too, under certain conditions:

$$T_s = \bar{v}_m \left(\frac{n_{\mathrm{syn}}}{n} + \frac{1}{m}\right) \approx \bar{v}_m \left(1 + \frac{1}{m}\right).$$

These conditions are: firstly, that the original data constitute a simple random sample - therefore, T_s would not be valid if the data originate from a complex survey design - and secondly, that the original data are large enough to support a large sample assumption. The overriding advantage of T_s is that, assuming its conditions do indeed hold, it allows valid variance estimates to be obtained from $m = 1$ synthetic data set.

The large sample assumption facilitates the use of a normal distribution for the sampling distribution (or the posterior distribution) of $\hat{Q}$ when T_s is used to estimate the variance. The notion is that, in large samples, b_m can be replaced with $\bar{v}_m$. It is difficult to assess, however, when a large sample assumption is reasonable, because it also depends on the specific analysis being undertaken on the synthetic data, that is, it depends on the analysis's sufficient statistic(s).

The estimators T_p and T_s assume that $n_{\text{syn}} = n$ (or that n_{syn} is constant across the m synthetic data sets in the case of T_s). When using count models as opposed to multinomial models, n_{syn} is stochastic and this assumption is violated. However, in a simulation study unreported here, the effect of varying n_{syn} was found to have a negligible effect on the validity of inferences, for example, confidence intervals still achieved the nominal coverage. Nevertheless, in some cases, new estimators may be required; such estimators may introduce weights $w_1 \ldots, w_m$ that relate to $n_{\text{syn}}^{(1)}, \ldots, n_{\text{syn}}^{(m)}$, the sample sizes of the m synthetic data sets.

Averaging the $m > 1$ Data Sets Before Analysing Them. When faced with multiple categorical data sets, analysts (and attackers) may either pool or average the data sets *before* analysing them. This is feasible only with contingency tables, as they have the same structure across the $m > 1$ data sets. There are several advantages to doing so. Firstly, it means that analysts only have to undertake their analyses once rather than multiple times, thus leading to reduced computational time. Note, although averaging leads to non-integer "counts", standard software such as the glm function in R can typically cope with this and still allow models to be fit. Secondly, model-fitting in aggregated data is often hampered by the presence of zero counts, but either averaging or pooling reduces the proportion of zero counts, since it only takes one non-zero across the $m > 1$ data sets to produce a non-zero when averaged or pooled.

When the NBI is used, for a given original count $f_i = N$ $(i = 1, \ldots, K)$, the corresponding mean synthetic cell count $\bar{f}_i^{\text{syn}}$ has mean and variance,

$$E(\bar{f}_i^{\text{syn}}) = N \quad \text{and} \quad \text{Var}(\bar{f}_i^{\text{syn}}) = \frac{1}{m}(N + \sigma N^2), \tag{1}$$

as the synthetic data sets are independent.

Thus, for a given original count, the variance of the corresponding mean synthetic count is inversely proportional to m, and linearly related to σ. This means that the minimum obtainable variance when σ alone is tuned - which is achieved as $\sigma \to 0$ and the NBI tends towards its limiting distribution, the Poisson - is N/m. On the other hand, increasing m can essentially take the variance to zero. If m is too large, though, the original counts are simply returned when averaged, which, of course, renders the synthesis worthless. This, perhaps, suggests the suitability of m as a tuning parameter in cases where the original counts are large and relatively low risk, such that a relatively small variance suffices.

4   Introducing the $\tau_3(k, d)$ and $\tau_4(k, d)$ Metrics

When multiple synthetic data sets are generated and the mean synthetic count calculated - which is no longer always an integer - it becomes more suitable to consider the proportion of synthetic counts *within a certain distance of* original counts of k. To allow this, the metrics $\tau_3(k)$ and $\tau_4(k)$ can be extended to $\tau_3(k,d)$ and $\tau_4(k,d)$, respectively:

$$\tau_3(k,d) := p(|f^{\text{syn}} - k| \le d \mid f = k), \quad \tau_4(k,d) := p(f = k \mid |f^{\text{syn}} - k| \le d).$$

The metric $\tau_3(k,d)$ is the probability that a cell count of size k in the original data is synthesized to within d of k; and $\tau_4(k,d)$ is the probability that a cell count within d of k in the synthetic data originated from a cell of k. Unlike k, $d > 0$ does not need to be an integer. By extending the $\tau_1(k)$ metric, such that $\tau_1(k,d)$ is the proportion of synthetic counts within d of k, it follows that $\tau_3(k,d)\tau_2(k) = \tau_4(k,d)\tau_1(k,d)$.

The $\tau_3(k)$ and $\tau_4(k)$ metrics are then special cases of $\tau_3(k,d)$ and $\tau_4(k,d)$, respectively (the case where $d = 0$). For small k, these $\tau(k,d)$ metrics are intended primarily as risk metrics, because they are dealing with uniques or near uniques. However, when d is reasonably large, $\tau_3(k,d)$ and $\tau_4(k,d)$ are, perhaps, better viewed as utility metrics, because they are dealing with the proportion of uniques that are synthesized to much larger counts (which impacts utility).

When $m > 1$ is sufficiently large, tractable expressions for the $\tau_3(k,d)$ and $\tau_4(k,d)$ metrics can be obtained via the Central Limit Theorem (CLT), as the distribution of each mean synthetic count can be approximated by a normal distribution, with mean and variance as given in (1). That is, given an original count $f_i = N$ $(i = 1, \ldots, K)$, when m is large, the distribution of the corresponding mean synthetic cell count $\bar{f}_i^{\text{syn}}$ is given as:

$$\bar{f}_i^{\text{syn}} \mid f_i = N, \sigma, m \sim \text{Normal}(N, (N + \sigma N^2)/m).$$

This can be used to approximate $\tau_3(k,d)$ and $\tau_4(k,d)$:

$$\begin{aligned}
\tau_3(k,d) &= p(|\bar{f}^{\text{syn}} - k| \le d \mid f = k), \\
&= p(\bar{f}^{\text{syn}} < k + d \mid f = k) - p(\bar{f}^{\text{syn}} < k - d \mid f = k), \\
&= \Phi\left(\frac{(k+d)-k}{\sqrt{(k+\sigma k^2)/m}}\right) - \Phi\left(\frac{(k-d)-k}{\sqrt{(k+\sigma k^2)/m}}\right) \\
&= 2\Phi\left(\frac{d}{\sqrt{(k+\sigma k^2)/m}}\right) - 1,
\end{aligned} \tag{2}$$

$$\begin{aligned}
\tau_4(k,d) &= p(f = k \mid |\bar{f}^{\text{syn}} - k| \le d) \\
&= \frac{\tau_3(k,d) \cdot \tau_2(k)}{\sum_{i=0}^{\infty} p(|f^{\text{syn}} - k| \le d \mid f = i) \cdot p(f = i)} \\
&= \frac{\left[2\Phi\left(d/\sqrt{(k+\sigma k^2)/m}\right) - 1\right] \cdot \tau_2(k)}{\sum_{i=1}^{\infty} \left[\Phi\left((k+d-i)/\sqrt{(i+\sigma i^2)/m}\right) - \Phi\left((k-d-i)/\sqrt{(i+\sigma i^2)/m}\right)\right] \cdot \tau_2(i)}
\end{aligned} \tag{3}$$

where Φ is which is used to denote the cumulative distribution function (CDF) of the standard normal distribution.

5   Empirical Study

The data set synthesized here was constructed with the intention of being used as a substitute to the English School Census, an administrative database held by the Department for Education (DfE). It was constructed using publicly available data sources such as English School Census published data and 2011 census output tables. The data - along with a more detailed description of its origin - is available at [1]. While the data is constructed from public sources, it shares relevant features present in large administrative databases that serve to illustrate risk and utility in synthetic data and, specifically, the role that m plays in relation to the risk-utility trade-off. The framework developed here could be equally applied to any categorical data set.

The data comprises 8.2×10^6 individuals observed over $p = 5$ categorical variables. The local authority variable has the greatest number of categories with 326; while sex has the fewest with 4. When aggregated, the resulting contingency table has $K = 3.5 \times 10^6$ cells, 90% of which are unobserved, that is, have a count of zero.

The function rNBI from the R package **gamlss.dist** [16] was used to generate multiple synthetic data sets using the (σ, α)-synthesis mechanism described in Sect. 2. This was done for a range of σ, 0, 0.1, 0.5, 2 and 10, and 50 synthetic data sets were generated for each. This allowed comparisons to be drawn for a range of m, for example, taking the first five data sets gives $m = 5$, taking the first ten gives $m = 10$, etc.

5.1   Measuring Risk

Evaluating risk in synthetic data, particularly in synthetic categorical data, is not always straightforward. Attempting to estimate the risk of re-identification [12] is not possible, because the ability to link records is lost when a microdata set is aggregated, synthesized and disaggregated back to microdata again.

The $\tau_3(1, d)$ and $\tau_4(1, d)$ metrics (that is, setting $k = 1$), introduced in Sect. 4, were used as risk metrics. Figure 1 in the Appendix shows that either increasing m or decreasing σ increases $\tau_3(1, d)$ and $\tau_4(1, d)$ and hence risk. There is an initial fall in the $\tau_3(1, 0.1)$ curves as m increases initially, suggesting lower not higher risk. However, this is just owing to the small d: for example, when $d = 0.1$, the only way to obtain a mean synthetic count within 0.1 of k when, say $m = 5$, is by obtaining a one in each of the five synthetic data sets, compared to just once when $m = 1$.

When m is large, the $\tau_3(k, d)$ and $\tau_4(k, d)$ metrics can be approximated analytically through (2), which relies on the CLT. There is uncertainty in both the empirical values (owing to simulation error) and the analytical values (owing to the normal approximation), though the divergences between the empirical and analytical values are small.

In general, then, increasing m or decreasing σ increases risk. This is also shown visually in Fig. 2 (Appendix), which demonstrates how m and σ can be used in tandem to adjust risk. Here, $\tau_3(1, 0.1)$ is used as the z-axis (risk) but any $\tau_3(k, d)$ or $\tau_4(k, d)$ would give similar results.

5.2   Measuring Utility

As saturated models are used, increasing m (for a given σ) causes the mean synthetic counts to tend towards the original counts. This can be seen in the Hellinger and Euclidean distances given in Fig. 3 (Appendix), which show an improvement in general utility when either increasing m or reducing σ.

These measures are equally relevant to risk, too, hence Fig. 3 reiterates that risk increases with m. It is fairly trivial, however, that reducing simulation error increases risk and utility. It is more useful to gain an insight into the *rate* at which risk and utility increase with m, that is, the shape of the curves. For example, Fig. 3, shows that increasing m has greater effect when $\sigma = 1$ than when $\sigma = 0.1$.

The utility of synthetic data can also be assessed for specific analyses by, for example, comparing regression coefficient estimates obtained from a model fit to both the observed and synthetic data. While such measures only assess the synthetic data's ability to support a particular analysis, they nevertheless can be a useful indicator to, for example, the required m needed to attain a satisfactory level of utility.

Here, the estimand of interest is the slope parameter from the logistic regression of age Y (aged $\leq 9 = 0$, $\geq 10 = 1$) on language X. A subset of the data were used, as just two of the language variable's seven categories were considered, while the age variable was dichotomised. When estimated from the original data, β_1 - which is a log marginal odds ratio - was equal to -0.0075 with a 95% confidence interval of $(-0.0151, -0.0001)$. Note that, in order to estimate this, it was assumed that the original data constituted a simple random sample drawn from a much larger population. It is hugely doubtful whether such an assumption would be reasonable in practice, but the purpose here was just to evaluate the ability of the synthetic data to produce similar conclusions to the original data.

The analysis was undertaken in the two ways described in Sect. 3. Firstly, the $m > 1$ synthetic data sets were analysed separately and variance estimates were obtained through the estimator T_p. Secondly, the $m > 1$ synthetic data sets were pooled into one data set prior to the analysis and variance estimates were obtained through the estimator T_s.

As can be seen in Fig. 4, the estimates from T_p were noticeably larger than those from T_s, for small m. This was worrying for the validity of T_s - and the confidence intervals subsequently computed using T_s - especially since the sampling distribution of T_p was not approximated by a normal distribution, but by a t-distribution with ν_p degrees of freedom, thus widening confidence intervals further. This suggests that the large sample approximation that T_s relies on was not reasonable in this case.

The confidence interval computed from the original data set was compared with the confidence intervals computed from the synthetic data sets via the confidence interval overlap metric [5,15]. This metric is a composite measure that takes into account both the length and the accuracy of the synthetic data confidence interval. Yet whether these factors are weighted appropriately is open to debate. Valid confidence intervals estimated from synthetic data, that is, confidence intervals that achieve the nominal coverage, are longer than the corresponding confidence intervals estimated from the original data, because synthetic data estimates are subject to the uncertainty present in the original data estimates, plus have additional uncertainty from synthesis. However, a synthetic data confidence interval, say, one that is $x\%$ narrower than the original data confidence interval (hence clearly invalid) would yield roughly the same overlap as, say, a confidence interval that is $x\%$ wider. Moreover, either an infinitely wide or infinitely small synthetic data confidence interval would achieve an overlap of 0.5.

The confidence interval overlap results are presented in Table 1 in the Appendix. The top frame gives the overlap values from when the data sets are analysed separately, and the bottom frame gives the results from when the data sets are pooled. It can be seen that increasing m broadly results in an increase in the overlap; and that the overlap tends towards 1 as the original and synthetic data confidence intervals converge. The confidence intervals computed using T_s are less robust as those using T_p, which is evident in the zero overlap when $m = 20$ and $\sigma = 10$. This is because, unlike the variance estimator T_p, T_s only considers the within-synthesis variability $\bar{v}_m$, not the between-synthesis variability b_m.

5.3   Tuning m and σ in Relation to the Risk-Utility Trade-Off

The plots in Fig. 5 (Appendix) show how m and σ can be tuned in tandem to produce synthetic data sets that sit favourably within the risk-utility trade-off. These trade-off plots, though, depend on the metrics used to measure risk and utility. Here, risk was measured by either $\tau_4(1, 0.5)$ or $\tau_4(1, 0.75)$, and utility by either confidence interval overlap (using T_p) or Hellinger distance. The Hellinger distances were standardised onto the interval of [0,1] (by dividing by the largest Hellinger distance observed and then subtracting from 1, so that 1 and 0 represent maximum and minimum utility, respectively).

It is possible to strictly dominate synthetic data sets over others, that is, obtain lower risk *and* greater utility values. For example, looking at the top-left plot, synthetic data sets generated with $m = 50$, $\sigma = 2$ have higher risk but lower utility than when $m = 20$, $\sigma = 0.5$. These visual trade-offs are plotted using the empirical results, so are subject to variation from simulation; the confidence interval overlap values, in particular, can be volatile, especially when σ is large.

The intention is that the synthesizer produces such plots before releasing the data. Furthermore, as many metrics can be expressed analytically when using saturated models, they can be produced before the synthetic data is even generated.

6   Discussion

The setting of the synthesis mechanism's tuning parameters is a policy decision, and therefore is subjective. The general notion is that the synthesizer decides on an acceptable level of risk and maximises utility based on this; a larger m would necessitate a larger σ to maintain a given level of risk. As many metrics can be expressed as functions of the synthesis mechanism's tuning parameters, these functions' partial derivatives may be useful to determine the *rate* at which risk and utility change; for example, there may be a point where any further increases in m lead to a disproportionately small improvement in utility.

In addition to m, the synthesizer could also increase or decrease $E(n_{\mathrm{syn}})$, the expected sample size of each synthetic data set. A single synthetic data set $(m = 1)$ with $E(n_{\mathrm{syn}}) = n$ contains roughly the same number of records as two synthetic data sets $(m = 2)$ each with $E(n_{\mathrm{syn}}) = n/2$. To generate a synthetic data set with an expected sample size of $n/2$, the synthesizer simply takes draws from NBI distributions with means exactly half of what they were previously. Reducing $E(n_{\mathrm{syn}})$ should reduce risk, as fewer records are released, but inevitably reduces utility, too; once again, it calls for an evaluation with respect to the risk-utility trade-off.

Moreover, there are further tuning parameters that could be incorporated into this synthesis mechanism. One way would be to use a three-parameter distribution. When using a two-parameter count distribution, the synthesizer can increase the variance but cannot control how the variability manifests itself. The use of a three-parameter count distribution would allow the synthesizer to control the skewness, that is, they could change the shape of the distribution for a given mean and variance.

There are, of course, disadvantages to generating $m > 1$ synthetic data sets with the most obvious being the increased risk. Nevertheless, the potential benefits warrant further exploration, especially in relation to the risk-utility trade-off: does the gain in utility outweigh the increase in risk?

Organisations are taking a greater interest in making data - such as administrative data - available to researchers, by producing their own synthetic data. For this to be successful, organisations need to guarantee the protection of individuals' personal data - which, as more data becomes publicly available, becomes ever more challenging - while also producing data that are useful for analysts. Therefore, there needs to be scope to fine tune the risk and utility of synthetic data effectively, and integrating m as a tuning parameter into this *a priori* framework helps to achieve this.

Acknowledgements. This work was supported by the Economic and Social Research Council (ESRC) via the funding of a doctoral studentship.

Appendix

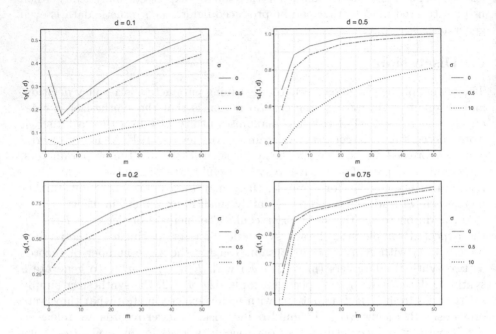

Fig. 1. The left hand plots give the empirical values of $\tau_3(1, d)$ for $d = 0.1$ and 0.2; the right hand plots give the empirical values of $\tau_4(1, d)$ for $d = 0.5$ and 0.75.

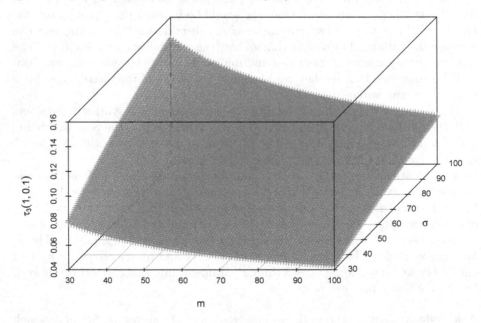

Fig. 2. The expected $\tau_3(1, 0.1)$ values for m and σ greater than 30.

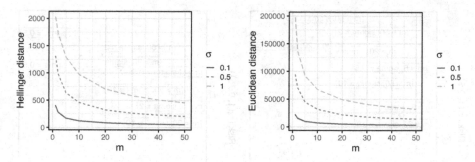

Fig. 3. The Hellinger and Euclidean distances as m increases, for various values of σ. These plots have been created using the cell counts rather than the cell probabilities, though the two are proportional.

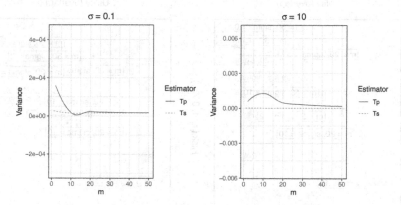

Fig. 4. The values of the estimators T_p and T_s. For small m, T_p is larger than T_s, before converging for larger m. The estimator T_s remains fairly constant across m.

Table 1. The confidence interval overlap results from when: (i) the data sets were analysed separately and T_p was used to estimate confidence intervals; and (ii) the data sets were pooled and T_s was used to estimate confidence intervals.

	$m = 2$	$m = 5$	$m = 10$	$m = 20$	$m = 30$	$m = 40$	$m = 50$
The overlap when the data sets were analysed separately and T_p used							
$\sigma = 0$	0.883	0.901	0.950	0.992	0.990	0.994	0.983
$\sigma = 0.1$	0.533	0.692	0.822	0.898	0.913	0.925	0.917
$\sigma = 0.5$	0.536	0.635	0.778	0.843	0.878	0.909	0.923
$\sigma = 2$	0.000	0.587	0.667	0.726	0.716	0.742	0.780
$\sigma = 10$	0.522	0.535	0.554	0.583	0.604	0.623	0.638
The overlap when the data sets were pooled and T_s used							
$\sigma = 0$	0.881	0.905	0.951	0.988	0.990	0.994	0.983
$\sigma = 0.1$	0.700	0.317	0.802	0.942	0.904	0.920	0.915
$\sigma = 0.5$	0.221	0.344	0.653	0.789	0.864	0.915	0.967
$\sigma = 2$	0.020	0.436	0.856	0.775	0.825	0.809	0.906
$\sigma = 10$	0.000	0.664	0.454	0.000	0.078	0.258	0.465

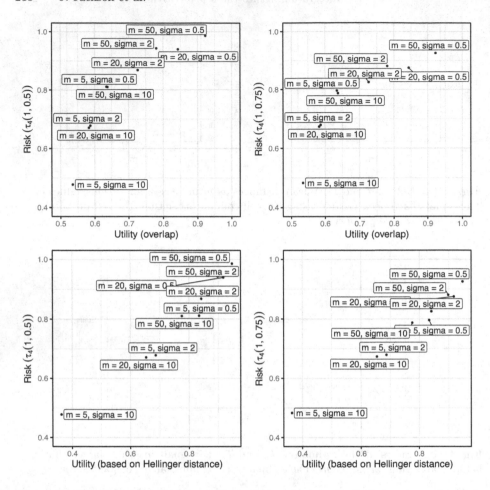

Fig. 5. Risk-utility trade-off plots to show where various synthetic data sets are located with respect to the risk-utility trade-off. The optimal position in each plot - that is, the lowest risk and the highest utility - is the bottom right corner. To measure risk, the metrics $\tau_4(1, 0.5)$ and $\tau_4(1, 0.75)$ were used. To measure utility, the confidence interval overlap and Hellinger distance were used.

References

1. Blanchard, S., Jackson, J.E., Mitra, R., Francis, B.J., Dove, I.: A constructed English School Census substitute (2022). https://doi.org/10.17635/lancaster/researchdata/533
2. Drechsler, J.: Synthetic Datasets for Statistical Disclosure Control: Theory and Implementation, vol. 201. Springer, New York (2011). https://doi.org/10.1007/978-1-4614-0326-5
3. Elliot, M., et al.: Functional anonymisation: personal data and the data environment. Comput. Law Secur. Rev. **34**(2), 204–221 (2018). https://doi.org/10.1016/j.clsr.2018.02.001. https://www.sciencedirect.com/science/article/pii/S0267364918300116

4. Jackson, J.E., Mitra, R., Francis, B.J., Dove, I.: Using saturated count models for user-friendly synthesis of large confidential administrative databases. J. Roy. Stat. Soc. Ser. A (Stat. Soc.) (2022, forthcoming). Preprint available at https://arxiv.org/abs/2107.08062
5. Karr, A.F., Kohnen, C.N., Oganian, A., Reiter, J.P., Sanil, A.P.: A framework for evaluating the utility of data altered to protect confidentiality. Am. Stat. **60**(3), 224–232 (2006)
6. Little, R.J.: Statistical analysis of masked data. J. Off. Stat. **9**(2), 407–426 (1993)
7. Nowok, B., Raab, G.M., Dibben, C., et al.: synthpop: bespoke creation of synthetic data in R. J. Stat. Softw. **74**(11), 1–26 (2016)
8. Raab, G.M., Nowok, B., Dibben, C.: Practical data synthesis for large samples. J. Privacy Confid. **7**(3), 67–97 (2016)
9. Raghunathan, T.E., Reiter, J.P., Rubin, D.B.: Multiple imputation for statistical disclosure limitation. J. Off. Stat. **19**(1), 1–16 (2003)
10. Reiter, J.P.: Inference for partially synthetic, public use microdata sets. Surv. Pract. **29**(2), 181–188 (2003)
11. Reiter, J.P.: Releasing multiply imputed, synthetic public use microdata: an illustration and empirical study. J. R. Stat. Soc. A. Stat. Soc. **168**(1), 185–205 (2005)
12. Reiter, J.P., Mitra, R.: Estimating risks of identification disclosure in partially synthetic data. J. Privacy Confid. **1**(1) (2009)
13. Rigby, R.A., Stasinopoulos, M.D., Heller, G.Z., De Bastiani, F.: Distributions for Modeling Location, Scale, and Shape: Using GAMLSS in R. CRC Press, Boca Raton (2019)
14. Rubin, D.B.: Statistical disclosure limitation. J. Off. Stat. **9**(2), 461–468 (1993)
15. Snoke, J., Raab, G.M., Nowok, B., Dibben, C., Slavkovic, A.: General and specific utility measures for synthetic data. J. R. Stat. Soc. A. Stat. Soc. **181**(3), 663–688 (2018)
16. Stasinopoulos, D.M., Rigby, R.A.: Generalized additive models for location scale and shape (GAMLSS) in R. J. Stat. Softw. **23**(7), 1–46 (2007)

Challenges in Measuring Utility for Fully Synthetic Data

Jörg Drechsler(✉)

Institute for Employment Research, Regensburger Str. 104,
90478 Nuremberg, Germany
joerg.drechsler@iab.de

Abstract. Evaluating the utility of the generated data is a pivotal step in any synthetic data project. Most projects start by exploring various synthesis approaches trying to identify the most suitable synthesis strategy for the data at hand. Utility evaluations are also always necessary to decide whether the data are of sufficient quality to be released. Various utility measures have been proposed for this purpose in the literature. However, as I will show in this paper, some of these measures can be misleading when considered in isolation while others seem to be inappropriate to assess whether the synthetic data are suitable to be released. This illustrates that a detailed validity assessment looking at various dimensions of utility will always be inevitable to find the optimal synthesis strategy.

Keywords: Confidence interval overlap · Confidentiality · Global utility · pMSE · Privacy

1 Introduction

The synthetic data approach for disclosure protection gained substantial popularity in recent years. While applications were mostly limited to the U.S. Census Bureau [1,12,13] a decade ago, more and more statistical agencies and other data collecting organizations are now exploring this idea as a possible strategy to broaden access to their sensitive data [2,4,16,29]. Recent developments in computer science, most notably the use of Generative Adversarial Networks (GANs, [10]) further stimulated the synthetic data movement and several start-up companies now offer synthetic data as a product, often with high flying promises regarding the unlimited usefulness of the data paired with claims of zero risk of disclosure. However, from an information theoretic stand point it is obvious that we can never have both, preservation of all the information from the original data while offering full protection of any sensitive information (except for the corner case in which the original data can be released without risk making the release of synthetic data pointless). Thus, all we can hope for is to find the optimal trade-off between utility and data protection, that is, we can try to maximize the utility for a desired level of data protection or maximize the level of protection for a level of utility that

© Springer Nature Switzerland AG 2022
J. Domingo-Ferrer and M. Laurent (Eds.): PSD 2022, LNCS 13463, pp. 220–233, 2022.
https://doi.org/10.1007/978-3-031-13945-1_16

is still deemed acceptable. Of course, in practice this is easier said than done. To fully utilize this optimization problem the data disseminating agency would need to know which levels of utility and disclosure protection the different stakeholders consider acceptable. Even more important, the agency needs reliable measures of utility and risk. Various metrics have been proposed in the literature for measuring the risk and utility for datasets that have undergone some procedure for statistical disclosure control [5, 23]. However, not all of them are suitable for synthetic data. Especially with fully synthetic data measuring the disclosure risk remains an open research question. Most risk measures that have been proposed in the literature try to estimate the risk of re-identification. Given that there is no one-to-one mapping between the original and the fully synthetic data, these measures cannot be meaningfully applied. The few proposals for measuring the risk of disclosure for fully synthetic data either rely on the opaque concept of perceived risk (the synthetic record looks too close to a real record), are computationally infeasible in practical settings [22] or make unrealistic assumptions regarding the knowledge of the attacker [25] (but see [26] for an interesting approach for measuring the risk of attribute disclosure).

However, even measuring the utility of the generated data can be more difficult than it might seem. A key challenge is that the data disseminating agencies typically only have limited knowledge for which purposes the data will be used (if they had this information they could simply publish all the analyses of interest as protecting the analysis output is typically much easier than protecting the full microdata). Thus, utility is typically measured by running a couple of analyses deemed to be of interest for the users and comparing the results from the synthetic data with those obtained for the original data. Alternatively, utility measures have been proposed that try to directly compare the synthetic data with the original data. In this paper, I will demonstrate that some of the measures that have been proposed in the literature can be misleading when considered in isolation while others seem to be inappropriate to assess whether the synthetic data are suitable to be released. The main conclusion based on this small assessment is that users of the synthetic data approach should always ensure that they evaluate several dimensions of utility before they decide which synthesis method works best for their data and whether the data are ready to be released.

2   Measuring the Utility

Utility measures are typically divided into two broad categories: narrow or analysis-specific measures and broad or global measures. The former focus on evaluating the utility by measuring how well the protected data preserve the results for a specific analysis of interest, while the letter try to directly compare the original and synthetic data providing a measure of similarity between the two datasets. Examples of analysis-specific utility measures are the confidence interval overlap measure proposed by [11] or the ratio of estimates (ROE) proposed by [27] for tabular data. The global utility is commonly assessed using distance measures such as the Kullback-Leibler divergence [11] or the propensity score mean squared error (pMSE) proposed in [30] and further developed in [24]. As pointed out by various authors [9,11,17], both types of measures have important drawbacks. While narrow measures provide useful information regarding the specific analysis considered, high utility based on these measures does not automatically imply high utility for other types of analyses. Since the data providers typically do not know which purposes the data will be used for later, it will be impossible to fully assess the utility of the synthetic data based on these measures. The global utility measures on the other hand are so broad that they might miss important weaknesses of the synthetic data. Furthermore, the measures are typically difficult to interpret, that is, it is difficult to decide whether the level of utility is acceptable or not. In practice, these measures are therefore mostly used to compare different synthesis approaches and not to decide whether the synthetic data offer enough utility to be released.

A final class of measures–termed *fit-for-purpose* measures here–can be considered to lie between the previous two. These measures typically only focus on specific aspects of the data, that is, they cannot be considered as global measures but also do not necessarily reflect statistics users might be interested in directly. Examples include plausibility checks such as ensuring only positive age values in the synthetic data, but also visual comparisons of univariate and bivariate distributions. Goodness-of-fit measures such as the χ^2-statistic for various cross-tabulations of the data or Kolmogoroff-Smirnov tests for continuous variables also belong to this group. As illustrated by [17], the pMSE can also be used for this purpose by simply including only the variables to be evaluated as predictors in the propensity model. Fit-for-purpose measures are typically the first step when assessing the utility of the generated data and we will illustrate the importance of these measures in the next section demonstrating that both, the global and the analysis specific measures of utility can be misleading when considered in isolation. Before we discuss the empirical evaluations, we provide further details regarding the utility measures used.

2.1 A Global Utility Measure: The pMSE

As mentioned above the pMSE has become a popular measure in recent years to assess the utility of the generated data. The procedure consists of the following steps:

1. Stack the n_{org} original records and the n_{syn} synthetic records adding an indicator, which is one if the record is from the synthetic data and zero otherwise.
2. Fit a model to predict the data source (original/synthetic) using the information contained in the data. Let p_i, $i = 1, \ldots, N$ with $N = n_{org} + n_{syn}$ denote the predicted value for record i obtained from the model.
3. Calculate the pMSE as $1/N \sum_N (p_i - c)^2$, with $c = n_{syn}/N$.

The smaller the pMSE the higher the analytical validity of the synthetic data. A downside of the pMSE is that it increases with the number of predictors included in the propensity model even if the model is correctly specified. To overcome this problem, [24] derived the expected value and standard deviation of the pMSE under the hypothesis that both, the original and the synthetic data are generated from the same distribution, that is, the synthesis model is correctly specified. Based on these derivations, the authors propose two utility measures: The pMSE ratio, which is the empirical pMSE divided by its expected value under the null and the standardized pMSE (S_pMSE), which is the empirical pMSE minus its expectation under the null divided by its standard deviation under the null.

2.2 Two Outcome-Specific Measure: The Confidence Interval Overlap and the Mean Absolute Standardized Coefficient Difference

The confidence interval overlap measure was first proposed by [30]. Paraphrasing from [6], its computation can be summarized as follows: For any estimate, we first compute the 95% confidence intervals for the estimand from the synthetic data, (L_s, U_s), and from the original data, (L_o, U_o). Then, we compute the intersection of these two intervals, (L_i, U_i). The utility measure is

$$I = \frac{U_i - L_i}{2(U_o - L_o)} + \frac{U_i - L_i}{2(U_s - L_s)}. \tag{1}$$

When the intervals are nearly identical, corresponding to high utility, $I \approx 1$. When the intervals do not overlap, corresponding to low utility, $I = 0$. The second term in (1) is included to differentiate between intervals with $(U_i - L_i)/(U_o - L_o) = 1$ but different lengths.

The mean absolute standardized coefficient difference (MASD) is implemented in the *synthpop* package as a utility measure for regression models. It computes the standardized difference for each regression coefficient as $z_j = (\bar{q}_m - \hat{Q})/(\sqrt{(v_{org}/m)})$, where $\bar{q}_m$ and $\hat{Q}$ denote the estimated coefficient from the synthetic and original data, respectively, v_{org} is the estimated variance of $\hat{Q}$ and m is the number of synthetic datasets. The MASD is then computed as $\sum_{j=1}^{p} |z_j|/p$, where p is the total number of regression coefficients in the model.

3 Misleading Utility Measures: An Illustration

For this small illustration, I use a subset of variables and records from the public use file of the March 2000 U.S. Current Population Survey (CPS). The data comprise eight variables measured on $N = 5,000$ heads of households (see Table 1 for details). Similar data are used in [7,8,19,20] to illustrate and evaluate various aspects of synthetic data. To simplify the modeling task I have removed some variables, subsampled the data, excluded some records, and recoded some variables compared to previous applications.

3.1 Synthesis Strategies

Overall we use four different synthesis strategies, three based on fully parametric models and one using a CART approach. We use the R package *synthpop* [15]

Table 1. Description of variables used in the empirical studies

Variable	Label	Range
Sex	*sex*	Male, female
Race	*race*	White, other
Marital status	*marital*	5 categories
Highest attained education level	*educ*	4 categories
Age (years)	*age*	15–90
Social security payments ($)	*ss*	0, 1–50,000
Household property taxes ($)	*tax*	0, 1–98,366
Household income ($)	*income*	1–582,896

to generate the synthetic data leaving most of the parameters at their default values. Specifically, we always synthesize all variables, keeping the size of the synthetic data the same as the size of the original data. We also keep the hyperparameters for the CART models at their default values and use standard options for the parametric variables: All continuous variables are synthesized using a linear regression model, while *sex* and *race* are synthesized using a logit model, and *marital* and *educ* are synthesized using multinomial regression. We always use the same synthesis order relying on the order in which the variables appear in the dataset with the minor adjustment that synthesis always starts with the variable *sex*. This adjustment was necessary as *synthpop* currently forces the synthesis for the first variable to be based on sampling when generating fully synthetic data (according to the maintainers of *synthpop* this issue will be fixed in future versions of the package). Since simply sampling from the marginal distribution arguably can be risky for continuous variables as exact values from the original data will be revealed, we decided to start the synthesis with a binary variable for which sampling original values does not pose any risks. Based on the same concerns–releasing exact values for continuous variables–we also use the smoothing option for the CART models. This option fits a kernel density estimator to the original values in any leaf of the tree and samples synthetic values from this estimator instead of sampling original values directly. We always generate $m = 5$ synthetic datasets.

The three parametric synthesizers differ in the way they account for distributional aspects of the original data. The first synthesizer (which we label the *naive* synthesizer below) does not consider these aspects at all, running the synthesis models without preprocessing the data. The second synthesizer (*transform*) tries to address the skewness of the continuous variables by taking the cubic root of all continuous variables before the imputation. Figure 1 shows the distribution of income and age before and after the transformation. The transformations make the distribution more symmetric, which can help to make the assumptions of the linear model more plausible. The final synthesis model (*two-stage*) additionally accounts for the fact that *ss* and *tax* have large spikes at zero as illustrated in

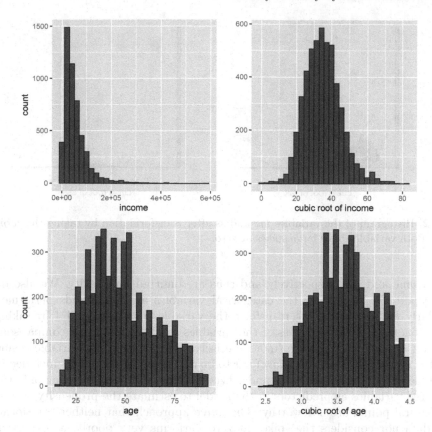

Fig. 1. Histogram of the variables *income* and *age* on the original scale and after transformation by taking the cubic root.

Fig. 2. To account for these spikes, we use the *semicont* option in *synthpop* which implements the two-stage imputation approach described in [18].

Looking at these synthesis strategies we would expect that the utility of the synthetic data would improve with each model, as the synthesis models better account for the properties of the original data. The only question should be, how these strategies perform relative to the CART synthesis model. We note that the synthesis models could certainly be further improved. The goal of this small exercise is not to find the best way of synthesizing the CPS data. We only use these synthetic datasets to highlight some caveats when relying on commonly used utility metrics.

3.2 Results for the Fit-for-Purpose Measures

Figures 3 and 4 provide visual comparisons of the distribution of the original and synthetic data for the variables *tax* and *income* generated using the *compare* function in *synthpop*. The findings for *age* and *ss* are comparable to the findings

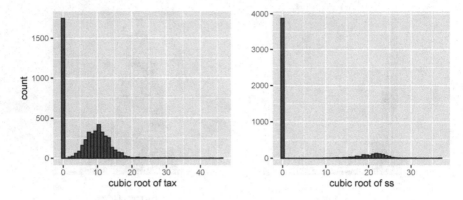

Fig. 2. Histogram of the variables *tax* and *ss* after transformation by taking the cubic root. Both variables have large spikes at zero.

for *income* and *tax*, respectively and thus are omitted for brevity. We also do not report the results for the categorical variables as all methods performed similarly well for those. We transform the variables in Figs. 3 and 4 by taking the cubic root as the skewness of the variables would make visual comparisons difficult on the original scale (more precisely, we transform the variables using $f(x) = sign(x)|x|^{1/3}$ to also allow for x to be negative). The numbers included in the title of each figure are the standardized pMSEs computed by using only the depicted variables as predictors in the model to estimate the propensity scores.

Several points are noteworthy: The *naive* approach that neither transforms the data nor considers the spikes at zero performs very poorly, as expected. Both variables have a considerable amount of negative values in the synthetic data despite the fact that they all only contain positive values in the original data. The spread of the synthetic data is also considerably larger than for the original data. This is especially true for *tax* due to its large spike at zero. Moving to the second synthesis approach (*transform*), which transformed the continuous variables to bring them closer to normality, we see that this approach helped to improve the quality of the synthetic data. Especially for *income* the distribution is much better preserved. This is also reflected in the substantial reduction in the standardized pMSE from over 260 to 7.66. However, the problems from not modeling the spike at zero are still obvious for *tax*.

This problem is also taken into account in the *two-stage* synthesis strategy where we see the positive impacts of separately modeling the spike. With this approach the spike is well preserved and the distributions in the synthetic data never differ substantially from the distributions in the original data. The results for the CART synthesizer are compatible with the results for the *two-stage* synthesis. We see some minor deviations in the distributions between the original and the synthetic data, but overall the distributions are well preserved. In terms of the standardized pMSE the *two-stage* approach outperforms the CART synthesis for *income*. Interestingly, the pMSE for *tax* is much small for the CART

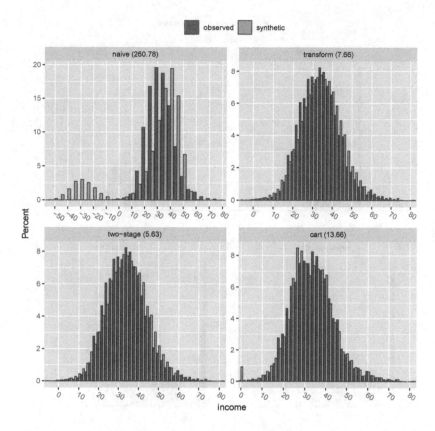

Fig. 3. Histogram of the variable *income* based on the original and synthetic data for various synthesis methods. The numbers in parentheses are the standardized pMSEs.

approach, although the CART synthesis does not preserve the spike at zero as well as the *two-stage* approach. We speculate that this is due to *synthpop* not using the variable directly to compute the measure. Instead, a categorical variable is derived by grouping the units into five equal sized bins using quantiles. The *S_pMSE* is computed using class membership as a predictor. This is obviously a crude measure as it ignores heterogeneity within the bins.

3.3   Results for the Outcome Specific Measures

We focus on one linear regression example to illustrate that weaker performance regarding the preservation of the marginal distributions does not necessarily imply worse results for a specific analysis task. We assume the analyses of interest is a linear regression of log(income) on the other variables contained in the dataset. This model is only for illustrative purposes and we do not claim that the model specification is appropriate. Results for the different synthesis methods are depicted in Fig. 5. The numbers in parentheses are the average confidence interval

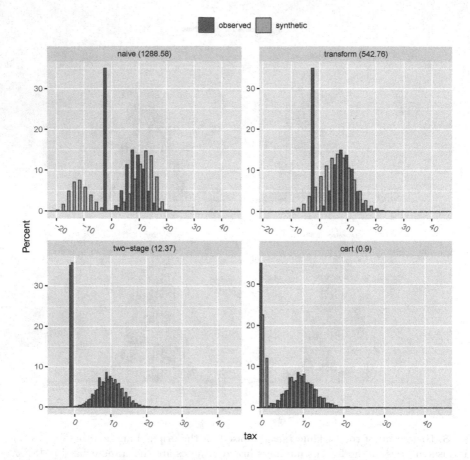

Fig. 4. Histogram of the variable *tax* based on the original and synthetic data for various synthesis methods. The numbers in parentheses are the standardized pMSEs.

overlaps (CIO) computed as the average across all regression coefficients from the model and the mean absolute standardized coefficient difference (MASD). We note that the CART synthesis model offers the lowest utility for both measures (note that larger values are better for CIO, while smaller values indicate higher utility for MASD). The CART model did not capture the relationship between marital status and income correctly. Even more problematically, the *sex* variable has the wrong sign in the synthetic data. For the CIO measure the utility order matches the order from the previous section, that is, *two-stage* offers higher utility than *transform*, which has higher utility than *naive*. Interestingly, the MASD measure indicates a reversed order with the *naive* approach offering the highest utility.

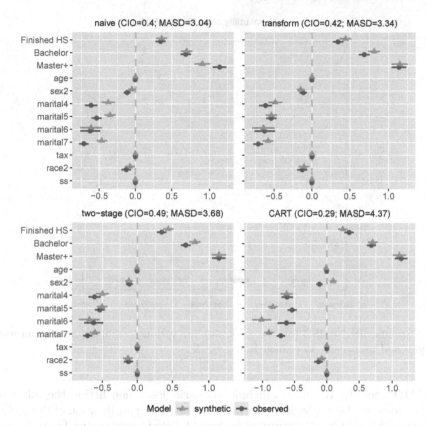

Fig. 5. Comparison of results of a regression of log(income) on the other variables included in the CPS data for various synthesis methods. The lines indicate the length of the 95% confidence intervals. The numbers in parentheses are the average confidence interval overlap (CIO) and the mean absolute standardized coefficient difference (MASD). Both are computed by averaging across all regression coefficients.

The results illustrate that utility can be high for certain analysis goals even if some aspects of the data are poorly captured but also that aggregated measures which average results across various estimates can potentially be misleading and visualizations such as those shown in Fig. 5 might be better suited to identify strengths and weaknesses of the synthetic data.

3.4 Results for the Global Utility Measures

To estimate the standardized pMSE for the entire dataset, we need to specify a model for estimating the individual membership propensities p_i. The most common choice in the propensity score literature is the logit model, but any model can be used as long is it returns predicted probabilities of class membership. In our application, we use the two models available in *synthpop*: the logit model

Global utility based on S_pMSE

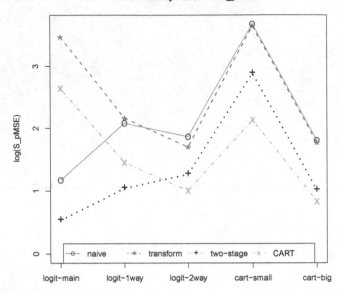

Fig. 6. Standardized pMSE (on the log-scale) for various combinations of synthesis strategies and propensity score models.

and CART models. We always include all variables, when fitting the different models. However, for the logit model we vary the exact specification of the model evaluating three different settings: The first, labeled *main* in the figure below, only includes all main effects. The second (*1-way*) additionally includes all one-way interactions between the variables. The final model (*2-way*) also includes all two-way interactions. For the CART models we do not need to specify the model, as the method will automatically identify the best splits of the data. However, an important tuning parameter with CART models as implemented in *rpart* [28], the library used in this application, is the complexity parameter *cp*. Any split that does not decrease the overall lack of fit by a factor of *cp* is not attempted. Thus, smaller values of *cp* generally imply that larger trees are grown. We evaluate two settings: In the first setting (*CART_small*), we use the default settings of *synthpop*, which presumably use the default values from the *rpart* package, which is $cp = 0.01$. In the second setting (*CART_big*), we use a very small value of $cp = 10^{-7}$.

Results are presented in Fig. 6. A couple of things are noteworthy: First, the results confirm that like other global utility measures, the *S_pMSE* cannot be used to assess whether the data are ready to be released. It can vary substantially depending on which model is used to estimate the propensity score. For example, for the *naive* synthesis approach, the *S_pMSE* changes from 39.76 to 6.06 when switching from *CART_small* to *CART_big*. Second, the measure is unable to detect the substantial improvements in the synthetic data when switching from the *naive* approach to the *transform* approach. In fact, the 2-way logit model is the only model that suggests that transforming the variables before the synthesis improves the results. All other models indicate that the *naive* synthesis offers at least similar utility, with the *main* model suggesting substantial quality improvements without transformation. Finally, while two of the logit models (*main* and *1-way*) suggest that a careful parametric synthesis approach (the *two-stage* approach) should be preferred over CART synthesis, the two CART based propensity score models always prefer the CART synthesis strategy.

The results indicate that the global utility measure is not a reliable indicator for deciding which synthesis strategy should be preferred. Results are highly dependent on the model used to estimate the propensity scores and the approach sometimes seems incapable of detecting major differences in the quality of the synthesis models.

4   Conclusions

Given the large variety of synthesis strategies that have been proposed in recent years, picking the most suitable synthesis method is a difficult task. In this situation it seems tempting to rely on global utility measures that return only one number and just pick the strategy that achieves the highest utility according to this score. As I have shown in this paper, things unfortunately are not that easy. The small illustration included in this paper demonstrates fundamental weaknesses of one popular global utility metric: the standardized pMSE. I showed that results for this metric are highly dependent on the model used to estimate the propensity score. Maybe even more worrying, the metric was unable to detect important differences in the utility for most of the model specifications.

On the other hand, I also showed (perhaps unsurprisingly) that utility can still be relatively high for certain types of analyses even if some distributional features of the data are poorly preserved. This implies that a thorough assessment of the utility is inevitable when deciding which synthesis method to pick and whether the data are ready to be released. This assessment should start by evaluating if the data are fit for purpose in various dimensions. If the data disseminating agency already has some information for which types of analyses the data will be used later, it will also be useful to compute outcome specific utility measures for a large variety of analyses to better understand, which analysis models still cause problems and use this information to refine the synthesis models. The findings from this paper seem to indicate that decisions based on global utility measures should better be avoided.

From a user's perspective, verification servers can also be an important alternative tool to increase confidence in the results obtained from the synthetic data. These servers hold both the synthetic and the original data. Researchers can submit their analysis of interest to the server, it runs the analysis on both datasets, and reports back some fidelity measure how close the results from the synthetic data are to the results based on the original data. However, some care must

232   J. Drechsler

be taken, as even fidelity measures might spill sensitive information. Developing such measures is currently an active area of research [3,14,21,31].

References

1. Abowd, J.M., Stinson, M., Benedetto, G.: Final report to the Social Security Administration on the SIPP/SSA/IRS public use file project. Technical report, Longitudinal Employer-Household Dynamics Program, U.S. Bureau of the Census, Washington, DC (2006)
2. Australian Bureau of Statistics: Methodological news, Dec 2021 (2021). https://www.abs.gov.au/statistics/research/methodological-news-dec-2021. Accessed 17 May 2022
3. Barrientos, A.F., et al.: Providing access to confidential research data through synthesis and verification: an application to data on employees of the US federal government. Ann. Appl. Statist. **12**(2), 1124–1156 (2018)
4. Bowen, C.M.K., et al.: A synthetic supplemental public use file of low-income information return data: methodology, utility, and privacy implications. In: Domingo-Ferrer, J., Muralidhar, K. (eds.) PSD 2020. LNCS, vol. 12276, pp. 257–270. Springer, Cham (2020). https://doi.org/10.1007/978-3-030-57521-2_18
5. Domingo-Ferrer, J., et al.: Statistical Disclosure Control. Wiley Series in Survey Methodology. Wiley, Hoboken (2012)
6. Drechsler, J.: Synthetic Datasets for Statistical Disclosure Control: Theory and Implementation. Lecture Notes in Statistics, vol. 201. Springer, New York (2011). https://doi.org/10.1007/978-1-4614-0326-5
7. Drechsler, J., Reiter, J.P.: Accounting for intruder uncertainty due to sampling when estimating identification disclosure risks in partially synthetic data. In: Domingo-Ferrer, J., Saygın, Y. (eds.) PSD 2008. LNCS, vol. 5262, pp. 227–238. Springer, Heidelberg (2008). https://doi.org/10.1007/978-3-540-87471-3_19
8. Drechsler, J., Reiter, J.P.: Sampling with synthesis: a new approach for releasing public use census microdata. J. Am. Statist. Assoc. **105**, 1347–1357 (2010)
9. Drechsler, J., Hu, J.: Synthesizing geocodes to facilitate access to detailed geographical information in large-scale administrative data. J. Surv. Statist. Methodol. **9**(3), 523–548 (2021)
10. Goodfellow, I.J., et al.: Generative Adversarial Networks. arXiv:1406.2661 [cs, stat] (2014)
11. Karr, A.F., Kohnen, C.N., Oganian, A., Reiter, J.P., Sanil, A.P.: A framework for evaluating the utility of data altered to protect confidentiality. Am. Statist. **60**, 224–232 (2006)
12. Kinney, S.K., Reiter, J.P., Reznek, A.P., Miranda, J., Jarmin, R.S., Abowd, J.M.: Towards unrestricted public use business microdata: the synthetic longitudinal business database. Int. Statist. Rev. **79**(3), 362–384 (2011)
13. Machanavajjhala, A., Kifer, D., Abowd, J.M., Gehrke, J., Vilhuber, L.: Privacy: Theory meets practice on the map. In: IEEE 24th International Conference on Data Engineering, pp. 277–286 (2008)
14. McClure, D.R., Reiter, J.P.: Towards providing automated feedback on the quality of inferences from synthetic datasets. J. Priv. Confident. **4**(1), 1–7 (2012)
15. Nowok, B., Raab, G.M., Dibben, C.: synthpop: bespoke creation of synthetic data in R. J. Statist. Softw. **74**, 1–26 (2016)

16. Nowok, B., Raab, G.M., Dibben, C.: Providing bespoke synthetic data for the UK longitudinal studies and other sensitive data with the synthpop package for R 1. Statist. J. IAOS **33**(3), 785–796 (2017)
17. Raab, G.M., Nowok, B., Dibben, C.: Guidelines for producing useful synthetic data. arXiv preprint arXiv:1712.04078 (2017)
18. Raghunathan, T.E., Lepkowski, J.M., van Hoewyk, J., Solenberger, P.: A multivariate technique for multiply imputing missing values using a series of regression models. Surv. Methodol. **27**, 85–96 (2001)
19. Reiter, J.P.: Releasing multiply-imputed, synthetic public use microdata: an illustration and empirical study. J. R. Statist. Soc. Ser. A **168**, 185–205 (2005)
20. Reiter, J.P.: Using CART to generate partially synthetic, public use microdata. J. Official Statist. **21**, 441–462 (2005)
21. Reiter, J.P., Oganian, A., Karr, A.F.: Verification servers: enabling analysts to assess the quality of inferences from public use data. Comput. Statist. Data Anal. **53**(4), 1475–1482 (2009)
22. Reiter, J.P., Wang, Q., Zhang, B.: Bayesian estimation of disclosure risks for multiply imputed, synthetic data. J. Priv. Confid. **6**(1), 1–18 (2014)
23. Shlomo, N., Skinner, C.: Measuring risk of re-identification in microdata: state-of-the art and new directions. J. R. Statist. Soc. Ser. A. **64**, 855–867 (2022)
24. Snoke, J., Raab, G.M., Nowok, B., Dibben, C., Slavkovic, A.: General and specific utility measures for synthetic data. J. R. Statist. Soc. Ser. A (Statist. Soc.) **181**(3), 663–688 (2018)
25. Stadler, T., Oprisanu, B., Troncoso, C.: Synthetic data-a privacy mirage. arXiv e-prints arXiv-2011 (2020)
26. Taub, J., Elliot, M.: The synthetic data challenge. In: Joint UNECE/Eurostat Work Session on Statistical Data Confidentiality, The Hague, The Netherlands (2019)
27. Taub, J., Elliot, M., Sakshaug, J.W.: The impact of synthetic data generation on data utility with application to the 1991 UK samples of anonymised records. Trans. Data Priv. **13**(1), 1–23 (2020)
28. Therneau, T., Atkinson, B., Ripley, B.: rpart: Recursive Partitioning and Regression Trees (2015). https://CRAN.R-project.org/package=rpart, r package version 4.1-10
29. de Wolf, P.P.: Public use files of Eu-SILC and Eu-LFS data. In: Joint UNECE/Eurostat Work Session on Statistical Data Confidentiality, Helsinki, Finland, pp. 1–10 (2015)
30. Woo, M.J., Reiter, J.P., Oganian, A., Karr, A.F.: Global measures of data utility for microdata masked for disclosure limitation. J. Priv. Confid. **1**, 111–124 (2009)
31. Yu, H., Reiter, J.P.: Differentially private verification of regression predictions from synthetic data. Trans. Data Priv. **11**(3), 279–297 (2018)

Comparing the Utility and Disclosure Risk of Synthetic Data with Samples of Microdata

Claire Little[1], Mark Elliot[1(✉)], and Richard Allmendinger[2]

[1] School of Social Sciences, University of Manchester, Manchester M13 9PL, UK
{claire.little,mark.elliot}@manchester.ac.uk
[2] Alliance Manchester Business School, University of Manchester, Manchester M13 9PL, UK
richard.allmendinger@manchester.ac.uk

Abstract. Most statistical agencies release randomly selected samples of Census microdata, usually with sample fractions under 10% and with other forms of statistical disclosure control (SDC) applied. An alternative to SDC is data synthesis, which has been attracting growing interest, yet there is no clear consensus on how to measure the associated utility and disclosure risk of the data. The ability to produce synthetic Census microdata, where the utility and associated risks are clearly understood, could mean that more timely and wider-ranging access to microdata would be possible.

This paper follows on from previous work by the authors which mapped synthetic Census data on a risk-utility (R-U) map. The paper presents a framework to measure the utility and disclosure risk of synthetic data by comparing it to samples of the original data of varying sample fractions, thereby identifying the sample fraction which has equivalent utility and risk to the synthetic data. Three commonly used data synthesis packages are compared with some interesting results. Further work is needed in several directions but the methodology looks very promising.

Keywords: Data synthesis · Microdata · Disclosure risk · Data utility

1 Introduction

Many statistical agencies release randomly selected Census samples to researchers (and sometimes publicly), usually with sample fractions under 10% and with other forms of statistical disclosure control (SDC) [14] applied. However, as noted by Drechsler et al. [8], intruders (or malicious users) are becoming more sophisticated and agencies using standard SDC techniques may need to apply them with higher intensity, leading to the released data being of reduced quality for statistical analysis.

An alternative to SDC is data synthesis [21,38], which takes original data and produces an artificial dataset with the same structure and statistical properties as the original, but that (in the case of fully synthetic data) does not contain any

© Springer Nature Switzerland AG 2022
J. Domingo-Ferrer and M. Laurent (Eds.): PSD 2022, LNCS 13463, pp. 234–249, 2022.
https://doi.org/10.1007/978-3-031-13945-1_17

of the original records. As the data is synthetic, attributes which are normally suppressed, aggregated or top-coded (such as geographical area or income) may then be included, allowing more complete analysis. As no synthetic record should correspond to a real individual, fully synthetic data should present very low disclosure risk – the risk of re-identification is not meaningful, although there is still likely to be a residual risk of attribution [41]. Interest in synthetic data is growing, yet there is no clear consensus on how to measure the associated utility and disclosure risk of the data, such that users may have an understanding of how closely a synthetic dataset relates to the original data.

This paper follows on from previous work [20], which mapped synthetic Census data on the risk-utility (R-U) map, and presents a framework to measure the utility and disclosure risk of synthetic data by comparing it against random samples of the original data of varying sample fractions, thereby identifying the sample fraction equivalence of the synthetic dataset. For instance, a particular synthetic dataset might be equivalent in terms of utility to a 20% sample of the original data, and in terms of disclosure risk to a 10% sample. Since Census microdata tends to be released with sample fractions between 1% to 10%, the ability to determine how a synthetic dataset compares in terms of utility and disclosure risk would allow data producers a greater understanding of the appropriateness of their synthetic data. To test our framework, we performed experiments using four Census microdata sets with synthetic data generated using three state-of-the-art data synthesis methods (*CTGAN* [45], *Synthpop* [25] and *DataSynthesizer* [29]).

Section 2 provides a brief introduction to the data synthesis problem, particularly in terms of microdata, and an introduction to the data synthesis methods. Sect 3 outlines the design of the study, describing the methods and the Census data used. Sect 4 provides the results whilst Sect. 5 considers the findings, and Sect. 6 concludes with directions for future research.

2 Background

2.1 Data Synthesis

Rubin [38] first introduced the idea of synthetic data as a confidentiality protection mechanism in 1993, proposing using multiple imputation on all variables such that none of the original data was released. In the same year, Little [21] proposed an alternative that simulated only sensitive variables, thereby producing partially synthetic data. Rubin's idea was slow to be adopted, as noted by Raghunathan et al. [32], who along with Reiter [35–37], formalised the synthetic data problem. Further work (e.g. [8,9,34]) has involved using non-parametric data synthesis methods such as classification and regression trees (CART) and random forests and more recently deep learning methods such as Generative Adversarial Networks [12] have also been used to generate synthetic data.

There are two competing objectives when producing synthetic data: high data utility (i.e., ensuring that the synthetic data is useful, with a distribution close to the original) and low disclosure risk. Balancing this trade-off can be difficult, as, in general, reducing disclosure risk comes with a concomitant cost for utility. This trade-off can be visualised using the R-U confidentiality map developed by Duncan et al. [10]. There are multiple measures of utility, ranging from comparing summary statistics, correlations and cross-tabulations, to considering data performance using predictive algorithms. However, for synthetic data there are fewer that measure disclosure risk. As noted by Taub et al. [41], much of the SDC literature focuses on re-identification risk, which is not meaningful for synthetic data, rather than the attribution risk, which is relevant. The Targeted Correct Attribution Probability (TCAP) [11,41] can be used to assess attribution risk.

2.2 Synthetic Census Microdata

Since Census microdata is predominantly categorical, it requires synthesis methods that can handle categorical data. CART, a non-parametric method developed by Breiman et al. [2], can handle mixed type (and missing) data, and can capture complex interactions and non-linear relationships. CART is a predictive technique that recursively partitions the predictor space, using binary splits, into relatively homogeneous groups; the splits can be represented visually as a tree structure, meaning that models can be intuitively understood (where the tree is not too complex). Reiter [34] used CART to generate partially synthetic microdata, as did Drechsler and Reiter [8], who replaced sensitive variables in the data with multiple imputations and then sampled from the multiply-imputed populations. Random forests, developed by Breiman [3], is an ensemble learning method and an extension to CART in that the method grows multiple trees. Random forests were used to synthesise a sample of the Ugandan Census [9] and to generate partially synthetic microdata [4].

Synthpop, an open source package written in the R programming language, developed by Nowok et al. [25], uses CART as the default method of synthesis (other options include random forests and various parametric alternatives). Since CART is a commercial product, *Synthpop* uses an open source implementation of the algorithm provided by the *rpart* package [43]. *Synthpop* synthesises the data sequentially, one variable at a time; the first is sampled, then the following are synthesised using the previous variables as predictors. Whilst an advantage of *Synthpop* is that it requires little tuning and performs very quickly, a disadvantage is that it (and tree-based methods in general) can struggle computationally with variables that contain many categories. As suggested by Raab et al. [31] methods to deal with this include aggregation, sampling, changing the sequence order of the variables and excluding variables from being used as predictors. *Synthpop* has been used to produce synthetic longitudinal microdata [26] and synthetic Census microdata [30,42].

Another method that can process mixed type data is *DataSynthesizer*, developed by Ping et al. [29], a Python package that implements a version of the PrivBayes [46] algorithm. PrivBayes constructs a Bayesian network that models the correlations in the data, allowing approximation of the distribution using a set of low-dimensional marginals. Noise is injected into each marginal to ensure differential privacy and the noisy marginals and Bayesian network are then used to construct an approximation of the data distribution. PrivBayes then draws samples from this to generate a synthetic dataset. *DataSynthesizer* allows the level of differential privacy to be set by the user, or turned off. It has been used to generate health data [33] and in exploratory studies [7,13,24].

In the field of deep learning [19], Generative Adversarial Networks (GANs) have been generating much research interest and have been used for various applications, although as detailed by Wang et al. [44] these are predominantly in the image domain. GANs, developed by Goodfellow et al. [12], typically train two neural network (NN) models: a generative model that captures the data distribution and generates new data samples, and a discriminative model that aims to determine whether a sample is from the model distribution or the data distribution. The models are trained together in an adversarial zero-sum game, such that the generator goal is to produce data samples that fool the discriminator into believing they are real and the discriminator goal is to determine which samples are real and which are fake. Training is iterative with (ideally) both models improving over time to the point where the discriminator can no longer distinguish which data is real or fake. From a data synthesis perspective GANs are interesting in that the generative model does not access the original (or training) data at all, and starts off with only noise as input; in theory this might reduce disclosure risk.

GANs for image generation tend to deal with numerical, homogeneous data; in general, they must be adapted to deal with Census microdata, which is likely to be heterogeneous, containing imbalanced categorical variables, and skewed or multimodal numerical distributions. Several studies have done this by adapting the GAN architecture, these are often referred to as tabular GANs (e.g. [5,6,28,47]). *CTGAN*, or Conditional Tabular GAN, developed by Xu et al. [45] uses "mode-specific normalisation" to overcome non-Gaussian and multimodal distribution problems, and employs oversampling methods and a conditional generator to handle class imbalance in the categorical variables. In their study *CTGAN* outperformed Bayesian methods, and other GANs, for generating mixed type synthetic data.

National statistical agencies have released synthetic versions of microdata using forms of multiple imputation. The United States Census Bureau releases a synthetic version of the Longitudinal Business Database (SynLBD) [18], the Survey of Income and Program Participation (SIPP) Synthetic Beta [1] and the OnTheMap application [22]. Whilst governmental organisations have not so far released synthetic microdata created using deep learning methods, research in this area is ongoing (e.g. [15,16]).

3   Research Design

To determine the sample equivalence, the risk and utility of generated synthetic Census data was compared to the risk and utility of samples of the original

Census data (of various sample fractions).[1] Four different Census microdata sets were used to demonstrate results on datasets from different underlying population data structures. Three state-of-the-art data synthesis methods were used to generate the synthetic data, each using the default parameters. To obtain consistent results, multiple datasets were generated (using different random seeds) for the sample and synthetic data, and the mean of the utility and risk metrics for each calculated.[2]

3.1   Data Synthesisers

The methods used were *Synthpop* [25], *DataSynthesizer* [29] and *CTGAN* [45]. These were selected as they are established, open-source methods that should produce good quality data. Whilst the focus of these experiments was on evaluating the resulting utility and risk of the generated data, rather than the individual methods, for *DataSynthesizer* the differential privacy parameter was varied in order to understand how the use of differential privacy affects the quality and risk of the synthetic data and how such differentially private synthetic datasets compare to samples. For each parameter setting, five fully synthetic datasets were generated, each using a different random seed.

Synthpop. Version 1.7-0 of *Synthpop* was used with default parameters. As described in Sect. 2.2, *Synthpop* allows the sequence order of the variables to be set by the user, however there is no default for this (other than the ordering the data is in). Since the Census microdata used for these experiments is predominantly categorical, with many variables containing many categories, and it is known that *Synthpop* can struggle with variables containing many categories, the visit sequence was set such that variables were ordered by the minimum to maximum number of categories, with numerical variables first (and a tie decided by ordering alphabetically). Moving variables with many categories to the end of the sequence is suggested by Raab et al. [31].

DataSynthesizer. Version 0.1.9 of *DataSynthesizer* (described in Sect. 2.2) was used with Correlated Attribute mode (which implements the PrivBayes [46] algorithm). Default parameters were used, whilst differing the Differential Privacy (DP) parameter. DP is controlled by the ϵ parameter and a value of zero turns DP off. Four different values were used (DP = off, ϵ = 0.1, 1, and 10). Lower values of ϵ tend to be used in practise, but the range of values aims to understand the effect at both the higher and lower end, as well as turning off DP altogether.

---

[1] Note that, for calculating the risk and utility, the sample data was treated in the same way as the synthetic data, namely by comparing against the original data. However, for simplicity, in the metric descriptions only synthetic data is mentioned.

[2] The project code is available here: https://github.com/clairelittle/psd2022-comparing-utility-risk.

Table 1. Census data summary

Dataset	Sample size	#Total variables	#Categorical	#Numerical
Canada 2011	32149	25	21	4
Fiji 2007	84323	19	18	1
Rwanda 2012	31455	21	20	1
UK 1991	104267	15	13	2

CTGAN. Version 0.4.3 of *CTGAN* was used for all experiments. *CTGAN* as described in Sect. 2.2, is a Conditional GAN implemented in Python. There are many hyperparameters that might be altered for a GAN; the default values were used, with the number of epochs set at 300.

3.2   Data

Four Census microdatasets were used, each from a different continent. Each dataset contains individual records, pertaining to adults and children. The variables include demographic information such as age, sex and marital status (i.e., variables that are often considered key identifiers) and a broad selection of variables pertaining to employment, education, ethnicity, family, etc. Each dataset contained the same key variables, and target variables that broadly cover the same overall themes. The purpose of using multiple datasets was not to directly compare the countries, but rather to determine whether any patterns uncovered during the experiments were replicated on similar (but not identical) datasets. Table 1 describes the data in terms of sample size and features.

The data was minimally preprocessed and missing values were retained. Three of the datasets (UK [27], Canada [23] and Rwanda [23]) were subsetted on a randomly selected geographical region; this was to reduce data size and also to naturally reduce the categories for some of the variables. The entire Fiji sample [23] was used. Appendix A contains a summary of the datasets.

Creating the Census Data Samples. For each Census dataset, random samples (without replacement) of increasing sizes were drawn (0.1%, 0.25%, 0.5%, 1%, 2%, 3%, 4%, 5%, 10%, 20%, 30%, 40%, 50%, 60%, 70%, 80%, 90%, 95%, 96%, 97%, 98%, 99%). A focus was placed on those sample fractions closer to zero since released Census samples tend to be relatively small, and closer to 100% to map the data as it became closer in size to the original. For each sample fraction 100 datasets were generated (with different random seeds), and the results of the individual risk and utility measures averaged.

3.3   Measuring Disclosure Risk Using TCAP

Elliot [11] and Taub et al. [41] introduced a measure for the disclosure risk of synthetic data called the Correct Attribution Probability (CAP) score. The disclosure risk is calculated using an adaptation used in Taub et al. [40] called the

Targeted Correct Attribution Probability (TCAP). TCAP is based on a scenario whereby an intruder has partial knowledge about a particular individual. That is, they know the values for some of the variables in the dataset (the keys) and also that the individual is in the original dataset[3], and wish to infer the value of a sensitive variable (the target) for that individual. The TCAP metric is then the probability that those matched records yield a correct value for the target variable (i.e. that the adversary makes a correct attribution inference).

TCAP has a value between 0 and 1; a low value would indicate that the synthetic dataset carries little risk of disclosure whereas a TCAP score close to 1 indicates a higher risk. However, a baseline value can be calculated (this is essentially the probability of the intruder being correct if they drew randomly from the univariate distribution of the target variable) and therefore a TCAP value above the baseline might indicate some disclosure risk. Given this, we have chosen to scale the TCAP value between the baseline and a value of 1. This does create the possibility of a synthetic dataset receiving a negative TCAP score (which can still be plotted on the R-U map) but that simply indicates a risk level below that of the baseline. We refer to the scaled TCAP value as the marginal TCAP; i.e. it is the increase in risk above the baseline.

For each Census dataset, three target and six key variables were identified and the corresponding TCAP scores calculated for sets of 3, 4, 5 and 6 keys. The overall mean of the TCAP scores was then calculated as the overall disclosure risk score. Where possible, the selected key/target variables were consistent across each country. Full details of the target and key variables are in Appendix B.

3.4 Evaluating Utility

Following previous work by Little et al. [20] and Taub et al. [42], the utility of the synthetic and sample data was assessed using multiple measures. The confidence interval overlap (CIO) and ratio of counts/estimates (ROC) were calculated. This was to provide a more complete picture of the utility, rather than relying upon just one measure. The propensity score mean squared error (pMSE) [39] was not used as, whilst it is suitable for analysing the synthetic data it is not suited to the analysis of sample data as it is structurally tied to the original data (since the sample data is a subset of the original data).

To calculate the CIO (using 95% confidence intervals), the coefficients from regression models built on the original and synthetic datasets are used. The CIO, proposed by Karr et al. [17], is defined as:

$$\text{CIO} = \frac{1}{2} \left\{ \frac{\min(u_o, u_s) - \max(l_o, l_s)}{u_o - l_o} + \frac{\min(u_o, u_s) - \max(l_o, l_s)}{u_s - l_s} \right\}, \quad (1)$$

where u_o, l_o and u_s, l_s denote the respective upper and lower bounds of the confidence intervals for the original and synthetic data. This can be summarised

---

[3] This is a strong assumption, which has the benefit of then dominating most other scenarios, the one possible exception is a presence detection attack. However, for Census data, presence detection is vacuous, and the response knowledge assumption is sound by definition.

by the average across all regression coefficients, with a higher CIO indicating greater utility (maximum value is 1 and a negative value indicating no overlap). For each synthetic Census dataset, two logistic regressions were performed, with the CIO for each calculated. The mean of these two results (where a negative, or no overlap was counted as zero) was taken as the overall CIO utility score for that dataset. Details of the regression models for each dataset are presented in Appendix C.

Frequency tables and cross-tabulations are evaluated using the ROC, which is calculated by taking the ratio of the synthetic and original data estimates (where the smaller is divided by the larger one). Thus, given two corresponding estimates (for example, the number of records with sex = female in the original dataset, compared to the number in the synthetic dataset), where y_{orig} is the estimate from the original data and y_{synth} is the corresponding estimate from the synthetic data, the ROC is calculated as:

$$\text{ROC} = \frac{\min(y_{orig}, y_{synth})}{\max(y_{orig}, y_{synth})}. \tag{2}$$

If $y_{orig} = y_{synth}$ then the ROC $= 1$. Where the original and synthetic datasets are of different sizes (as is the case when calculating the ROC for the various sample datasets) the proportion, rather than the count can be used. The ROC was calculated over univariate and bivariate cross-tabulations of the data, and takes a value between 0 and 1. For each variable the ROC was averaged across categories to give an overall score.

To create an overall utility score for comparing against the overall disclosure risk score (marginal TCAP), the mean of the ROC scores and the CIO was calculated – a score closer to zero indicates lower utility; a score closer to 1 indicates higher utility.[4] The results for the synthetic and sample data were plotted on the R-U map for each country separately.

4   Results

Fully synthetic datasets of the same size as the original were generated, and no post-processing was performed on the data. For *Synthpop* and *CTGAN* five different models were created (using different random seeds) and a synthetic dataset generated for each. The mean utility and risk over the five datasets is plotted on the R-U map. For *DataSynthesizer* five different models for each value of ϵ were created and a point (the mean over the 5) is plotted for each of a series of values for ϵ. Figure 1 illustrates the R-U map for the UK 1991 Census data. The sample fractions form a curved line, with a point representing each increasing sample fraction (from left to right). The utility starts to drop quite steeply once the sample fraction drops below about 3%. Both the risk and utility of the data at 100% (i.e. the whole original sample) is necessarily 1. The synthetic datasets are

---

[4] We recognise that averaging different utility metrics may not be optimal and in future work we will consider an explicitly multi-objective approach to utility optimisation.

plotted alongside the different sample fractions to illustrate how they compare with the sample data. Considering Fig. 1 the *Synthpop* point falls almost on the sample curve, meaning it has utility and disclosure risk equivalence of between a 10% and 20% sample of the original data. Plots for the other three Census datasets are contained in Appendix D.

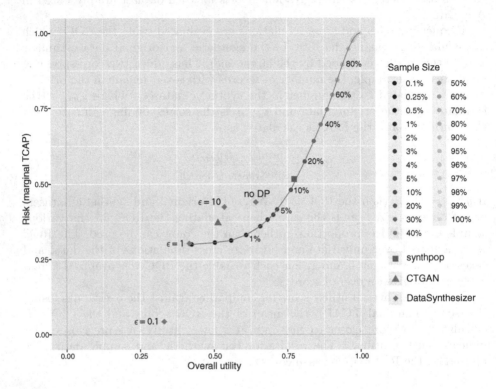

Fig. 1. Risk-Utility map plotting the mean synthetic data and sample fraction results for UK 1991 Census data. DP stands for Differential Privacy.

Table 2 provides detail on the utility and risk (mean over five datasets) of the synthetic data for each of the Census datasets together with the associated sample fraction equivalence.[5] A synthetic dataset with higher sample fraction equivalence for utility but lower sample fraction equivalence for risk would be optimal from the point of view of synthetic data producers. For all four data sets *Synthpop* has higher sample fraction equivalence for utility than for risk. *CTGAN* has mixed results with two outcomes of higher risk equivalence than utility (although all equivalences are low compared to the other methods). The effect of different ϵ values can be observed for *DataSynthesizer*. For $\epsilon = 0.1$ the risk and utility equivalence is less than a 0.1% sample (across all Census datasets); and all but the UK dataset have a negative value for risk (meaning

---

[5] Standard deviation not included for clarity as this was generally small, <0.01.

Table 2. Synthetic data risk and utility (mean of 5 datasets), and comparable sample equivalence for each of the Census datasets. (Note that the values of ϵ indicated here are per dataset created, not for the whole synthesis process which will be five times as large given that m = 5)

Data	Synthesizer	Overall utility	Risk (marginal TCAP)	Sample Equiv. for Utility	Sample Equiv. for Risk
UK 1991	*CTGAN*	0.514	0.371	0.25%–0.5%	2%–3%
	Synthpop	0.774	0.516	10%–20%	10%–20%
	DataSynthesizer:				
	$\epsilon = 0.1$	0.330	0.043	<0.1%	<0.1%
	$\epsilon = 1$	0.416	0.303	<0.1%	0.1%–0.25%
	$\epsilon = 10$	0.536	0.424	0.25%–0.5%	5%–10%
	No DP	0.643	0.440	1%–2%	5%–10%
Canada 2011	*CTGAN*	0.495	0.165	0.25%–0.5%	<0.1%
	Synthpop	0.830	0.294	20%–30%	2%–3%
	DataSynthesizer:				
	$\epsilon = 0.1$	0.342	−0.102	<0.1%	<0.1%
	$\epsilon = 1$	0.425	0.011	0.1%–0.25%	<0.1%
	$\epsilon = 10$	0.521	0.126	0.25%–0.5%	<0.1%
	No DP	0.688	0.231	3%–4%	1%–2%
Fiji 2007	*CTGAN*	0.469	0.439	0.1%–0.25%	3%–4%
	Synthpop	0.816	0.555	20%–30%	10%–20%
	DataSynthesizer:				
	$\epsilon = 0.1$	0.301	−0.173	<0.1%	<0.1%
	$\epsilon = 1$	0.360	0.233	<0.1%	<0.1%
	$\epsilon = 10$	0.477	0.414	0.1%–0.25%	2%–3%
	No DP	0.727	0.526	5%–10%	5%–10%
Rwanda 2012	*CTGAN*	0.430	0.412	0.5%–1%	0.25%–0.5%
	Synthpop	0.752	0.437	20%–30%	1%–2%
	DataSynthesizer:				
	$\epsilon = 0.1$	0.203	−0.404	<0.1%	<0.1%
	$\epsilon = 1$	0.259	−0.045	<0.1%	<0.1%
	$\epsilon = 10$	0.373	0.230	0.1%–0.25%	<0.1%
	No DP	0.720	0.413	10%–20%	0.25%–0.5%

the risk is below the baseline). Considering Figs. 1 and 2 the *DataSynthesizer* points have a curvilinear relationship with each other, although where they fall in relation to the sample fractions equivalence varies between the four different Census datasets.

5   Discussion

The initial results are very interesting in several respects. Firstly the risk-utility relationship for sample data is curvilinear. With risk dropping fast at first as the sample reduces before utility declines rapidly with smaller sample fractions. This is of course ideal and if repeated on larger trial would be a vindication of the use of sampling as a default disclosure control for Census microdata. The curve also indicates a sweet spot sample fraction is around 2–3%, below this level there is little risk benefit and a large decrease in utility. There is a big caveat to place on this finding, which we will come to shortly. Second, the results are mixed when comparing synthetic data and samples, with outcomes appearing to vary by country. *Synthpop* generally performs well with the datasets it produced in each country (other than UK) falling to the right of the risk utility curve for the samples. *CTGAN* produced two results on the curve and two to the left of the curve as did *DataSynthesiser*. This would need a more thorough investigation before conclusions could be reached but the driver is presumably variations in data structure.

Third, the impact of varying ϵ in data synthesiser was also curvilinear but in the opposite less favourable direction (utility decreasing first). Simply switching the differential privacy option on (but with a high value of ϵ) causes a substantial decrease in utility with little appreciable impact on risk. The often advised level of $\epsilon = 0.1$ produces datasets that are right down in the left hand corner, with little utility and no risk. This result if validated through larger scale studies would vindicate that impression analysts have about the impact of differential privacy. The above findings must be strongly caveated on three points.

1. The experiments were conducted using samples of microdata. The experimental samples were in fact sub-samples. The results may not generalise to full population data (i.e. we should not assume that sub sample to sample relationships will be replicated in sample to population ones). The true test will be to compare synthetic populations with microdata samples.
2. The study underestimates the risk of samples relative to synthetic data in general. While we might reasonably assume that synthetic data do not contain identification risk, this is not true for samples (by virtue of them being drawn from real data).
3. The risk measure employed here uses a response knowledge attribution disclosure. This is sound for Census data but for other datasets, presence detection might be a significant risk that would need to be taken account of. In further work we will be examining this issue further.

6   Conclusion

This paper has introduced the notion of sample fraction equivalence risk and utility. With experiments using Census data from four countries, we have demonstrated a mechanism for comparison of data synthesis and sampling for microdata. A second subsidiary aim was to bring differential privacy into the same evaluation framework.

The results of the experiments are quite compelling and illustrate the value of the approach. In future work we will be aiming to extend this initial study in three ways: (i) to run experiments on full population data, (ii) to assess other disclosure control methods for sample fraction equivalence, and (iii) to integrate a measure of re-identification disclosure risk into the framework.

Acknowledgement. The authors wish to acknowledge IPUMs International and the statistical offices that provided the underlying data making this research possible: Statistics Canada; Bureau of Statistics, Fiji; National Institute of Statistics, Rwanda; and the Office for National Statistics, UK.

Appendices

Appendix A

A brief summary of the Census microdata:

Canada 2011: Subsetted on the province of Manitoba, containing 32,149 records (3.47% of the total available dataset which was a 2.78% sample of the 2011 Census). Downloaded from IPUMs [23], courtesy of Statistics Canada.

Fiji 2007: The entire 10% sample (n = 84,323) of the 2007 Fiji Census. Downloaded from IPUMs [23] courtesy of the Bureau of Statistics, Fiji.

Rwanda 2012: Subsetted on the Karongi region, containing 31,455 records (3.03% of the total available, a 10% sample of the 2012 Census). Downloaded from IPUMs [23] courtesy of the National Institute of Statistics, Rwanda.

UK 1991: Subsetted on the region of West Midlands, containing 104,267 records (9.34% of total, a 2% sample of the 1991 Individual Sample of Anonymised Records for the British Census). Downloaded from UK Data Service [27].

Appendix B

Summary of TCAP key/target variables. The six key variables are listed together; the first 3 were used in the case of 3 keys, first 4 for 4 keys, etc.

Canada 2011: For target variables (RELIG, CITIZEN and TENURE) the key variables were: AGE, SEX, MARST (marital status), MINORITY (part of a visible minority), EMPSTAT (labour force status), BPL (birthplace).

Fiji 2007: For target variables (RELIGION, WORKTYPE and TENURE) the key variables were: PROVINCE (of residence), AGE, SEX, MARST (marital status), ETHNIC (part of a visible minority), CLASSWKR (employment status).

Rwanda 2012: For target variables (RELIGION, EMPSECTOR and OWN-ERSH (tenure)) the key variables were: AGE, SEX, MARST (marital status), CLASSWK (employment status), URBAN (urban/rural area), BPL (birthplace).

UK 1991: For target variables (LTILL (long-term illness), FAMTYPE and TENURE) the key variables were: AREAP, AGE, SEX, MSTATUS (marital status), ETHGROUP (ethnic group), ECONPRIM (economic status).

Appendix C

Description of regression models used to calculate the CIO. For each dataset two logistic regressions were performed using marital status and housing tenure as the targets (a binary target was created). Eight predictors were used, these were the same for both models (with tenure/marital status removed accordingly):

Canada Predictors: ABIDENT (aboriginal identity), AGE, CLASSWK, DEGREE, EMPSTAT, SEX, URBAN, TENURE/MARST.

Fiji Predictors: AGE, CLASSWKR, ETHNIC, RELIGION, EDATTAIN (educational level attained), SEX, PROVINCE, TENURE/MARST.

Rwanda Predictors: AGE, DISAB1, EDCERT (highest educational qualification), CLASSWK, LIT (languages spoken), RELIG, SEX, TENURE/MARST.

UK Predictors: AGE, ECONPRIM, ETHGROUP, LTILL, QUALNUM, SEX, SOCLASS, TENURE/MSTATUS.

Appendix D

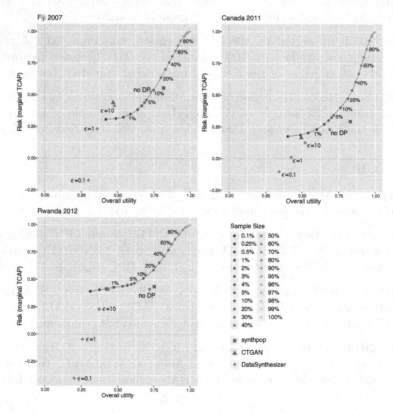

Fig. 2. Risk-Utility map plotting the mean synthetic data and sample fraction results for Fiji 2007, Canada 2011 and Rwanda 2012 Census data.

References

1. Benedetto, G., Stinson, M.H., Abowd, J.M.: The creation and use of the SIPP synthetic beta. Technical report, November, U.S. Census Bureau (2018)
2. Breiman, L., Friedman, J., Stone, C.J., Olshen, R.A.: Classification and regression trees. Wadsworth International Group, Belmont, California (1984). https://doi.org/10.1201/9781315139470
3. Breiman, L.: Random forests. Mach. Learn. **45**(1), 5–32 (2001). https://doi.org/10.1023/A:1010933404324
4. Caiola, G., Reiter, J.P.: Random forests for generating partially synthetic, categorical data. Trans. Data Privacy **3**(1), 27–42 (2010)
5. Camino, R., Hammerschmidt, C., State, R.: Generating multi-categorical samples with generative adversarial networks. arXiv preprint arXiv:1807.01202 (2018)
6. Chen, H., Jajodia, S., Liu, J., Park, N., Sokolov, V., Subrahmanian, V.S.: Fake tables: using GANs to generate functional dependency preserving tables with bounded real data. In: IJCAI, pp. 2074–2080 (2019). https://doi.org/10.24963/ijcai.2019/287
7. Dankar, F.K., Ibrahim, M.K., Ismail, L.: A multi-dimensional evaluation of synthetic data generators. IEEE Access **10**, 11147–11158 (2022). https://doi.org/10.1109/ACCESS.2022.3144765
8. Drechsler, J., Reiter, J.P.: Sampling with synthesis: a new approach for releasing public use census microdata. J. Am. Stat. Assoc. **105**(492), 1347–1357 (2010). https://doi.org/10.1198/jasa.2010.ap09480
9. Drechsler, J., Reiter, J.P.: Sampling with synthesis: a new approach for releasing public use census microdata. Comput. Stat. Data Anal. **55**(12), 3232–3243 (2011). https://doi.org/10.1016/j.csda.2011.06.006
10. Duncan, G.T., Keller-McNulty, S.A., Stokes, S.L.: Database security and confidentiality: examining disclosure risk vs. data utility through the R-U confidentiality map. Technical report, National Institute of Statistical Sciences (2004)
11. Elliot, M.: Final report on the disclosure risk associated with the synthetic data produced by the SYLLS team. Technical report, University of Manchester (2014). http://hummedia.manchester.ac.uk/institutes/cmist/archive-publications/reports/2015-02%20-Report%20on%20disclosure%20risk%20analysis%20of%20synthpop%20synthetic%20versions%20of%20LCF_%20final.pdf
12. Goodfellow, I., et al.: Generative adversarial nets. In: Advances in Neural Information Processing Systems, vol. 27. Curran Associates, Inc. (2014). https://proceedings.neurips.cc/paper/2014/file/5ca3e9b122f61f8f06494c97b1afccf3-Paper.pdf
13. Hittmeir, M., Ekelhart, A., Mayer, R.: Utility and privacy assessments of synthetic data for regression tasks. In: 2019 IEEE International Conference on Big Data (Big Data), pp. 5763–5772 (2019). https://doi.org/10.1109/BigData47090.2019.9005476
14. Hundepool, A., et al.: Statistical Disclosure Control. Wiley Series in Survey Methodology. Wiley, Hoboken (2012). https://doi.org/10.1002/9781118348239
15. Joshi, C.: Generative adversarial networks (GANs) for synthetic dataset generation with binary classes (2019). https://datasciencecampus.ons.gov.uk/projects/generative-adversarial-networks-gans-for-synthetic-dataset-generation-with-binary-classes/
16. Kaloskampis, I., Joshi, C., Cheung, C., Pugh, D., Nolan, L.: Synthetic data in the civil service. Significance **17**(6), 18–23 (2020). https://doi.org/10.1111/1740-9713.01466

17. Karr, A.F., Kohnen, C.N., Oganian, A., Reiter, J.P., Sanil, A.P.: A framework for evaluating the utility of data altered to protect confidentiality. Am. Stat. **60**(3), 224–232 (2006). https://doi.org/10.1198/000313006X124640

18. Kinney, S.K., Reiter, J.P., Reznek, A.P., Miranda, J., Jarmin, R.S., Abowd, J.M.: Towards unrestricted public use business microdata: the synthetic longitudinal business database. Int. Stat. Rev. **79**(3), 362–384 (2011). https://doi.org/10.1111/j.1751-5823.2011.00153.x

19. LeCun, Y., Bengio, Y., Hinton, G.: Deep learning. Nature **521**(7553), 436–444 (2015). https://doi.org/10.1038/nature14539

20. Little, C., Elliot, M., Allmendinger, R., Samani, S.S.: Generative adversarial networks for synthetic data generation: a comparative study. In: Joint UNECE/Eurostat Expert Meeting on Statistical Data Confidentiality (2021). https://unece.org/sites/default/files/2021-12/SDC2021_Day2_Little_AD.pdf

21. Little, R.J.A.: Statistical analysis of masked data. J. Off. Stat. **9**(2), 407–426 (1993)

22. Machanavajjhala, A., Kifer, D., Abowd, J., Gehrke, J., Vilhuber, L.: Privacy: theory meets practice on the map. In: 2008 IEEE 24th International Conference on Data Engineering, pp. 277–286 (2008). https://doi.org/10.1109/ICDE.2008.4497436

23. Minnesota Population Center. Integrated Public Use Microdata Series, International: Version 7.3 [dataset]. IPUMS: IPUMs Census Data, Minneapolis (2020). https://doi.org/10.18128/D020.V7.2

24. Nixon, M.P., Barrientos, A.F., Reiter, J.P., Slavković, A.: A latent class modeling approach for generating synthetic data and making posterior inferences from differentially private counts (2022). https://doi.org/10.48550/ARXIV.2201.10545

25. Nowok, B., Raab, G.M., Dibben, C.: Synthpop: bespoke creation of synthetic data in R. J. Stat. Softw. **74**(11), 1–26 (2016). https://doi.org/10.18637/jss.v074.i11

26. Nowok, B., Raab, G.M., Dibben, C.: Providing bespoke synthetic data for the UK longitudinal studies and other sensitive data with the synthpop package for R. Stat. J. IAOS **33**(3), 785–796 (2017). https://doi.org/10.3233/SJI-150153

27. Office for National Statistics, Census Division, University of Manchester, Cathie Marsh Centre for Census and Survey Research: Census 1991: Individual Sample of Anonymised Records for Great Britain (SARs) (2013). https://doi.org/10.5255/UKDA-SN-7210-1

28. Park, N., Mohammadi, M., Gorde, K., Jajodia, S., Park, H., Kim, Y.: Data synthesis based on generative adversarial networks. Proc. VLDB Endow. **11**(10), 1071–1083 (2018). https://doi.org/10.14778/3231751.3231757

29. Ping, H., Stoyanovich, J., Howe, B.: DataSynthesizer: privacy-preserving synthetic datasets. In: Proceedings of the 29th International Conference on Scientific and Statistical Database Management. ACM, New York (2017). https://doi.org/10.1145/3085504.3091117

30. Pistner, M., Slavković, A., Vilhuber, L.: Synthetic data via quantile regression for heavy-tailed and heteroskedastic data. In: Domingo-Ferrer, J., Montes, F. (eds.) PSD 2018. LNCS, vol. 11126, pp. 92–108. Springer, Cham (2018). https://doi.org/10.1007/978-3-319-99771-1_7

31. Raab, G.M., Nowok, B., Dibben, C.: Guidelines for producing useful synthetic data (2017). https://doi.org/10.48550/ARXIV.1712.04078

32. Raghunathan, T.E., Reiter, J.P., Rubin, D.B.: Multiple imputation for statistical disclosure limitation. J. Off. Stat. **19**(1), 1–16 (2003)

33. Rankin, D., Black, M., Bond, R., Wallace, J., Mulvenna, M., Epelde, G.: Reliability of supervised machine learning using synthetic data in health care: model to preserve privacy for data sharing. JMIR Med. Inform. **8**(7), e18910 (2020). https://doi.org/10.2196/18910

34. Reiter, J.: Using CART to generate partially synthetic public use microdata. J. Off. Stat. **21**(3), 441–462 (2005)

35. Reiter, J.P.: Satisfying disclosure restrictions with synthetic data sets. J. Off. Stat. **18**(4), 531 (2002)

36. Reiter, J.P.: Inference for partially synthetic, public use microdata sets. Surv. Methodol. **29**(2), 181–188 (2003)

37. Reiter, J.P.: Releasing multiply imputed, synthetic public use microdata: an illustration and empirical study. J. R. Stat. Soc. Ser. A Stat. Soc. **168**(1), 185–205 (2003). https://doi.org/10.1111/j.1467-985X.2004.00343.x

38. Rubin, D.B.: Statistical disclosure limitation. J. Off. Stat. **9**(2), 461–468 (1993)

39. Snoke, J., Raab, G.M., Nowok, B., Dibben, C., Slavkovic, A.: General and specific utility measures for synthetic data. J. Roy. Stat. Soc. Ser. A Stat. Soc. **181**(3), 663–688 (2018). https://doi.org/10.1111/rssa.12358

40. Taub, J., Elliot, M.: The synthetic data challenge. Joint UNECE/Eurostat Work Session on Statistical Data Confidentiality (2019). https://unece.org/fileadmin/DAM/stats/documents/ece/ces/ge.46/2019/mtg1/SDC2019_S3_UK_Synthethic_Data_Challenge_Elliot_AD.pdf

41. Taub, J., Elliot, M., Pampaka, M., Smith, D.: Differential correct attribution probability for synthetic data: an exploration. In: Domingo-Ferrer, J., Montes, F. (eds.) PSD 2018. LNCS, vol. 11126, pp. 122–137. Springer, Cham (2018). https://doi.org/10.1007/978-3-319-99771-1_9

42. Taub, J., Elliot, M., Sakshaug, J.W.: The impact of synthetic data generation on data utility with application to the 1991 UK samples of anonymised records. Trans. Data Priv. **13**(1), 1–23 (2020)

43. Therneau, T., Atkinson, E., Ripley, B.: Package 'rpart' (2019). https://cran.r-project.org/package=rpart

44. Wang, L., Chen, W., Yang, W., Bi, F., Yu, F.R.: A state-of-the-art review on image synthesis with generative adversarial networks. IEEE Access **8**, 63514–63537 (2020). https://doi.org/10.1109/ACCESS.2020.2982224

45. Xu, L., Skoularidou, M., Cuesta-Infante, A., Veeramachaneni, K.: Modeling tabular data using conditional GAN. In: Advances in Neural Information Processing Systems, Vancouver, Canada, vol. 32 (2019). https://proceedings.neurips.cc/paper/2019/file/254ed7d2de3b23ab10936522dd547b78-Paper.pdf

46. Zhang, J., Cormode, G., Procopiuc, C.M., Srivastava, D., Xiao, X.: PrivBayes: private data release via Bayesian networks. ACM Trans. Database Syst. **42**(4), 1–41 (2017). https://doi.org/10.1145/3134428

47. Zhao, Z., Kunar, A., Birke, R., Chen, L.Y.: CTAB-GAN: effective table data synthesizing. In: Proceedings of 13th Asian Conference on Machine Learning, vol. 157, pp. 97–112. PMLR (2021). https://proceedings.mlr.press/v157/zhao21a.html

Utility and Disclosure Risk for Differentially Private Synthetic Categorical Data

Gillian M. Raab[✉]

Scottish Centre for Administrative Data Research, University of Edinburgh,
Edinburgh, Scotland, UK
gillian.raab@ed.ac.uk

Abstract. This paper introduces two methods of creating differentially private (DP) synthetic data that are now incorporated into the *synth**pop*** package for **R**. Both are suitable for synthesising categorical data, or numeric data grouped into categories. Ten data sets with varying characteristics were used to evaluate the methods. Measures of disclosiveness and of utility were defined and calculated. The first method is to add DP noise to a cross tabulation of all the variables and create synthetic data by a multinomial sample from the resulting probabilities. While this method certainly reduced disclosure risk, it did not provide synthetic data of adequate quality for any of the data sets. The other method is to create a set of noisy marginal distributions that are made to agree with each other with an iterative proportional fitting algorithm and then to use the fitted probabilities as above. This proved to provide useable synthetic data for most of these data sets at values of the differentially privacy parameter ϵ as low as 0.5. The relationship between the disclosure risk and ϵ is illustrated for each of the data sets. Results show how the trade-off between disclosiveness and data utility depend on the characteristics of the data sets.

1 Introduction

Differential privacy (DP)[1] [9] is considered by theoretical computer scientists to be the most rigorous system of protecting the privacy of individuals in data released to the public. Formally, the release of a statistic that has been altered to comply with ϵ-DP limits, restricts the absolute value of the log-likelihood-ratio of obtaining this result with the complete data to the that from data without any one individual. Small values of ϵ lead to greater protection of privacy by increasing the distance between the original and the ϵ-DP result, while large values do little to preserve privacy and give results closer to those that would be found from the original data. The initial development of DP focussed on privacy without reference to utility, although it was recognised that DP results could provide answers far from the original, expecially for statistics based on small data sets.

---

[1] This acronym will also be used for "differentially private".

© Springer Nature Switzerland AG 2022
J. Domingo-Ferrer and M. Laurent (Eds.): PSD 2022, LNCS 13463, pp. 250–265, 2022.
https://doi.org/10.1007/978-3-031-13945-1_18

In contrast, the generation of synthetic data, as initially implemented in our *synthpop* package for **R** [21], provides a tool to obtain and evaluate the best possible utility but without any formal privacy guarantee, although the fact that no synthetic record corresponds to a real individual gives some reassurance. Additional privacy protection for the output from the *synthpop* package is afforded by the use of statistical disclosure control measures (sdc) that are available as part of the package. This includes the removal of "replicated uniques" defined as records that are unique in the synthetic data, but are also present and unique in the original data, as well as other methods such as smoothing and top and bottom coding of numeric data.

The original conception of DP was that a DP mechanism would be designed to answer individual queries, or a series of queries. When a series of queries are answered for the same data set, the person receiving these DP answers would have increased the disclosure risk to the sum of all the ϵs in the individual queries. This appeared to rule out the possibility of using it to generate synthetic data for which there would be no limit on the number of queries. The few early attempts to create DP synthetic data seemed to result in poor utility, but the last ten years have seen many developments of practical DP methodology, much of it encouraged by endeavours to apply DP to outputs from the 2020 US Census [2,7, 10,12]. These initiatives are not universally accepted as a good thing. There have been criticisms that DP distorts tasks such as redistricting voter areas [15] as well as claims that DP does not prevent the leakage of confidential information [19].

The computer science literature uses different language and conventions from the statistical literature. This makes it difficult for statisticians to evaluate, especially as some of the DP algorithms are complex, see e.g. [13]. In this paper we introduce two easy-to-understand models for creating synthetic data for grouped data that are now incorporated into the development version of *synthpop*, currently on github[2]. Since its introduction there have been several modifications of the original DP definition, such as ϵ-δ-DP and other variants [22]. In this paper we use only the original definition and use only the Laplace methanism for adding DP noise. The models are evaluated on 10 data sets with different characteristics, and the utility and disclosure risk of the synthetic data are assessed. These methods could be developed further and they may not work as well as those developed by teams working with the US Census, but they can provide an experience of creating DP synthetic data to a wider group of people who can gain experience with the method.

2 Methods for Creating Synthetic Data Sets

2.1 Without DP Guarantee

Many methods of creating synthetic data are based on original proposals in 1993 [16,26] that began to be implemented and developed from 2003 [25]. There is already a large literature for creating completely synthetic data,

---

[2] It can be installed from https://github.com/bnowok/synthpop.

without reference to DP; see Drechslerś monograph [8] for a comprehensive review. While the original developments in this field built on ideas from multiple imputation, it is simpler to consider the methodology as generating data from a model fitted to the real data [23]. Synthetic data are simulated from this model using either the fitted parameters, or a sample or samples from the posterior distribution of the fitted parameters. The parameters may be obtained by a fit to the joint distribution of all the variables or, more commonly, by defining the joint distribution in terms of a series of conditional distributions.

Synthesis Methods Implemented in *synthpop*. Synthesising from conditional distributions was the only method available in the original version of the *synthpop* package. It provides great flexibility by allowing a choice of model for each conditional distribution, by changing the order of the conditional distributions and by defining the predictors to be used for each conditional distribution. This allows the synthesiser to improve the quality of the synthetic data using tools to evaluate its utility [24] that are now part of the package.

From Version 1.5 [20] in 2018, *synthpop* includes two methods based on a log-linear fit to the joint distribution for data when all variables are categorical. The first (*catall*) fits a saturated model by selecting a sample from a multinomial distribution with probabilities calculated from the complete cross-tabulation of all the variables in the data set. This is very close to a method recently proposed by Jackson et al. [14]. Jackson et al. also use a saturated model, but they generate data from a Poisson distribution. They present interesting results of different ways in which the synthetic data can be made less disclosive by simulating from overdispersed distributions, such as the negative binomial or the Poisson inverse Gamma. The *catall* procedure can be made exactly equivalent to the Poisson method if the number of records in the synthetic data is itself sampled from a Poisson distribution. When the cross-tabulation contains cells with zero counts these will be reproduced as zero cells in the synthetic data unless a small positive quantity (α) is added to each cell in the table. The *synthpop* package allows for this by spreading a total count, defined by the parameter *nprior*, across all the cells in the table that are not structural zeros, so that $\alpha = nprior/k$ is added to each cell where k is the number of cells in the table that are not structural zeros (e.g. a cell for the type of qualification for people with no qualifications). The second method *ipf* fits log-linear models to a set of margins defined by the user using the method of iterative proportional fitting[3] implemented in the package *mipfp* in **R**. As for *catall* the parameter α can be set to allow non-structural zeros to appear in the synthetic data. Both *catall* and *ipf* are only feasible for a small number of variables because of the need to create very large cross-tabulations of all the variables. *synthpop* currently prints a warning if the total cells exceed 10^8, though a user can attempt to increase this.

Other Methods.   More recently, the computer science and machine learning literature has introduced many new methods for creating synthetic data. A

---

[3] Also known as the RAS algorithm and as raking in Computer Science.

web search for "synthetic data solutions" generates links to many firms offering synthetic data services[4, 5]. One can even find a link to a list of firms who can provide synthetic data for businesses[6]. Many of the machine learning algorithms are based on Generalized Adversarial Networks (GANs) [11]. These methods involve an iterative process where a 'Generator' fits a model to the data and a 'Discriminator' attempts to distinguish between the real data and the model. Feedback to the 'Generator' is used to improve the fit of the data at the next iteration.

Compatible Generative Models. Users of synthetic data can use it to calculate any summary statistic or to estimate any statistical model. The results obtained and the standard errors calculated for them [23] will only be unbiassed if the model used in generating the synthetic data is at least as complex as that used in the analysis. We will refer to this aspect as the use of a generative model that is compatible with the analysis. Even more important is the condition that the original data is well represented by the generative model. The synthetic data may then appear to provide a good fit for an analysis compatible with the synthesis model, when an analysis of the original data would have indicated a lack-of-fit. If the analysis being used is incompatible with the generative model creating the synthetic data, results may be biassed compared to those from the original data; i.e. synthetic data will have poor utility. Synthesis methods that appear to provide the most useful synthetic data are adaptive methods where the fitting procedure explores the relationships between variables to obtain the best fit. Such methods include a full conditional model where all conditional distributions are fitted by an adaptive CART model (the default model in *synthpop*) as well as the many different implementations of GANs.

The method *catall* is compatible with any analysis of categorical data. Synthesis with *ipf* is only compatible with analyses for which the sufficient statistics are members of the set of margins used to contrain the iterative fitting. If the data have significant interactions that are not included in the set of margins, the results of any incompatible analyses will be biassed. Poor utility will be found in these cases when evaluated from a fit to an incompatible model. However, unlike synthetic data created by adaptive methods, knowledge of the marginals will allow the user of synthetic data to know which analyses can be trusted. For example, one could create synthetic data with *ipf* with all two-way marginals and their interactions with the outcome of interest. Then a logistic regression of the outcome on all other variables and their two-way interactions would be a compatible analysis model.

---

[4] Hazy https://hazy.com/, Accessed 18 May 2022.

[5] Mostly AI https://mostly.ai/ebook/synthetic-data-for-enterprises, Accessed 18 May 2022.

[6] https://research.aimultiple.com/synthetic-data/, Multiple AI: In-Depth Synthetic Data Guide, Accessed 18 May 2022.

2.2   Adapting Methods for Synthetic Data to Make Them DP

Background. For a comprehensive review of practical methods for DP data synthesis see Bowen and Liu, 2020 [5]. This section will provide a brief summary, focussing mainly on the methods for categorical data used in this paper.

The earliest real application of DP synthetic data was the data set that sits behind the US Census Bureau's online tool 'On the Map'[7] that allows visualisation of where people work and where they live. The technique used was to add noise to the raw data for this example by either simulating from data with a Multinomial-Dirchilet distribution or by adding Laplace noise to the counts [4,17]. The values of ϵ had to be larger than was desirable and data had to be modified to make it appear plausible.

A more promising approach to creating synthetic data is to add DP noise to the parameters of the joint distribution, rather than to the raw data. The ϵ for the whole dataset is the sum of the individual values for each parameter. Shlomo [27] has recently illustrated this approach for synthesising data with a multivariate Normal distribution.

Methods Used in the NIST Challenge. In recent years several groups have developed methods and software for creating DP synthetic data. These initiatives have been encouraged by a series of challenges, with substantial prizes, promoted by the National Institute of Standards and Technology (NIST)[8]. Participating teams had to provide synthetic versions of the data for each challenge where the versions satisfied DP for a given set of values of ϵ and δ. The teams had to provide code that would convince the judges that their data satisfied ϵ-δ differential privacy. The software used by many of the teams who entered can be accessed via the NIST web site. Those submissions that passed this hurdle were then compared using utility metrics. Bowen and Snoke [6] have published a detailed evaluation of the synthetic data that were submitted and describe the methods used by the teams.

Two main types of methods were employed in the challenge. The first was those based on DP GANs. GANS can be easily made DP by providing feedback from the Discriminator to the Generator via the results of DP queries. The total ϵ for the method is the sum of those used at each iteration and the iterations have to stop once this DP budget is exhausted. The second method, used by many teams, was to generate synthetic data from a model fitted to the margins of the data. Each margin is first made DP and a model is fitted that assumes that only the interactions defined by the chosen margins are present in the population from which the data can be considered to have been generated. After noise has been added to the margins to make them DP, they no longer sum to the same totals for lower order marginals. Several methods have been used to

---

[7] see https://lehd.ces.census.gov/applications/, Accessed 16th March 2022.

[8] https://www.nist.gov/ctl/pscr/open-innovation-prize-challenges/past-prize-challenges, where you will find details of the winning methods and links to some of the software used.

make the margins agree and ensure they are non-negative. Once this is achieved synthetic data can be generated from a model fitted to the DP margins. The resulting synthetic data are DP because once the DP margins are created, any data derived from them without further input from the original data will also be DP[9]. The choice of margins cannot be based on an analysis of the data to be synthesised as this would incur an addition to the privacy budget. The NIST teams used analyses of other similar data sets to inform their choices or, in some cases, used part of their privacy budget to identify the margins to fit. In the evaluation the teams who used marginal methods scored more highly than those using GANs. One such team who scored highly on all challenges [18] has provided a detailed description of their methodology including their model choice and the post-processing methods used.

DP Methods in *synth*pop. The methods *catall* and *ipf* can be made DP. The first by a method similar to the method used by Abowd and Villhuber [4] in 2008. The second is similar to the marginal models used by the NIST teams. Note that the choice of margins to use for DP *ipf* cannot be made from an analysis of the original data, unless the analyses to determine the choice of margins contributes to the privacy budget. In this preliminary investigation the choice of margins is all two-way interactions in all cases. The process of making these two methods DP, as implemented in the latest version *synth*pop[10], involves the following steps:

1. Determine the value of ϵ to use.
2. Create cross-tabulations of all the variables (*catall*) or of the selected margins (*ipf*).
3. If the parameter *nprior* > 0 add $nprior/n_{cells}$ to each cell in every table, where n_{cells} is the number of cells over which the prior is to be spread.
4. Add Laplace noise with dispersion parameter $1/\epsilon$ (*catall*) or M/ϵ for (*ipf*), where M is the number of margins fitted, to each cell of the table.
5. Set any negative counts to zero or a small positive value.
6. Rescale the counts to become probabilities that sum to unity.
7. For *catall* create the synthetic data as a multinomial distribution from this probability vector with the selected sample size.
8. For *ipf* use the *Ipfp* algorithm from the *mipfp* package[11] for **R** to obtain a fit to the probabilities calculated from the noisy marginals, and then generate synthetic data as for *catall*.

Note that when iterative proportional fitting is applied to the probabilities from incompatible margins, rather than compatible counts, the margins are adjusted as part of the process so they become compatible with each other. Convergence for iterative proportional fitting is always slow compared to the Newton-Raphson algorithms, but it is usually slow but sure. It becomes even

---

[9] This is described as robustness to post-processing.
[10] See footnote 2. It will be made available on CRAN after Version 1.7.0.
[11] See https://CRAN.R-project.org/package=mipfp.

slower when the initial margins are incompatible, and in a few cases it may fail to converge in a large number of iterations; see Table 3 for details. For the case when the margins are compatible the eventual convergence is guaranteed, but is not clear if this is the case when margins are not compatible. Steps 5 and 6 are post-processing steps that can have a considerable influence on disclosiveness and utility.

3   Measures of Utility and Disclosure Risk for Synthetic Categorical Data

3.1   Disclosure Risk

For DP synthetic data ϵ provides a measure of disclosure risk. It is desirable to have another measure of disclosure risk that can be calculated for non-DP synthetic data and that can be compared to ϵ for DP synthetic data. One such measure is the percentage of unique records in the synthetic data that are also unique in the original data, designated as ru (replicated uniques)[12]. Below we introduce the notation used and define replicated uniques for the case when all variables are categorical. The measure ru will depend on the disclosiveness of the original data. In the extreme, if there are no unique records in the original data then ru will always be zero. This measure relates to the expected behaviour of an intruder who knows the characteristics of a unique individual in the data base and attempts to identify them. A more nuanced version of this type of measure has been discussed by Taub et al. [29], but this relates to a particular target variable for which the value might be determined from a set of key variables. To obtain a measure for a data set this would need to be averaged over a selection of targets. Replicated uniques were used by Jackson et al. [14] in their evaluation of syntheses from saturated models.

N	Number of observations in the original data
k	Total cells in the cross-tabulation of all variables
$y_1, y_2, ..., y_k$	Counts of original data in each cell of the cross-tabulation
$s_1, s_2, ..., s_k$	Counts of synthetic data in each cell of the cross-tabulation
$100 \, \Sigma_{i=1,k}(y_i = 1)/N$	Percentage of unique records in the original data ($p1$)
$100 \, \Sigma_{i=1,k}(y_i = 1)/k$	Percentage of empty cells in the cross-tabulation ($p0$)
$100 \, \Sigma_{i=1,k}(s_i = 1 \ \& \ y_i = 1)/k$	Percentage of uniques in the synthetic data that are also unique in the original (replicated uniques ru)

3.2   Utility

Raab et al. [24] have carried out an extensive review of utility measures that have been proposed for synthetic data. The measures are computed by first attempting to discriminate the synthetic data from the original according to some method. All the measures evaluated were found to be highly correlated with each other,

---

[12] Also sometimes termed "correct matches".

and in some cases even identical. The model used to discriminate between the real
and synthetic data is more important than the choice of measure. Some of the
measures have expected values that can be calculated, or obtained by simulation,
when the generative model is the correct one that could have produced the
original data. The utility measure that we will use here is the propensity score
mean-square-error ($pMSE$) whose expected value for a correct generative model
is discussed in [28]. A standardised version can be calculated as the ratio of
$pMSE$ to its expectation (S_pMSE). For non-DP synthesis this value will have
an expected value of 1.0 when the generative model is correct, with higher values
indicating poor utility. Experience of using this measure for many synthetic data
sets suggests that synthetic data with values of S_pMSE below 10 provide useful
results that agree well with analysis of the original data, and those under 30
generally produce usable results.

   When the model used to evaluate utility consists of a tabulation, the $pMSE$
has a simple form [24] that is identical to the utility measure proposed by Voas
and Williams [30]. The formula for this and its standardised version for table m,
designated as U_m, is shown below. These formulae assume that a synthetic data
set of the same size as the original has been produced. A summary measure for
the whole data set can be obtained by averaging U_m over a set of M marginals
to give U_M. When the synthetic data are created by a method like ipf that
constrains a set of marginals, then the expected value of U_m for each constrained
marginal will be 1.0.

M	Number of margins selected
$m_1, m_2, ... m_M$	Number of cells in the mth margin
$yi_1, yi_2, ..., yi_m$	Counts of original data in each cell of the ith margin
$si_1, si_2, ..., si_m$	Counts of synthetic data in each cell of the ith margin
$pMSE_m = \Sigma_j[(yi_j - si_j)^2/\{(yi_j + si_j)/2\}]$	$pMSE$ for the ith margin
df_m	Degrees of freedom for the mth marginal
$U_m = pMSE_m/df_m$	S_pMSE standardised utility for the mth marginal
$U_M = \Sigma_m U_m/M$	Average standardised utility for the set of M marginals

4   Data Sets Used for the Evaluation

Table 1. Features of the data sets.

	S3	S5	S7	P3	P5	P7	P7x	Ps3	Ps5	Ps7
N	5,000			1,035,201				13,309		
k	60	2,160	77,760	70	2,100	94,500	3,420,900	70	2,100	94,500
$p0$	0	6.23	98.20	5.71	37.62	94.49	96.62	34.29	73.19	99.10
$p1$	0	6.74	35.84	0.00	0.02	0.12	4.51	0.03	0.96	2.66

Data sets used to evaluate the DP synthesis methods were subsets of variables
from two sources. The data set (SD2011) is a sample of 5000 records from a sur-

vey on the quality of life in Poland in 2011[13]. In all cases any numeric variables were grouped into 5 classes[14]. The data used to synthesise was a subset of the first 3, first 5 and first 7 variables in this data set, labelled S3, S5 and S7. The second data set was supplied to teams from National Statistics Agencies (NSOs) as part of the evaluation of a guide to synthetic data for NSOs [1]. It consisted of an extract of just over a million records from the American Community Survey, made available by the IPUMS project[15]. As well as demographic data it included an area identifier that divided the area for the survey into 181 Public Use Microdata Areas (PUMAs). Data extracts from this source were 3, 5 and 7 demographic variables (P3, P5 and P7), then one with 7 variables including PUMA (P7x) and finally three smaller data sets from one PUMA with 13 thousand records (Ps3, Ps5 and Ps7). This is important since better utility can be obtained for this type of data by stratifying it into subsets by area. If each subset is made DP the overall ϵ is the maximum of those used for each subset, since a person can only be in one PUMA. Features of the data sets are summarised in Table 1 and the variables in each are given in the Appendix A.1.

The number of cells in the cross-tabulation of all the variables ranged from 60 to over 3 million. Extracts with 7 variables had the most sparse tables (high $p0$). The % of unique records is a disclosure measure for the original data that is an upper limit for ru from synthetic data sets. For a given sample size $p0$ and $p1$ increase with the number of cells in the table and they are greatest for small sample sizes. Two of the three data sets with just three variables have no unique records, and the third has only a very small number of unique records. Those datasets with 7 variables are all very sparse with two having over 98% of the cells in the cross-tabulation as zeros.

5   Results

5.1   Utility and Disclosure Risk for Non-DP Synthesis

Each of the 10 datasets was synthesised using *synthpop* by the method *catall* and then by *ipf* with all two-margins. Results for disclosiveness and utility, averaged over 10 replications, are shown in Table 2. Synthesis by *catall* reduces the disclosure risk, as assessed by ru, to an average of about 37% of $p1$ and to a greater extent (varying by data set) for *ipf* synthesised from two-way margins. As expected all the two-way and three-way utility measures are centred on 1.0 for *catall*. This is also true for synthesis by *ipf* for utility evaluated from two-way margins, but for utility evaluated from three-way margins we find the expected lack-of-fit from the incompatible generating model. There was very little evidence of three-way interactions in the SD2011 data sets, but more for the PUMS data. For synthesis by *ipf* we show the utility for the three-way marginal that gave

---

[13] Available as a data set as in the *synthpop* package.

[14] The only numeric variables selected from SD2011 were Age and Income, and from the PUMS data only Age.

[15] https://www.ipums.org/.

the worst utility in an evaluation of a two-way synthesis. This choice of worst marginal was carried forward into the DP syntheses reported below. This will allow the loss of utility due to an incompatible margin to be compared to that due to the addition of DP noise. For the PUMS data the variables with the most evidence of a three way interaction were age and sex with either marital status or group quarters.

Table 2. Non DP synthesis results for the 10 data sets, all results are average for 10 syntheses.

	S3	S5	S7	P3	P5	P7	P7x	Ps3	Ps5	Ps7
Original data										
$p1$	0.00	6.74	35.84	0.00	0.02	0.12	4.51	0.03	0.96	2.66
catall saturated model										
% replicated uniques ru		2.44	13.16		0.01	0.04	1.66	0.01	0.34	0.95
(ru) as a % of $p1$		36	37		37	36	37	28	36	36
Mean two-way utility	0.97	1.02	1.00	0.93	1.01	0.92	0.99	1.10	1.02	0.99
Mean.three-way.utility	0.92	1.03	0.99	0.96	0.98	0.97	1.02	1.05	1.01	1.00
ipf with all two-way margins										
% replicated.uniques(ru)		1.67	6.41		0.00	0.03	1.00	0.00	0.28	0.63
(ru) as a % of $p1$		30	13		25	17	22	15	22	24
Mean two-way utility	0.97	0.96	1.04	1.09	0.98	0.98	1.04	0.97	0.98	1.07
Mean three-way utility	1.48	1.76	1.77	4.41	20.58	10.34	10.26	1.54	2.14	1.67
Worst.three-way.utility	1.48	1.92	2.37	4.41	81.74	81.40	82.14	1.54	5.52	1.20

5.2   Utility and Disclosure Risk for DP Synthesis

The results for DP synthesis by *catall* confirmed the expectation that this method would not prove useful; similar results were found by Bowen and Liu [5]. Even for some larger values of ϵ utilities often exceeding 100 times their value for non-DP synthesis. The disclosure, measured by ru did fall to low levels, especially for data sets with very sparse cross-tabulations, but this was not associated with acceptable utility for any of the data sets. The process of adding noise leads to many negative values. These have to be set to either zero or a small value to allow the probabilities to be used in generating the synthetic data. Exactly how steps 5 and 6 distort the data for these sparse tables depends on the relative proportion of 0s and 1s in the cross-tabulation. For small ϵ applied to sparse tables step 5 increases the sum of the noisy counts and step 6 reduces large counts. The US Census Bureau have noted the contribution of this type of post-processing to their area counts [10,12], but the top-down solution they have used [3], including some agreed fixed totals, does not appear applicable to synthetic data generation. Post-processing by steps 5 and 6 are less of a problem for *ipf* because counts in the margins are larger. Some teams using marginal approaches

Replicated uniques % of uniques in original vs. epsilon for 7 data sets

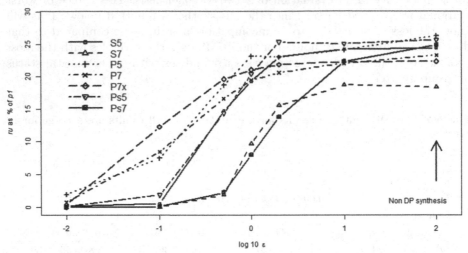

Fig. 1. Percentage of replicated uniques for *ipf* DP and non-DP syntheses expressed as a % of the percentage of uniques in the original data set. Results for each of 7 data sets with DP method of adding Laplace noise to the two-way margins.

in the NIST challenge appear to have used the method described here [31], but better methods could be explored.

Table 3 displays the disclosure risk and utility for the 10 data sets and six different choices of ϵ synthsised by *ipf*. In a few cases with small values of ϵ the fitting procedure failed to converge in 5,000 iterations.[16] Figure 1 plots *ru* as a % of *p1* against $log_{10}(\epsilon)$, with the non-DP synthesis being plotted at 3, equivalent to an ϵ of 100. Disclosure risk, as measured by replicated uniques decreases with ϵ as we would expect. Two data sets (S7 and Ps7) each with 7 variables and a relatively small sample size, show a good reduction in *ru* by an ϵ of 1, and better for lower values. To evaluate utility, we first consider two-way margins that will not be affected by the lack-of-fit found for three-way margins in the non-DP synthesis. For the data sets analysed here ϵ values of 1 and above give satisfactory utility, judged by having average values below 30. Other data sets require a lower ϵ of 0.1 to get an improvement in %*ru*, and hence higher (worse) utility measures. An ϵ of 0.5 gives satisfactory utility for most of the data sets. For the lower values of ϵ used here (0.1 and 0.01) utility is too poor to be acceptable. Additional exploratory checks illustrated the failure of the synthetic data to show the same relationships between variables as was found in the original data. The process of adding noise to the margins has the effect of bringing them all closer to a constant value and thus inevitably weakening the relationships found in the original. Looking at the results for utility from three-way tables, we can see that the results are limited by the utility that can be achieved by non-DP synthesis with the same models.

---

[16] It is possible that the added noise produced margins that are impossible to reconcile. These cases do not correspond to useable syntheses.

Table 3. Disclosure measures and utility for differentially private *ipf* synthesis with 6 values of ϵ and 10 data sets. All figures are averages over 10 independent syntheses. Rows marked with * indicate that one or more of the *ipf* fits failed to converge in 5,000 iterations

	S3	S5	S7	P3	P5	P7	P7x	Ps3	Ps5	Ps7
ϵ 0.01										
% *ru*	0.00	0.01*	0.00*	0.00	0.00	0.00	0.02	0.00	0.00	0.00
ru as a % of *p1*		0.18	0.01		1.95	0.75	0.41	0.00	0.08	0.00
Mean two-way utility	245.71	325.45	365.47	83.03	412.38	1942.99	1692.89	227.55	767.38	1240.11
Mean three-way utility	139.81	135.90	133.32	62.22	258.10	949.60	786.60	131.96	365.65	543.98
Worst three-way utility	139.81	221.15	159.31	62.22	501.79	1236.90	1211.84	131.96	501.07	441.72
ϵ 0.1										
% *ru*	0.00	0.04	0.02*	0.00	0.00	0.01	0.55	0.00	0.02	0.00
ru as a % of *p1*		0.56	0.05		7.46	8.45	12.24	0.00	1.88	0.14
Mean two-way utility	12.63	99.01	172.29	5.47	23.05	128.11	104.59	16.48	81.11	261.94
Mean three-way utility	10.27	56.77	79.06	9.51	32.04	78.55	61.82	11.46	49.17	135.13
Worst three-way utility	10.27	75.29	94.35	9.51	102.79	136.49	159.38	11.46	58.53	149.41
ϵ 0.5										
% *ru*	0.00	0.99	0.68*	0.00	0.00	0.02	0.88	0.00	0.14	0.06
ru as a % of *p1*		14.69	1.90		18.64	16.58	19.57	10.00	14.30	2.26
Mean two-way utility	1.69	14.59	31.67	1.38	3.51	19.66	15.65	2.85	15.16	46.60
Mean three-way utility	1.90	8.91	20.85	4.61	21.67	21.20	17.45	2.89	9.82	28.59
Worst three-way utility	1.90	9.70	24.30	4.61	81.96	91.95	92.17	2.89	18.52	7.38
ϵ 1										
% *ru*	0.00	1.29	3.46	0.00	0.00	0.02	0.95	0.00	0.20	0.21
ru as a % of *p1*		19.20	9.66		23.08	19.41	20.99	10.00	20.39	7.99
Mean two-way utility	1.15	5.48	15.21	1.14	2.04	8.32	6.58	2.09	5.34	22.03
Mean three-way utility	1.40	4.21	9.51	4.34	20.72	14.47	12.77	2.06	4.69	14.50
Worst three-way utility	1.40	4.89	11.82	4.34	80.48	84.96	82.32	2.06	9.18	2.78
ϵ 2										
% *ru*	0.00	1.55	5.55	0.00	0.00	0.02	0.98	0.01	0.24	0.37
ru as a % of *p1*		23.03	15.49		22.96	20.57	21.79	17.50	25.31	13.76
Mean two-way utility	1.08	2.84	5.86	0.85	1.29	3.90	3.32	1.36	3.31	11.86
Mean three-way utility	1.31	2.65	4.37	3.97	20.27	11.87	11.16	1.65	3.27	8.16
Worst three-way utility	1.31	3.16	5.57	3.97	79.48	82.24	80.19	1.65	7.10	1.66
ϵ 10										
% *ru*	0.00	1.63	6.68	0.00	0.00	0.03	1.00	0.01	0.24	0.59
ru as a % of *p1*		24.15	18.64		24.02	21.96	22.18	35.00	25.00	22.26
Mean two-way utility	1.15	1.15	1.64	0.91	1.07	1.32	1.54	1.05	1.25	2.55
Mean three-way utility	1.40	1.82	2.05	4.06	20.02	10.38	10.20	1.52	2.28	2.58
Worst three-way utility	1.40	2.14	2.53	4.06	78.29	80.65	80.81	1.52	5.97	1.26

In many cases, especially for the worst three-way interaction this lack-of-fit to the generative model is more damaging to utility than the DP adjustment. All of the evaluations were carried out with a very small value of a prior α added to each cell of the tables[17]. Increasing α can itself contribute to preventing disclosure [14],

---

[17] A count determined by the parameter *nprior* is distributed equally over the table or margin entries except those defined as structural zeros. This is important for non-DP synthesis so as to prevent them remaining as zeros. The default value of 1 for *nprior* was used.

as used in the Dirchilet-Multinomial synthesiser [17]. Using this approach without any DP adjustment gave some improvement in ru for large α but at the expense of greatly damaged utility. When α was adjusted for DP syntheses, its influence was much less than that of the DP parameter.

6   Discussion and Future Work

We have introduced a very simple modification of two methods to allow them to produce DP synthetic data. The saturated model is useful for non-DP data synthesis, well suited to large administrative data bases of count data. But efforts, so far, to make this model differentially private find poor utility, even for large values of ϵ that provide little improvement in disclosure.

In contrast, the ipf method, based on margins, seems to be easily adapted to provide DP synthetic data. This model can be fitted in *synthpop* with code like this example. This code creates a synthetic data object (`synipf`) by the default method of fitting all two-way interactions. The percentage of replicated uniques can then be found with the next line, and the third line calculates the average two-way utility for all two-way interactions. The default utility measure, $pMSE$, is used but there is a choice of a further 15 measures available.

```
synipf <- syn(S7, method= "ipf", ipf.priorn = 0, ipf.epsilon = 1)

ru_S7_1 <- replicated.uniques(synipf, S7)$per.replications

util_S7_2way <- mean(utility.tables(synipf,SD20113,"twoway", plot = FALSE)$tabs[,2])
```

This is just a preliminary version of DP synthesis with *synthpop*. It has some clear limitations. In particular, it is limited to data sets with a relatively small number of variables because of the need to store the complete cross tabulation. The teams using marginal models for the NIST challenge have used graphical models to define the parameters of the models. It may be possible to make their open-source routines, usually written in *Python*, accessible within R. A further limitation is that the methods apply only to categorical variables. Non-DP versions of *catall* and *ipf* can be used for numeric data by asking the program to categorise the variables, and then select from the groups in the original data at the end of the synthesis. For DP synthesis a method that did not access the original data again would be required. Many other modifications could be attempted. In particular options should be made available to add noise to the margins by methods other than a Laplace distribution. Examples are trimmed $\epsilon - \delta$ Gausian noise and the Exponential mechanism, see [18] for other choices.

Further investigation of the disclosure risks posed by synthetic data, both DP and non-DP, would be helpful. These should include realistic evaluations of the behaviour of those attempting to find confidential information about data subjects. A disclosure from synthetic data that turns out to be false can also cause harm, both to the data subject and to the reputation of the data holder. This underlines the importance of ensuring that everyone who has access to synthetic data is aware that it is not real.

Inference from synthetic data will always be limited by the model used in creating it. George Box's caveat that "all models are wrong, but some are useful" needs to be borne in mind when using results from synthetic data whether DP or not. There is a powerful argument that no important decisions should be taken using analyses of synthetic data. Confirmatory analyses on the real data, perhaps via a validation server, should be carried out. But synthetic data can still have an important role in widening access to confidential data and in providing realistic data sets for training.

Acknowledgements. This work would not have been possible without the work of Beata Nowok, the main author of *synthpop*. Any errors found in the new DP routines on github (see note 2) are entirely my responsibility. The ESRC/UKRI provided support for the Administrative data Research Centre and the Scottish Longitudinal Study. I would also like to thank two anonymous referees for helpful comments on an earlier version of this paper.

A   Appendix

A.1   Details of the Variables in Data Sets

Tables 4 and 5 give details of the variables selected from each of the two data sets. See section **data** for how each of the data set were created. The two Age variables were each grouped into 5 categories.

Table 4. Variables selected from the SD2011 data set.

	Variable	Number of missing values	Number of distinct values
1	Sex	0	2
2	Age grouped	0	5
3	Placesize	0	6
4	Education level	7	4
5	Social and professional group	33	9
6	Income grouped	683, 603 not applicable	5
7	Marital status	9	6

Table 5. Variables selected from the IPUMS data set.

	Variable	Number of missing values	Number of distinct values
1	Public use microdata area	0	181
2	Year	0	7
3	Group quarters	0	5
4	Sex	0	2
5	Age	0	73
6	Marital status	0	6
7	Race	0	9
8	Hispanic	0	5

References

1. Synthetic data for official statistics: a starter guide. United Nations, Geneva. UNECE: High Level Group for the Modernisation of Official Statisics, (2022, forthcoming)
2. Abowd, J.M.: The U.S. census bureau adopts differential privacy. In: 24th ACM SIGKDD International Conference on Knowledge Discovery and Data Mining (2018). https://doi.org/10.1145/3219819.3226070. Accessed May 2022
3. Abowd, J.M., et al.: The 2020 census disclosure avoidance system TopDown algorithm (2022). https://arxiv.org/abs/2204.08986
4. Abowd, J.M., Vilhuber, L.: How protective are synthetic data? In: Domingo-Ferrer, J., Saygın, Y. (eds.) PSD 2008. LNCS, vol. 5262, pp. 239–246. Springer, Heidelberg (2008). https://doi.org/10.1007/978-3-540-87471-3_20
5. Bowen, C.M., Liu, F.: Comparative study of differentially private data synthesis methods. Stat. Sci. **35**(2), 280–307 (2020)
6. Bowen, C.M., Snoke, J.: Comparative study of differentially private synthetic data algorithms from the NIST PSCR differential privacy synthetic data challenge. J. Priv. Confid. **11**(1), 12704 (2021)
7. Cole, D., Sautmann, V., (eds.) Handbook on Using Administrative Data for Research and Evidence-based Policy, Chap. 6 Designing Access with Differential Privacy, pp. 173–239 (2020). https://admindatahandbook.mit.edu/book/v1.0/diffpriv.html. Accessed on 19 May 2022
8. Drechsler, J.: Synthetic Data Sets for Statistical Disclosure Control: Theory and Implementation. Springer, New York (2011). https://doi.org/10.1007/978-1-4614-0326-5
9. Dwork, C., McSherry, F., Nissim, K., Smith, A.: Calibrating noise to sensitivity in private data analysis. In: Halevi, S., Rabin, T. (eds.) TCC 2006. LNCS, vol. 3876, pp. 265–284. Springer, Heidelberg (2006). https://doi.org/10.1007/11681878_14
10. Garfinkel, S.: Differential privacy and the 2020 us census. MIT Case Studies in Social and Ethical Responsibilities of Computing (Winter 2022) (2022). https://mit-serc.pubpub.org/pub/differential-privacy-2020-us-census
11. Goodfellow, I., et al.: Generative adversarial networks. In: Advances in Neural Information Processing Systems, vol. 3, pp. 2672–2680 (2014). https://arxiv.org/abs/1406.2661
12. Hawes, M.B.: Implementing differential privacy: seven lessons from the 2020 united states census. Harv. Data Sci. Rev. **2**(2) (2020). https://hdsr.mitpress.mit.edu/pub/dgg03vo6, https://hdsr.mitpress.mit.edu/pub/dgg03vo6
13. Hay, M., Machanavajjhala, A., Miklau, G., Chen, Y., Zhang, D.: Principled evaluation of differentially private algorithms using DPBENCH. In: Proceedings of the 2016 International Conference on Management of Data (2016). https://dl.acm.org/doi/10.1145/2882903.2882931
14. Jackson, J., Mitra, R., Francis, B., Dove, I.: Using saturated count models for user-friendly synthesis of categorical data. J. Roy. Statist. Soc. Serues A (2022, accepted). https://arxiv.org/abs/2107.08062v2)
15. Kenny, C.T., Kuriwaki, S., McCartan, C., Rosenman, E., Simko, T., Imai, K.: The use of differential privacy for census data and its impact on redistricting: the case of the 2020 US. Census. Sci. Adv. **7**(7), 1–17 (2021). https://imai.fas.harvard.edu/research/DAS.html
16. Little, R.J.A.: Statistical analysis of masked data. J. Off. Stat. **9**(2), 407–26 (1993)

17. Machanavajjhala, A., Kifer, D., Abowd, J., Gehrke, J., Vilhuber, L.: Privacy: theory meets practice on the map. In: 2008 IEEE 24th International Conference on Data Engineering, pp. 277–286 (2008)
18. McKenna, R., Miklau, G., Sheldon, D.: Winning the NIST contest: a scalable and general approach to differentially private synthetic data. J. Priv. Confidentiality **11**(3), 1–30 (2021). https://journalprivacyconfidentiality.org/index.php/jpc/article/view/407
19. Muralidhar, K., Domingo-Ferrer, J., Martínez, S.: ϵ-differential privacy for microdata releases does not guarantee confidentiality (let alone utility). In: Privacy in Statistical Databases 2020 (2020)
20. Nowok, B., Raab, G.M., Dibben, C.: synthpop: Generating Synthetic Versions of Sensitive Microdata for Statistical Disclosure Control, R package version 5.0-0 (2018). https://CRAN.R-project.org/package=synthpop
21. Nowok, B., Raab, G.M., Dibben, C.: synthpop: Generating Synthetic Versions of Sensitive Microdata for Statistical Disclosure Control (2021). https://CRAN.R-project.org/package=synthpop, R package version 1.7-0
22. Pejó, B.: Guide to Differential Privacy Modifications: A Taxonomy of Variants and Extensions. Springer Briefs in Computer Science Serries. Springer International Publishing AG, Cham (2022). https://doi.org/10.1007/978-3-030-96398-9_12
23. Raab, G., Nowok, B., Dibben, C.: Practical data synthesis for large samples. J. Priv. Confidentiality **7**, 67–97 (2017). https://journalprivacyconfidentiality.org/index.php/jpc/article/view/407
24. Raab, G.M., Nowok, B., Dibben, C.: Assessing, visualizing and improving the utility of synthetic data. Available as a vignette for the Synthpop package at https://cran.r-project.org/web/packages/synthpop/vignettes/utility.pdf. Accessed 1 May 2022
25. Raghunathan, T.E., Reiter, J.P., Rubin, D.B.: Multiple imputation for statistical disclosure limitation. J. Off. Stat. **19**(1), 1–17 (2003)
26. Rubin, D.B.: Discussion: Statistical disclosure limitation. J. Off. Stat. **9**(2), 461–468 (1993)
27. Shlomo, N.: Integrating differential privacy in the statistical disclosure control toolkit for synthetic data production. In: Domingo-Ferrer, J., Muralidhar, K. (eds.) PSD 2020. LNCS, vol. 12276, pp. 271–280. Springer, Cham (2020). https://doi.org/10.1007/978-3-030-57521-2_19
28. Snoke, J., Raab, G., Nowok, B., Dibben, C., Slavkovic, A.: General and specific utility measures for synthetic data. J. Roy. Stat. Soc. Ser. A **181**(3), 663–688 (2018)
29. Taub, J., Elliot, M., Pampaka, M., Smith, D.: Differential correct attribution probability for synthetic data: an exploration. In: Domingo-Ferrer, J., Montes, F. (eds.) PSD 2018. LNCS, vol. 11126, pp. 122–137. Springer, Cham (2018). https://doi.org/10.1007/978-3-319-99771-1_9
30. Voas, D., Williamson, P.: Evaluating goodness-of-fit measures for synthetic microdata. Geog. Environ. Model. **5**, 177–200 (2001)
31. Zhang, J., Cormode, G., Procopiuc, C., Srivastava, D., Xiao, X.: PrivBayes: private data release via Bbayesian networks. In: Proceedings of the 2014 ACM SIGMOD International Conference on Management of Data, SIGMOD 2014, pp. 1423–1434. ACM (2014)

Machine Learning and Privacy

Membership Inference Attack Against Principal Component Analysis

Oualid Zari[1]([✉]), Javier Parra-Arnau[2,3], Ayşe Ünsal[1], Thorsten Strufe[2], and Melek Önen[1]

[1] EURECOM, Biot, France
oualid.zari@eurecom.fr
[2] Karlsruhe Institute of Technology, Karlsruhe, Germany
[3] Universitat Politècnica de Catalunya, Barcelona, Spain

Abstract. This paper studies the performance of membership inference attacks against *principal component analysis* (PCA). In this attack, we assume that the adversary has access to the principal components, and her main goal is to infer whether a given data sample was used to compute these principal components. We show that our attack is successful and achieves high performance when the number of samples used to compute the principal components is small. As a defense strategy, we investigate the use of various differentially private mechanisms. Accordingly, we present experimental results on the performance of Gaussian and Laplace mechanisms under *naive* and *advanced compositions* against MIA as well as the utility of these differentially-private PCA solutions.

Keywords: Membership inference attack · Principal component analysis · Differential privacy · Laplace mechanism · Gaussian mechanism

1 Introduction

Over the past decade, machine learning (ML) algorithms have found application in a vast and rapidly growing number of systems for analyzing and classifying usually privacy-sensitive data.

In order to analyze and interpret such data, PCA [18] is employed as one of the most commonly used unsupervised ML algorithm. PCA is used for summarizing the information content in databases by reducing the dimensionality of the data while preserving as much variability as possible. The output of this statistical tool is a set of *principal components* whose size is usually much smaller than the total number of attributes of the underlying data.

The increasing popularity of ML algorithms, including PCA, opened the door for attackers especially when ML techniques are deployed in critical applications. This work focuses on a particular type of attack named *Membership Inference Attack* (MIA) against PCA, where an adversary is assumed to intercept the principal components computed over some dataset and infer whether a data

© Springer Nature Switzerland AG 2022
J. Domingo-Ferrer and M. Laurent (Eds.): PSD 2022, LNCS 13463, pp. 269–282, 2022.
https://doi.org/10.1007/978-3-031-13945-1_19

sample was part of this dataset or not. The membership prediction is yield by comparing the reconstruction error; the distance between the original target sample and its PCA projection against a threshold. In this paper, we study the effectiveness of MIA against PCA and show that it achieves high performance when the number of samples used by PCA is small.

Furthermore, to cope with such attacks that take advantage of the leakage of principal components, we propose to study the use of *differentially private* mechanisms and evaluate the privacy budget affects the success rate of the attack as well as the utility of the PCA under differential privacy (DP).

Our main contributions are summarized as follows.

1. We study, for the first time, the impact of MIA against PCA whereby the adversary has access to the principal components.
2. We propose the use of differentially-private PCA algorithms to cope with MIA and analyze the impact of the privacy budget on both utility and the success rate of MIA for both vector and scalar queries under the *so-called* naive and advanced composition approaches.
3. The experimental results present a comparison between the aforementioned different approaches under Gaussian and Laplace mechanisms for protecting the PCA against MIA.

2   Background

2.1   Principal Component Analysis

Given a set $D = \{x_n \in \mathbb{R}^d : n = 1 : N\}$ of N raw data samples corresponding to N individuals of dimension d, we denote the data matrix where each column is a data sample by $X = [x_1, \ldots, x_N]$. We assume that data X has zero mean, which can be ensured by centering the data. The standard PCA algorithm is to find a k–dimensional subspace that approximates each sample x_n. This problem can be formulated as follows:

$$\min_{\Pi_k} \mathcal{L} = \frac{1}{N} \sum_{n=1}^{N} \mathcal{L}_n = \frac{1}{N} \sum_{n=1}^{N} \|x_n - \Pi_k x_n\|_2^2 \tag{1}$$

where $\mathcal{L}$ denotes the average reconstruction error and Π_k is an orthogonal projector which is used for approximating each sample x_n by $\hat{x}_n = \Pi_k x_n$. The solution to this problem can be achieved via singular value decomposition (SVD) of the sample covariance matrix, which is defined by $A = \frac{1}{N} X X^T = \frac{1}{N} \sum_{n=1}^{N} x_n x_n^T$. A is a symmetric positive semi-definite matrix, hence its singular value decomposition is equivalent to its spectral decomposition. SVD of A yields $A = \sum_{i=1}^{d} \lambda_i v_i v_i^T$, where $\lambda_1 \geq \lambda_2 \geq \ldots, \lambda_d \geq 0$ and $v_1, v_2, \ldots, v_d$ denote the eigenvalues and their corresponding eigenvectors of A, respectively. Let us denote the matrix whose columns are the top k eigenvectors by $V_k = [v_1, \ldots, v_k]$. The orthogonal projector $\Pi_k = V_k V_k^T$ is a solution to the problem in (1). PCA uses V_k to project the samples into the low k–dimensional subspace $Y = V_k^T X$.

2.2 Membership Inference Attacks

The goal of an MIA is to infer whether or not a target sample is included in the training dataset. When an adversary learns whether or not a target sample was used to release any statistics or to train a machine learning model, this refers to an information leakage. This attack could cause serious problems in terms of privacy if the training dataset contains privacy-sensitive information. An example that highlights the implications of such an attack is [7], which was able to identify individuals contributing their DNA to a health-related project.

3   Related Work

Since the introduction of MIA against deep neural network (DNN) models in [22], this attack has been extensively studied on DNNs and other ML models. The cited work formalized the attack as a binary classification problem and trained neural network (NN) classifiers to distinguish between training members and non-members. The authors demonstrates that the main factor contributing to the success of MIA on DNN models is overfitting. Subsequent works [13,15,21,23,27] further developed MIAs with different approaches against DNN of different architectures. The work in [23] revealed that by using suitable metrics, metric-based attacks result in similar attack performance when compared with NN-based attacks. Besides DNN, MIAs have also been investigated against logistic regression models [20,25], k-nearest neighbors [24,25], and decision tree models [25,27]. Our work extends the investigations of MIAs against machine learning models to PCA. As we shall elaborate later in Sect. 4.1, we propose, to this end, a new metric-based MIA against PCA. To the best of our knowledge, there is no previous work trying to perform MIA on PCA.

To mitigate MIAs, DP has been widely applied to various ML models [12,13,26,28]. In [1], the authors show how to train DNNs with DP by adding noise to the gradients or parameters during model training. In [19], the authors empirically evaluate MIAs using the proposal of [1]. They find that DP can partially mitigate the attack with an acceptable level of privacy budget. In our study, we investigate the effectiveness of DP PCA algorithms on mitigating our proposed attack.

4   Membership Inference Attacks Against PCA

The first part of this work focuses on the study of the impact of MIA targeting PCA. We aim to investigate how the sample size and the number of the intercepted principle components affect the performance of such attacks. In Sect. 4.1, we define the threat model and the actual MIA targeting PCA. This is followed by the experimental setup and the corresponding experimental results of Sects. 4.2 and 4.3, respectively.

4.1   Threat Model and Attack Methodology

In our setting, the curator computes the principal components V_k using the training dataset D, and sends these to a trusted party. We assume the adversary $\mathcal{A}$ intercepts some or all of those components by eavesdropping the communication channel. With them, the adversary aims to identify whether or not a certain sample z is included in D. In other words, the adversary's goal is to discover members of the training dataset.

Such an attack can of course occur in a distributed setting [2] where several parties may compute the principal components of their individual (and usually smaller [10]) training datasets and send those to an aggregator, which ultimately may compute the global principal components. Analogously to the non-distributed case, here $\mathcal{A}$ would compromise individual privacy by intercepting the principal components conveyed by each party.

To identify whether or not sample z was actually used for the computation of the principal components, $\mathcal{A}$ computes the reconstruction error $\mathcal{L}(z, V_k)$ of the target sample z based on the intercepted V_k, and then compares this error with some tunable decision threshold R. If the reconstruction error of the target sample is lower than the threshold, $\mathcal{A}$ predicts that z is a member of the training dataset D. Otherwise, $\mathcal{A}$ predicts that z is not a member of D. Our intuition is that samples from the training dataset are more likely to incur lower reconstruction error compared to other non-member samples.

4.2   Experimental Setup

We proceed with a detailed description of the datasets used in our experiments.

Datasets:[1] We assess the performance of the attack using two groups of datasets: (i) datasets including personal information, namely, UCI Adult [16] (for short, Adult), Census [4], and LFW [8]; and the image dataset MNIST [14], which is typically used in the literature of MIAs. As preprocessing, we standardize the datasets to unit variance before constructing our attack.

- The UCI Adult dataset includes 48,842 records with 14 attributes. It contains both numerical (e.g. age, hours per week, etc.) and categorical (e.g. working class, education, etc.) attributes. We employ the standard one-hot encoding approach to construct the numerical representation of the categorical attributes [9].
- Census: it contains 1080 records with 13 attributes of business statistics.
- Labeled Faces in the Wild (LFW): It includes 13,233 images of 5749 human faces collected from the Web. 1680 of the 5749 people pictured have at least two distinct images in this dataset. The resolution of the images is 25×18. In our evaluation, in order to balance the number of samples for each individual, we only take one picture of each individual in the dataset.

---

[1] Due to page limit constraint, we report only the results for Adult and LFW datasets. We refer the reader to the full version of this paper [29].

- MNIST: it includes 10 classes of handwritten digits formatted as grayscale 28×28 pixel images. The dataset is used to predict the class of the digit represented in the image. The total number of samples is 70,000.

Performance Metric: As an evaluation metric of the attack's success, we use the area under the receiver operating characteristic(ROC) curve (AUC) metric, which indicates the relationship between true positive and false-negative rates over several decision thresholds R that the adversary can use to construct the attack. In all experiments, we choose equal-sized samples for both members and non-members at random and report the mean of the results over 10 trials.

4.3   Experimental Results

We evaluate the success rate of the attack in terms of the number of principal components intercepted by the adversary, denoted by k. For this, we measure the attacker's performance through the AUC. Figure 1 shows the maximum AUC that the adversary can achieve by observing the top-k principal components. Recall that k may take values from 1 to d, where d is the number of attributes of the dataset. We report results for various number of samples N. The closer the AUC is to 0.5, the less successful the attack is as the adversary cannot distinguish between a member and a non-member.

We observe that the AUC increases with increasing k. This is justified by the fact that the attacker has access to more information and therefore is more likely to succeed in identifying the membership. We also observe that the AUC decreases with increasing N, perhaps, indicating that the sample covariance matrix A converges to the true covariance matrix of the dataset, which renders the reconstruction error of member and non-member samples of D indistinguishable. The same behaviour is observed with NNs when the training dataset is large [22].

The results for the MNIST and LFW datasets indicate that the AUC is always greater than 0.5 and reaches 0.9 when $N = 1,000$. As for the Census and Adult datasets, the corresponding AUC values are much lower (compared to the other datasets). This is mainly justified by the small dimension d of these datasets. We note that MIA against machine learning models trained using the Adult dataset is usually unsuccessful [21,22].

5   Differentially-Private PCA and MIA

In this section, we present PCA(DP-PCA) algorithms introduced in [6,17], and study their protection against MIAs with various privacy budget values and their utility. Accordingly, we first remind several preliminaries DP in Sect. 5.1. This is followed by the experimental results in Sect. 5.3.

5.1   Preliminaries on Differential Privacy

Definition 1 (Neighboring datasets). *Any two datasets that differ in one record are called neighbors. For two neighbor datasets* $\mathbf{x}$ *and* $\mathbf{x}'$*, the following equality holds:*

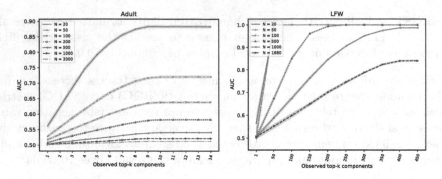

Fig. 1. Impact of the sample size N and the observed top-k components on the attack's performance. Shaded areas show 95% confidence intervals for the mean.

$$d(\mathbf{x}, \mathbf{x}') = 1,$$

where d denotes the Hamming distance.

Definition 2 ((ε, δ)-Differential privacy [5]). *A randomized mechanism $\mathcal{M}$ on a query function f satisfies ε-DP with $\varepsilon, \delta \geqslant 0$ if, for all pairs of neighbor databases $\mathbf{x}, \mathbf{x}'$ and for all $\mathcal{O} \subseteq \text{range}(\mathcal{M})$,*

$$\mathrm{P}\{\mathcal{M}(f(\mathbf{x})) \in \mathcal{O}\} \leqslant e^{\varepsilon}\, \mathrm{P}\{\mathcal{M}(f(\mathbf{x}')) \in \mathcal{O}\} + \delta.$$

We say that $\mathcal{M}$ satisfies *pure* DP if $\delta = 0$, and *approximate* DP otherwise.

Definition 3 (L_p-global sensitivity [5]). *Let $\mathcal{D}$ be the class of possible data sets. The L_p-global sensitivity of a query function $f \colon \mathcal{D} \to \mathbb{R}^d$ is defined as*

$$\Delta_p(f) = \max_{\forall \mathbf{x}, \mathbf{x}' \in \mathcal{D}} \|f(\mathbf{x}) - f(\mathbf{x}')\|_p,$$

where $\mathbf{x}, \mathbf{x}'$ are any two neighbor datasets.

Definition 4 (Laplace mechanism [5]). *Given any function $f \colon \mathcal{D} \to \mathbb{R}^d$, the Laplace mechanism mechanism is defined as follows:*

$$\mathcal{M}_L(\mathbf{x}, f(\cdot), \varepsilon) = f(\mathbf{x}) + (Y_1, \ldots, Y_d),$$

where Y_i are i.i.d. random variables drawn from a Laplace distribution with zero mean and scale $\Delta_1(f)/\varepsilon$.

Definition 5 (Gaussian mechanism [5]). *Given any function $f \colon \mathcal{D} \to \mathbb{R}^d$, the Gaussian mechanism mechanism is defined as follows:*

$$\mathcal{M}_G(\mathbf{x}, f(\cdot), \varepsilon) = f(\mathbf{x}) + (Y_1, \ldots, Y_d),$$

where Y_i are i.i.d. random variables drawn from a Gaussian distribution with zero mean and standard deviation $\Delta_2(f)\sqrt{2 \log(1.25/\delta)}/\varepsilon$.

Theorem 1 ([5]). *The Laplace mechanism satisfies $(\varepsilon, 0)$-DP.*

Theorem 2 ([5]). *For any $\varepsilon, \delta \in (0, 1)$, the Gaussian mechanism satisfies (ε, δ)-DP.*

Theorem 3 ([5]). *If each mechanism $\mathcal{M}_i$ in a k-fold adaptive composition $\mathcal{M}_1, \ldots, \mathcal{M}_k$ satisfies (ε', δ')-DP for $\varepsilon', \delta' \geq 0$, then the entire k-fold adaptive composition satisfies $(\varepsilon, k\delta' + \delta)$-DP for $\delta \geq 0$ and*

$$\varepsilon = \sqrt{2k \ln(1/\delta)}\varepsilon' + k\varepsilon'(e^{\varepsilon'} - 1). \tag{2}$$

5.2 Differentially Private PCA Approaches

As in the previous scenario where no privacy protection was implemented, the first step for the data curator is to compute the principal components of the covariance matrix A, which are to be shared with a trusted entity. However, to protect individual privacy against an adversary who may intercept some or all components of A, the curator now decides adding Laplace noise directly on the coefficients q_{ij} of A. In the context of DP, this approach is called *output perturbation*.

To protect the $\alpha \doteq d(d + 1)/2$ distinct[2] coefficients of A, we consider two strategies: (i) using a *joint* query function that simultaneously queries all such coefficients, and (ii) querying each coefficient *separately*. We shall refer to these procedures as *vector* and *scalar* queries, respectively.

For $i = 1, \ldots, d$, let attribute i take values in the interval $[l_i, u_i]$ after standardization, and denote by Λ_i the absolute difference $|l_i - u_i|$. Recall [17] that $\Delta_1(q_{ij}) = \Lambda_i \Lambda_j / N$, from which we can easily derive an upper bound on $\Delta_1(A)$ just by adding up the sensitivities of all distinct coefficients. Accordingly, the scale of the Laplace noise injected to each coefficient yields $\Delta_1(A)/\varepsilon$ in the vector case, and $\Delta_1(q_{ij})/\varepsilon_{ij}$ in the scalar case, where ε is the total privacy budget and ε_{ij} the fraction thereof assigned to the coefficient q_{ij}.

Using the standard sequential composition property, we can compute the total privacy cost of the scalar strategy by adding up all ε_{ij} for $i \geq j$. In our experiments, in order to compare the two approaches for a same total privacy budget, we shall assume $\varepsilon_{ij} = \varepsilon/\alpha$. Note that, in this case, the noise scales will coincide only if $\sum_{i \geq j} \Lambda_i \Lambda_j = \alpha \Lambda_i \Lambda_j$.

We shall also consider a variation of the scalar case that relies on the advanced (sequential) composition property. Notice that even though this property is defined in the context of approximate DP, Theorem 3 also applies if the mechanisms being composed satisfy pure ε-DP. With advanced composition, however, the total privacy cost can be estimated more tightly (compared to the standard property) when the number of coefficients is significantly large. Said otherwise, for the same privacy budget ε (and small δ) and for large α, the scale of the noise introduced with advanced composition can be reduced notably with respect to

---

[2] Recall that A is a symmetric matrix.

that injected with standard sequential composition. The noise scale yields in this case $\Lambda_i\Lambda_j/N\varepsilon'$, where ε' satisfies Eq. (2) for $k = \alpha$ and a given total privacy budget ε, δ.

Finally, the fourth protection approach we shall use in our experimental evaluation guarantees approximate DP through the *Gaussian* mechanism. More specifically, the algorithm in [6] queries all coefficients of A simultaneously and estimates $\Delta_2(A)$ to be $1/N$; the sensitivity bound follows after normalizing D so that each row has at most unit l_2 norm. Accordingly, the scale of the noise added to each coefficient yields $\sqrt{2\log(1.25/\delta)}/N\varepsilon$. Table 1 summarizes the four protection mechanisms we shall evaluate in the next subsection.

Table 1. Overview of the DP mechanisms aimed to protect PCA against MIA. Here, ε denotes the *total* privacy budget and ε' the *fraction* thereof assigned to each coefficient of A.

Approach	Privacy notion	noise scale
Laplace scalar query with naive composition	DP	$\alpha\Lambda_i\Lambda_j/N\varepsilon$
Laplace vector query	DP	$\sum_{i\geqslant j}\Lambda_i\Lambda_j/N\varepsilon$
Laplace scalar query with advanced composition	approx. DP	$\Lambda_i\Lambda_j/N\varepsilon'$
Analyze Gauss (AG) Algorithm [6]	approx. DP	$\sqrt{2\log(1.25/\delta)}/N\varepsilon$

5.3   Experimental Results

We first study the protection of DP mechanisms against our attack. Therefore, we implement the four aforementioned approaches and evaluate the AUC of the attack with various privacy budgets ε, ranging from 10^{-2} to 10^8. We would like to notice that this is not the usual range of values used in the literature. For example, in privacy-preserving data publishing, values of ϵ above 3 progressively seem to lose any meaningful guarantees [3]. However, for us, the fact that we will be using such large values is irrelevant, since we will empirically measure privacy leakage *not* through the ε itself, but through the effectiveness of an MIA. Finally, at the end of this section, we study the utility of the protected data provided by such approaches.

DP Mechanisms and AUC. Figure 2 shows the performance of the attack with respect to the k observed principal components when AG and Laplace vector query algorithms are used with various values of ε. In the case of the AG algorithm, ε varies from 0.01 to 1 and δ is set to $\delta = \frac{1}{N}$ whereas for the Laplace vector query algorithm, we select larger values of ε from 10^{-1} to 10^7. We also present the AUC of the attack in the non-private setting where DP-mechanisms are not adopted. Under AG, we observe that for all values of ε, the AUC of the attack is only marginally above 0.5 (random guess baseline). Hence, the AG algorithm mitigates the effectiveness of MIA. With larger ε values under the Laplace vector query approach, AUC starts to increase and gets closer to the non-private case. We also observe that for the Adult and Census datasets, for

$\varepsilon = 10^2$ the Laplacian vector query approach provides roughly the same level of protection than AG for $\varepsilon = 1$. For the LFW dataset, for $\varepsilon = 10^4$ the Laplace vector query approach provides the same protection as AG with $\varepsilon = 1$. Hence, even with a higher privacy budget ε, the Laplace vector query approach limits the success of the attack.

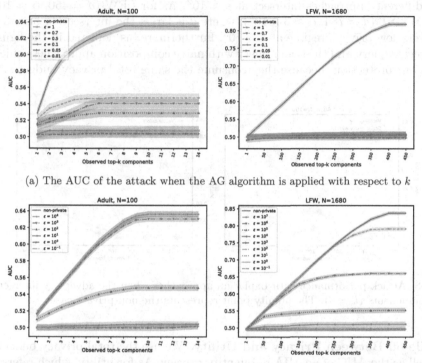

(a) The AUC of the attack when the AG algorithm is applied with respect to k

(b) The AUC of the attack when the Laplace vector query algorithm is applied w.r.t. k.

Fig. 2. The AUC of the attack when the AG algorithm (a) and the Laplace vector query approach (b) are applied with various values of ε. Shaded areas are the 95% confidence intervals for the mean.

Laplacian Approaches. Figure 3 compares the protection of the aforementioned Laplacian approaches for various levels of the total privacy budget based on the maximum AUC of the attack. We observe that the advanced composition approach achieves better protection than the naïve one in the low privacy regime (when ε is large). This observation can be explained through the noise scales injected by the two approaches. From Sect. 5.2, it is easy to verify that the algorithm based on the advanced composition will introduce less noise than that relying on the naive composition when $\varepsilon' < \varepsilon/\alpha$. In Fig. 4, we plot in the hashed area the set of points (ε, α) where this inequality holds. From the figure, we can see that, for a fixed α, increasing ε will ultimately result in less noise for naive composition. And the other way round, for a fixed ε, increasing the number of coefficients will, at some point, make the advanced mechanism introduce less noise. We thus justify the observation above by assuming that adding more noise leads to stronger protection against the MIA.

Specifically, for the Adult ($d = 14$, $\alpha = 105$) and Census ($d = 13$, $\alpha = 91$) datasets, where the dimension is relatively small, the AUCs corresponding to the two different approaches intersect at $\varepsilon \approx 10^2$. As for LFW ($d = 450$, $\alpha \approx 10^5$) and MNIST ($d = 784$, $\alpha \approx 3 \times 10^5$), where d is large, the intersection occurs at the very low privacy regime at $\varepsilon \approx 10^5$. Furthermore, as depicted in the figure, the vector query and the scalar query with naive composition approaches achieve the same protection, because they consume the same total privacy budget ε.

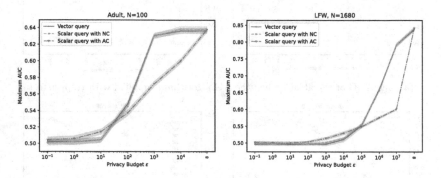

Fig. 3. Attack performance with Laplacian approaches when the adversary intercepts all components ($k = d$). The infinity point represents the non-private case

Trade-off Between Privacy and Utility. We use the total privacy budget ε as well as the AUC of an MIA to quantify privacy. As for utility, which refers to the accuracy of the principal components produced by the DP-PCA algorithms of Sect. 5.2, we adopt the metric introduced in [11]. In particular, we compute the percentage of captured energy of the principal components produced by

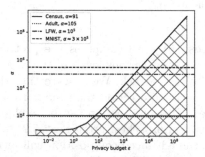

Fig. 4. The hashed area shows where naive composition introduces less noise than advanced composition.

those algorithms, $\hat{V}_k$, with respect to the principal components of non-private PCA (SVD), V_k. Accordingly, we measure utility as $q = \frac{tr(\hat{V}_k^T A \hat{V}_k)}{tr(V_k^T A V_k)}$, where A is the sample covariance matrix. We note that, for all the datasets, we select the reduced dimension k such that V_k have the captured energy of 90%.

Figure 5 and 6 show the utility of the DP-PCA algorithms as a function of the privacy budget ε, and of the AUC, respectively. We observe that the AG algorithm offers good utility for the Adult and Census datasets. However, AG has a low utility for the other datasets. The Laplacian PCA solutions show lower utility in comparison with AG for $\varepsilon \leq 1$. The vector and scalar query with naive composition approaches show almost the same utility, except for the MNIST and Census datasets, where the scalar query with naive composition achieves better utility than the vector query approach. Advanced composition provides better utility than the naive composition where ε and α are in the blank area of Fig. 4. In summary, the utility of the DP-PCA algorithms is influenced by the amount of noise added, as one would expect.

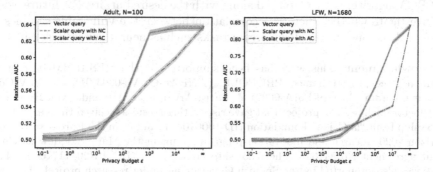

Fig. 5. Trade-off posed by the four DP-PCA algorithms described in Sect. 5.2, between the total privacy budget ε and data utility. Utility is measured as the percentage of captured energy w.r.t. SVD.

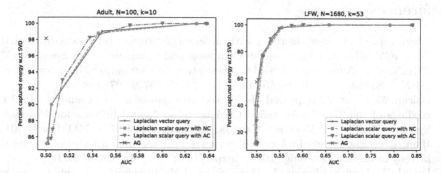

Fig. 6. Trade-off posed by the four DP-PCA algorithms described Sect. 5.2, between attack performance and data utility. We measure attack performance through AUC, and utility through the percentage of captured energy w.r.t. SVD.

On the other hand, the vector query approach outperforms the scalar query approach if the sensitivity of the coefficients is skewed. In order to enjoy better utility, the scalar query approach with advanced composition should be used rather than with naive composition when the privacy budget ε and the number of queries α are in the blank area of Fig. 4.

6   Conclusion

In this paper, we have implemented and evaluated the first membership inference attack against PCA, whereby an adversary has access to some or all principal components. Our attack sheds light on privacy leakage in PCA. Specifically, we have demonstrated that an MIA can be deployed successfully, with high performance, when the number of samples used by PCA is small. We have evaluated the protection of DP-PCA under different protection algorithms, privacy budgets, number of principal components intercepted, and number of covariance coefficients. Our work may be useful to assess the practical value of privacy when DP-PCA algorithms are employed along with the desired utility. For future work, to investigate whether there is a correlation between the vulnerable samples in PCA and the ones in the downstream tasks such as neural network classifiers.

Acknowledgment. This work has been supported by the MESRI-BMBF French-German joint project named PROPOLIS (ANR-20-CYAL-0004-01), the 3IA Côte d'Azur program (ANR19-P3IA-0002). J. Parra-Arnau is an Alexander von Humboldt postdoctoral fellow. The project that gave rise to these results received the support of a fellowship from "la Caixa" Foundation (ID 100010434) and from the European Union's Horizon 2020 research and innovation programme under the Marie Skłodowska-Curie grant agreement No 847648. The fellowship code is LCF/BQ/PR20/11770009. This work was also supported by the Spanish Government under research project "Enhancing Communication Protocols with Machine Learning while Protecting Sensitive Data (COMPROMISE)" (PID2020-113795RB-C31/AEI/10.13039/501100011033).

References

1. Abadi, M., et al.: Deep learning with differential privacy. In: Proceedings of the 2016 ACM SIGSAC Conference on Computer and Communications Security, CCS 2016, Vienna, Austria, pp. 308–318. Association for Computing Machinery (2016). ISBN: 9781450341394. https://doi.org/10.1145/2976749.2978318
2. Balcan, M.-F., et al.: Improved distributed principal component analysis. In: Proceedings of the 27th International Conference on Neural Information Processing Systems, NIPS 2014, Montreal, Canada, vol. 2, pp. 3113–3121. MIT Press (2014)
3. Blanco-Justicia, A., et al.: A critical review on the use (and misuse) of differential privacy in machine learning (2022). https://arxiv.org/abs/2206.04621
4. Brand, R., Domingo-Ferrer, J., Mateo-Sanz, J.M.: Reference data sets to test and compare SDC methods for protection of numerical microdata. Technical report. https://research.cbs.nl/casc/CASCrefmicrodata.pdf
5. Dwork, C., Roth, A.: The algorithmic foundations of differential privacy. Found. Trends Theor. Comput. Sci. **9**(3–4), 211–407 (2014). ISSN: 1551-305X

6. Dwork, C., et al.: Analyze gauss: optimal bounds for privacy-preserving principal component analysis. In: Proceedings of the Forty-Sixth Annual ACM Symposium on Theory of Computing, STOC 2014, pp. 11–20. Association for Computing Machinery, New York (2014). ISBN: 9781450327107. https://doi.org/10.1145/2591796.2591883
7. Homer, N., et al.: Resolving individuals contributing trace amounts of DNA to highly complex mixtures using high-density SNP genotyping microarrays. PLoS Genet. **4**, e1000167 (2008)
8. Huang, G.B., et al.: Labeled faces in the wild: a database for studying face recognition in unconstrained environments. Technical report 07-49, University of Massachusetts, Amherst, October 2007
9. Hundepool, A., et al.: Statistical Disclosure Control (2012). Ed. by S. Fischer-Hübner et al.
10. Imtiaz, H., Sarwate, A.D.: Differentially private distributed principal component analysis. In: 2018 IEEE International Conference on Acoustics, Speech and Signal Processing (ICASSP), pp. 2206–2210 (2018). https://doi.org/10.1109/ICASSP.2018.8462519
11. Imtiaz, H., Sarwate, A.D.: Symmetric matrix perturbation for differentially-private principal component analysis. In: 2016 IEEE International Conference on Acoustics, Speech and Signal Processing (ICASSP), pp. 2339–2343, March 2016. https://doi.org/10.1109/ICASSP.2016.7472095
12. Jayaraman, B., Evans, D.E.: Evaluating differentially private machine learning in practice. In: USENIX Security Symposium (2019)
13. Jayaraman, B., et al.: Revisiting membership inference under realistic assumptions. In: Proceedings on Privacy Enhancing Technologies 2021, pp. 348–368 (2021)
14. LeCun, Y., Cortes, C.: MNIST handwritten digit database (2010). http://yann.lecun.com/exdb/mnist/
15. Long, Y., et al.: Understanding membership inferences on well-generalized learning models. ArXiv, abs/1802.04889 (2018)
16. Blake, C.L., Newman, D.J., Merz, C.J.: UCI repository of machine learning databases (1998). http://www.ics.uci.edu/~mlearn/MLRepository.html
17. Parra-Arnau, J., Domingo-Ferrer, J., Soria-Comas, J.: Differentially private data publishing via cross-moment microaggregation. Inf. Fusion **53**, 269–288 (2020). ISSN: 1566-2535
18. Pearson, K.: On lines and planes of closest fit to systems of points in space. Philos. Mag. **2**(11), 559–572 (1901)
19. Rahman, M.A., et al.: Membership inference attack against differentially private deep learning model. Trans. Data Priv. **11**, 61–79 (2018)
20. Sablayrolles, A.: White-box vs black-box: bayes optimal strategies for membership inference. In: ICML (2019)
21. Salem, A., et al.: ML-Leaks: model and data independent membership inference attacks and defenses on machine learning models. CoRR, abs/1806.01246 (2018). http://arxiv.org/abs/1806.01246
22. Shokri, R., Stronati, M., Shmatikov, V.: Membership inference attacks against machine learning models. CoRR, abs/1610.05820 (2016). http://arxiv.org/abs/1610.05820
23. Song, L., Shokri, R., Mittal, P.: Membership inference attacks against adversarially robust deep learning models. In: 2019 IEEE Security and Privacy Workshops (SPW), pp. 50–56 (2019)
24. Tramèr, F., et al.: Truth serum: poisoning machine learning models to reveal their secrets. ArXiv, abs/2204.00032 (2022)

25. Truex, S., et al.: Demystifying membership inference attacks in machine learning as a service. IEEE Trans. Serv. Comput. **14**, 2073–2089 (2021)
26. Truex, S., et al.: Effects of differential privacy and data skewness on membership inference vulnerability. In: 2019 First IEEE International Conference on Trust, Privacy and Security in Intelligent Systems and Applications (TPS-ISA), pp. 82–91 (2019)
27. Yeom, S., et al.: Privacy risk in machine learning: analyzing the connection to overfitting. In: 2018 IEEE 31st Computer Security Foundations Symposium (CSF), pp. 268–282 (2018)
28. Ying, Z., Zhang, Y., Liu, X.: Privacy-preserving in defending against membership inference attacks. In: Proceedings of the 2020 Workshop on Privacy-Preserving Machine Learning in Practice (2020)
29. Zari, O., et al.: Membership inference attack against principal component analysis (2022). https://www.eurecom.fr/index.php/en/publication/6913

When Machine Learning Models Leak: An Exploration of Synthetic Training Data

Manel Slokom[1,2,3]([✉]), Peter-Paul de Wolf[2], and Martha Larson[3]

[1] Delft University of Technology, Delft, The Netherlands
m.slokom@tudelft.nl
[2] Statistics Netherlands, The Hague, The Netherlands
pp.dewolf@cbs.nl
[3] Radboud University, Nijmegen, The Netherlands
m.larson@cs.ru.nl

Abstract. We investigate an attack on a machine learning classifier that predicts the propensity of a person or household to move (i.e., relocate) in the next two years. The attack assumes that the classifier has been made publically available and that the attacker has access to information about a certain number of target individuals. That attacker might also have information about another set of people to train an auxiliary classifier. We show that the attack is possible for target individuals independently of whether they were contained in the original training set of the classifier. However, the attack is somewhat less successful for individuals that were not contained in the original data. Based on this observation, we investigate whether training the classifier on a data set that is synthesized from the original training data, rather than using the original training data directly, would help to mitigate the effectiveness of the attack. Our experimental results show that it does not, leading us to conclude that new approaches to data synthesis must be developed if synthesized data is to resemble "unseen" individuals to an extent great enough to help to block machine learning model attacks.

Keywords: Synthetic data · Propensity to move · Attribute inference · Machine learning

1 Introduction

Governmental institutions charged with collecting and disseminating information may use machine learning models to produce estimates, such as imputing missing values or inferring variables that cannot be directly observed. When such estimates are published, it is also useful to publish the machine learning model itself, so that researchers using the estimates can evaluate it closely, or

---

The views expressed in this paper are those of the authors and do not necessarily reflect the policy of Statistics Netherlands.

© Springer Nature Switzerland AG 2022
J. Domingo-Ferrer and M. Laurent (Eds.): PSD 2022, LNCS 13463, pp. 283–296, 2022.
https://doi.org/10.1007/978-3-031-13945-1_20

even produce their own estimates. Moreover, society also asks for more insight into the models that are used, e.g., to address possible discrimination caused by decisions based on machine learning models.

Unfortunately, machine learning models can be attacked in a way that allows an attacker to recover information about the data set that they were trained on [19]. For this reason, publishing machine learning models can lead to a risk that information in the training set is leaked. In this paper, we carry out a case study of an attribute inference attack on a machine learning classifier to better understand the nature of the risk. The classifier that we study predicts *propensity to move*, i.e., whether an individual or household will relocate their home within the next two years. The attack scenario assumes that the classifier has been released to the public, and that an attacker wishes to learn a sensitive attribute for a group of victims, i.e., target individuals. The attacker has non-sensitive information about these target individuals that is used for the attack and has scraped information about other people from the Web.

Our experimental investigation first confirms that a machine learning classifier is able to predict propensity to move for individuals in its training data set as well as for previously "unseen" individuals, reproducing [3]. We then attack this classifier and demonstrate that an attacker can learn sensitive attributes both for individuals in the training data as well as for previously "unseen" individuals. However, for "unseen" individuals the attack is somewhat less successful. We reason that data synthesis might potentially allow us to create data that we could use for training and that would be far enough from the original data, than any real individual would have the somewhat higher resistance to attack of an "unseen" individual. Based on this idea, we create a synthetic training set, train a machine learning classifier on that set, and repeat the attacks. Interestingly, the resulting classifier is just as susceptible to attack as the original classifier, which was trained on the original data. We relate this finding to the success of an attack that infers sensitive information from individuals using priors and not the machine learning model. Our findings point to the direction that future research must pursue in order to create synthetic data that could reduce the risk of attack when used to train machine learning models.

2   Threat Model

Our goal is to test whether a machine learning model trained on synthetic data can replace a machine learning model trained on original data. The idea is to release a machine learning model trained on synthetic data such that there is no leak of original data. The synthetic data serves as a replacement of the original data. In this section, we specify our goal more formally in the form of a threat model.

Inspired by [23], a threat model follows three main dimensions. First, the threat model describes the adversary by looking at the resources at the adversary's disposal and the adversary's objective. In other words, it specifies what the attacker is capable of and what the attacker's goal is. Second, it describes the vulnerability, including the opportunity that makes an attack possible. Then, the threat model specifies the nature of the countermeasures that can be taken to prevent the attack.

Table 1 provides the specifications of our threat model for each of the dimensions. As resources, we assume that the attacker has access to our released machine learning classifier. In addition to the ML model, the attacker has a subset of the data that is used to train an attacker model. The adversary's objective is to infer sensitive information about individuals. In our experiments, the attack model is trained using subset of data in addition to the released machine learning model that predicts propensity-to-move. The opportunity for attack is the possession of original data including sensitive attributes. Finally, the countermeasure that we are investigating is data synthesis.

Table 1. Threat model addressed by our approach

Component	Description
Adversary: Objective	Specific attributes about individuals
Adversary: Resources	The attacker has access to the released classifier and has a subset of data
Vulnerability: Opportunity	Possession of original data and inference of individuals' sensitive data
Countermeasure	Make access to original data and model unreliable

3   Background and Related Work

In this section, we give a brief overview on basic concepts and related work on predicting the propensity to move, on privacy in machine learning, and model inversion attribute inference attack.

3.1   Propensity to Move

The propensity to move is defined as desires, expectations, or plans to move to another dwelling [5]. Multiple factors come to play to understand and estimate the propensity to move in a population. In [5], the authors have grouped those factors into two categories: (1) *Residential satisfaction* which is defined as the satisfaction with the dwelling and its location or surroundings. Residential satisfaction is divided into housing satisfaction and neighborhood satisfaction. (2) *Household characteristics* which is related to demographic and socioeconomic characteristics of the household. The gender and age are indicators of a household are important demographic attributes. For instance, a male household has different mobility patterns than a female household. Also, education and income of the household are important socioeconomic attributes.

In [10], authors investigated the possible relationship between involuntary job loss and regional mobility. In a survey, the German socio-economic panel [10] looked at whether job loss increases the probability to relocate to a different region and whether displaced workers who relocate to another region after job loss have better labor market outcomes than those staying in the same area. They found that job loss has a strong positive effect on the propensity to relocate. In [17], the authors examined the residential moving behavior of older adults in the Netherlands. [17] used a data collected from the Housing Research Netherlands (HRN) to provide insights into the housing situation of the Dutch population and their living needs. A logistic regression model was used to assess the likelihood that respondents would report that they are willing to move in the upcoming two years. Among their key findings, they showed that older adults with a propensity to move are more often motivated by unsatisfactory conditions in the current neighborhood. Further results revealed that older adults are more likely to have moved to areas with little deprivation, little nuisance, and a high level of cohesion.

In [3], the authors studied the possibility of replacing a survey question about moving desires by a model-based prediction. To do so, they used machine learning algorithms to predict moving behavior from register data. The results showed that the models are able to predict the moving behavior about equally well as the respondents of the survey. In [4], the authors used data collected by the British Household Panel Survey. The data is conducted using a face to face interviews. They examined the reasons why people desire to move and how these desires affect their moving behavior. The results show that the reasons people report for desiring to move vary considerably over the life course. People are more likely to relocate if they desire to move for targeted reasons like job opportunities than if they desire to move for more diffuse reasons relating to area characteristics. In [18], the authors studied the social capital and propensity to move of four different resident categories in two Dutch restructured neighborhoods. They defined social capital as the benefit of cursory interactions, trust, shared norms, and collective action. Using a logistic regression model, they showed that (1) age, length of residency, employment, income, dwelling satisfaction, dwelling type and perceived neighborhood quality significantly predict residents' propensity to move and (2) social capital is of less importance than suggested by previous research.

3.2   Privacy in Machine Learning

In this section, we will discuss challenges and possible solutions in privacy preserving techniques. Existing works can be divided into three categories according to the roles of machine learning (ML) in privacy [19]: First, *making the ML model private*. This category includes making ML model (its parameters) and data private. Second, *using ML to enhance privacy protection*. In this category, the ML is used as a tool to enhance privacy protection of the data. Third, *ML based privacy attack*. The ML model is used as an attack tool of the attacker.

Based on the threat model, both data and the prediction model are important. Predicting and estimating the propensity to move requires access to models as well as to data. However, since the propensity to move data contains sensitive data such as income, gender, age, education level, the data is treated as sensitive and once collected from individuals it cannot be shared with third parties.

One possible solution is to generate synthetic data that captures the distribution of the original data and generates artificial, but yet realistic data. The synthetic data offers a replacement for the original data to enable model training, model validation and model explanation. In order to attempt to protect the machine learning model before release or sharing, we propose to train our model on the synthetic data instead of the original data. The goal is to test whether it is possible to release a machine learning model trained on synthetic data without leaking sensitive information.

Synthetic data generation is based on two main steps: First, we train a model to learn the joint probability distribution in the original data. Second, we generate a new artificial data set from the same learned distribution. In recent years, advances in machine learning and deep learning models have offered us the possibility to learn a wide range of data types.

Synthetic data was first proposed for Statistical Disclosure Control (SDC) [8]. The SDC literature distinguishes between two types of synthetic data [8]. First, *fully synthetic data sets* create an entirely synthetic data based on the original data set. Second, *partially synthetic data sets* contain a mix of original and synthetic values. It replaces only observed values for variables that bear a high risk of disclosure with synthetic values. In this paper, we are interested in fully synthetic data. For data synthesize, we used an open source and widely used R toolkit: *Synthpop*. We used a CART model for synthesize since it has been shown to perform well for other type of data [9]. Data synthesis is based on sequential modeling by decomposing a multidimensional joint distribution into conditional and univariate distributions. In other words, the synthesis procedure models and generates one variable at a time, conditionally to previous variables:

$$f_{x_1,x_2,..,x_n} = f_{x_1} \times f_{x_2|x_1} \times .. \times f_{x_n|x_1,x_2,..x_{n-1}} \tag{1}$$

Synthesis using CART model has two important parameters. First, the order in which variables are synthesized called *visiting.sequence*. This parameter has an important impact on the quality of the synthetic data since it specifies the order in which the conditional synthesize will be applied. Second, the *stopping rules* that dictate the number of observations that are assigned to a node in the tree.

3.3   Attribute Inference Attack

Privacy attacks in machine learning [6,22] include membership inference attacks [24], model reconstruction attacks such as attribute inference [29], model inversion attacks [11,12], and model extraction attacks [28]. Here, we focus on a form of model inversion attacks, namely, attribute inference attack.

Model inversion attacks try to recover sensitive features or the full data sample based on output labels and partial knowledge (subset of data) of some features [1,22]. [1] provided a summary of possible assumptions about adversary capabilities and resources for different model inversion attribute inference attacks. In [11,12], the authors introduced two types of model inversion attacks: Black-box attack and white-box attack. The difference between black-box attack and white-box attack lies in the amount of resources that are available for the adversary. In [1], the authors proposed two types of model inversion attacks: (1) confidence score-based model inversion attack and (2) label-only model inversion attack. The first attack assumed that the adversary has access to the target model's confidence scores, whereas the second assumed that the adversary has access to the target model's label predictions only. Other attacks such as [14] assumed that the attacker does not have access to target individuals non-sensitive features.

Attribute Inference Attack. An attribute inference attack or attribute disclosure occurs if an attacker is able to learn new information about a specific individual, i.e., the values of certain attributes. Examples from the Statistical Disclosure Control (SDC) literature include [8,16].

Here, we study attribute inference attack as prediction. An attacker trains a model to predict the value of an unknown sensitive attribute from a set of known attributes given access to raw or synthetic data [15,25]. We implemented our attribute inference attack using *adversarial robustness toolbox*[1]. In order to perform an attribute inference attack, we assume that the attacker has access to a subset of data, a marginal prior distribution representing possible values for the sensitive features in the training data, and the released ML model's predictions. Using this resources, an attacker is able to train a model to learn sensitive information. This attack is called black-box attack because the predictions of the model, but not the architecture or the weights are available to the attacker. Further details about our black-box attack will be discussed in Sect. 4.3.

In addition to black-box attack, we use two other attack models as baselines for comparison, namely, *random attack* and *baseline attack*. Both attacks assume that the attacker does not have access to the released ML model. First, the random attack has only access to the marginal prior distribution of the sensitive feature that is being targeted. Our random attack uses random classifier with a stratified strategy, i.e., it generates random predictions that respect the class distribution of the training data. Second, the baseline attack also access to the prior distribution of the sensitive feature. However, in addition it also uses a ML model, i.e., a random forest classifier, to infer sensitive attributes. Recall that only the black-box attack is related to our threat model defined in Sect. 2. The random and baseline attacks provide comparative conditions, which the black-box attack must outperform.

Measuring Success of Inference. Prior work on synthetic data disclosure risk [26] looked at either matching probability by comparing perceived match risk, expected match risk, and true match risk [20], or Bayesian estimation approach by assuming that an attacker seeks a Bayesian posteriori distribution [21]. In this paper, our black-box attack is considered successful if its accuracy outperforms the accuracy of a random attack. In other words, we assume that going beyond a random guess, can reveal sensitive information about individuals. This type of

---

[1] https://github.com/Trusted-AI/adversarial-robustness-toolbox.

measurement is similar to previous work on model inversion attribute inference attacks [11,12,14], which measure the difference between the adversary's predictive accuracy given the model and the best, i.e., ideal, accuracy that could be achieved without the model [29]. Methods for measurements of success are discussed in [2], who also covers the precise or probabilistic measures conventionally used in the SDC community, i.e., using matching or Bayesian estimate.

4   Experimental Setup

In this section, we describe our data sets, utility measures measured by applying different machine learning algorithms, and adversary resources.

4.1   Data Set

For our experiments, we used an existing data about someone's propensity-to-move. The data was collected by [3]. [3] linked several registers from the Dutch System of Social Statistical Datasets (SSD). The data set has around 150K individuals including 100K individuals drawn randomly from register data and 50K individuals are sampled from the Housing Survey 2015 (HS2015) respondents. The resulting data set has used in [3] has 700 variables containing for each individual: (1) "y01" the binary target variable indicating whether ($=1$) or not ($=0$) a person moved in year j where j = 2013, 2015. The target attribute "y01" is imbalanced and dominated by class 0. (2) time independent personal variables, (3) time dependent personal, household, and housing variables, (4) information about regional variables.

Feature Selection. Different from [3], we applied feature selection to reduce the number of features. Some features can be noise and potentially reduce the performance of the models. Also, reducing number of feature helps to reduce the complexity of synthesize and to better understand the output of the ML model. To do so, we applied *SelectKBest* from *Sklearn*[2]. We use chi2 method as a scoring function. We selected top $K = 30$ features with the highest scores. Our final data set contains 30 best features for a total of 150K individuals[3]. In addition to the 30 features, we added gender (binary), income (categorical with five categories), and age (categorical with seven categories) as sensitive features that will be used in our attribute inference attack later (Sect. 5.2). Gender, age, and income have balanced classes. Similar to [3], we found that the most important features are age (lft), time since latest change in household composition (inhehalgr3), and time since latest move or number of moves (rinobjectnummer).

---

[2] https://scikit-learn.org/stable/modules/generated/sklearn.feature_selection.Select KBest.html.

[3] We note that reducing the number of features does not have an impact on the success rate of the attack because there is a redundancy in some variables since they go until 17 years back [3].

Data Splits. As mentioned earlier, our propensity to move data was collected in 2013 and 2015. Following [3], we use the 2013 data to train our classifier and the 2015 data to test the classifier and to carry out the attacks. The 2015 data contains individuals who were present in the 2013 data set, and also new individuals. We split the 2015 data set into two parts "original individuals" (inclusive) and "new in 2015 individuals" (exclusive) in order to test our classifier and our attacks on individuals who were in the training set but also in the also on "unseen individuals".

4.2   Utility Measures

Machine Learning Algorithms. We selected a number of machine learning algorithms to predict propensity-to-move. The chosen machine learning techniques provide insight into the importance of the features and are easy to interpret and understand [3].

In our experiments in Sect. 5.1, we used: *decision tree* where a tree is created/ learned by splitting the source set into subsets based on an attribute value test. This process is repeated on each derived subset in a recursive manner. Extra trees and random forest are part of ensemble methods. In *random forest*, each tree in the ensemble is built from a sample drawn with replacement (i.e., a bootstrap sample) from the training set. *Extra trees* fits a number of randomized decision trees on various sub-samples of the data set and uses averaging to improve the predictive accuracy and control overfitting. *Naive Bayes* is a probabilistic machine learning algorithm based on applying Bayes' theorem with strong (naive) independence assumptions between the features. *KNN*, K-nearest neighbors, is a non-parametric machine learning algorithm. KNN uses proximity to make predictions about the grouping of an individual data point.

Metrics for Evaluating Performance of ML Models. Similar to [3] and since our target propensity-to-move attribute is imbalanced, we used: F1-score, as a harmonic mean of precision and recall score. Matthews Correlation Coefficient (MCC), and Area Under the Curve (AUC) that measures the ability of a classifier to distinguish between classes.

4.3   Adversary Resources

In Sect. 3.3, we provided description of our attack models. The attacker is interested to infer target individual sensitive features. Below, we briefly discuss different attack models used in our experiments along with different resources that are available for the attacker.

- *Random attack*: uses a subset of data and marginal prior distribution.
- *Baseline attack*: uses a subset of data, marginal prior distribution, and random forest classifier.
- *Black-box attack*: uses a subset of data, marginal prior distribution, released ML model, and random forest classifier.

In random attack model, a random classifier[4] randomly infers target individual's sensitive features i.e., gender, age, income. In baseline attack model, a random forest classifier[5] is trained on a subset of data and marginal prior distribution to predict sensitive features. Last but not least, a black-box attack model has access to the released ML model's predictions, in addition to having access to subset of data and marginal prior distribution. Then, a random forest classifier is trained to infer target individual's sensitive features.

Understanding the vulnerability of a model to attribute inference attack requires using right metric to evaluate different attack models. Since our sensitive target features (gender, age, income) are balanced [11], we used precision, recall to measure the effectiveness of the attacks. Precision measures the ability of the classifier not to label as positive a sample that is negative. Precision is the ratio of $tp/(tp + fp)$ where tp is the number of true positives and fp the number of false positives. Recall measures the ability of the classifier to find all the positive samples. Recall is the ratio of $tp/(tp + fn)$ where tp is the number of true positives and fn the number of false negatives. We also measure accuracy which is defined as the fraction of predictions that our classifier got right.

5 Experimental Results

Now, that we have defined our threat model including the adversary resources and capabilities, and utility measures to evaluate the quality of synthetic data and machine learning algorithms, we turn to discuss our experimental results.

5.1 Evaluation of Machine Learning Algorithms

Table 2 shows our results of classification performance of propensity to move, and confirms the results of [3]. As expected, all classifiers outperform the random baseline, with classifiers using trees generally the stronger performers. We also see that when the test set includes only individuals already present in the training set (*inclusive*), the performance is better than when it includes only "unseen" individuals (*exclusive*). Note that if the data for the *inclusive* individuals were identical in the training and test set, we would have expected very high classification scores. However, the data is not identical because it was collected on two different occasions with two years intervening, and individuals' situations would presumably have changed.

Reproducing Burger et al.,'s [3] *results* In Table 2, results show that all machine learning classifiers outperform random classifier. Overall we observe that our results are in line with [3] across different metrics. This confirms that we can still predict individuals moving behavior in the same level as in [3] even after reducing number of features.

---

[4] Random Classifier using Stratified strategy from https://scikit-learn.org/stable/modules/generated/sklearn.dummy.DummyClassifier.html.

[5] Random Forest Classifier: https://scikit-learn.org/stable/modules/generated/sklearn.ensemble.RandomForestClassifier.html.

Table 2. Classification performance of propensity-to-move measured in terms of AUC, MCC, and F1-score on **original data** and **synthetic data**. (Right) the data splitting is similar to [3]. The training set individuals and test set individuals are inclusive. (Left) A different data splitting where we train the model on individuals data from 2013, then, we test the model on different individuals from 2015.

Machine learning algorithms		Training and test individuals are exclusive			Training and test individuals are inclusive		
		AUC	MCC	F1-score	AUC	MCC	F1-score
Original data	Random	0.4962	−0.0105	0.2139	0.5014	0.0029	0.1633
	NaiveBayes	0.5656	−0.0328	0.5491	0.6815	0.2204	0.2992
	RandomForest	0.7061	0.3210	0.6322	0.7532	0.3121	0.4460
	DecisionTree	0.6372	0.2692	0.5376	0.6568	0.2292	0.3057
	ExtraTrees	0.7226	0.3197	0.6325	0.7597	0.3212	0.4525
	KNN	0.6304	0.2074	0.4104	0.6717	0.1744	0.2235
Synthetic data	Random	0.4991	−0.025	0.2261	0.5011	0.0022	0.1657
	NaiveBayes	0.5658	0.045	0.5451	0.6822	0.2029	0.2578
	RandomForest	0.7053	0.3282	0.6343	0.7467	0.3133	0.4471
	DecisionTree	0.6489	0.2598	0.4878	0.6618	0.2125	0.3078
	ExtraTrees	0.7188	0.3185	0.6321	0.7557	0.3138	0.4464
	KNN	0.6067	0.1152	0.1857	0.6542	0.1637	0.2070

In addition to reproducing [3], we looked at another prediction model where train and test individuals are exclusive/different. We found that it is also possible to predict moving behavior of new individuals from 2015 based on a classifier trained on different individuals from 2013.

Measuring the Utility of Synthetic Data. In order to evaluate the quality of synthetic data, we run machine learning algorithms on synthesized training set (2013 data). we used $TSTR$ [13] evaluation strategy where we train classifiers on 2013 synthetically generated data and we test on 2015 original data. Results in Table 2 show that the performance of machine learning algorithms trained on synthetic data is very close and comparable to the performance of machine learning algorithms trained on original data. This confirms that the synthetic training set can replace the original training set. In the remainder of the paper, we will focus on decision tree model. We will assume that we are releasing a decision tree model.

5.2   Model Inversion Attribute Inference Attack

In this section, we present the results of our experiments on attribute inference attack using the three attack models: (1) random attack, (2) baseline attack, (3) black-box attack (Sect. 4.3). Recall that we assume that the adversary can have access to three different subsets of data (Sect. 2).

1. **Inclusive individuals (2013):** the attacker has access to a subset of the data that is used from 2013 to train the released machine learning algorithm.
2. **Inclusive individuals (2015):** the attacker has access to a more recent subset of data from 2015, but for the same set of individuals that are used to train the released machine learning algorithm.
3. **Exclusive individuals (2015):** the attacker has access to a recent subset of data from 2015, but the individuals are different from individuals that are used to train the released machine learning algorithm.

Table 3 shows results of different attribute inference attacks for three type of sensitive features gender, age and income. We notice that attack always achieves better than random scores, which demonstrates the viability of the attack.

Table 3. Results of model inversion attribute inference attacks. Adversary resources can be either: **Inclusive individuals (2013)**, **Inclusive individuals (2015)**, or **Exclusive individuals (2015)**. ± represents the standard deviation over ten times of running the experiments. Numbers in gray represent the best inference results across conditions. Note that only black-box attack is related to threat model described in Sect. 2. An attack is considered successful if its score is higher than a score of random attack.

Adversary Resources	Released ML	Attack Models	Gender Accuracy	Gender Precision	Gender Recall	Age Accuracy	Age Precision	Age Recall	Income Accuracy	Income Precision	Income Recall
Inclusive individuals (2013)	Original	Random	0.500 ±0.00	0.500 ±0.00	0.500 ±0.00	0.1095 ±0.00	0.1129 ±0.00	0.1112 ±0.00	0.2086 ±0.00	0.2060 ±0.00	0.2077 ±0.00
		Baseline	0.6107 ±0.007	0.6103 ±0.007	0.6104 ±0.007	0.1472 ±0.003	0.1566 ±0.003	0.1407 ±0.001	0.1483 ±0.005	0.1590 ±0.005	0.2323 ±0.006
		Black-Box	0.6187 ±0.005	0.6181 ±0.005	0.6183 ±0.005	0.1482 ±0.004	0.1577 ±0.004	0.1412 ±0.001	0.1469 ±0.004	0.1576 ±0.005	0.2302 ±0.006
	Synthetic	Random	0.500 ±0.00	0.500 ±0.00	0.500 ±0.00	0.1164 ±0.00	0.1223 ±0.00	0.1213 ±0.00	0.1838 ±0.00	0.1889 ±0.00	0.1983 ±0.00
		Baseline	0.6262 ±0.00	0.6263 ±0.006	0.6264 ±0.006	0.1562 ±0.004	0.1561 ±0.004	0.1412 ±0.001	0.1509 ±0.003	0.1575 ±0.003	0.2189 ±0.004
		Black-Box	0.6298 ±0.005	0.6299 ±0.005	0.6300 ±0.005	0.1562 ±0.003	0.1561 ±0.003	0.1412 ±0.001	0.1492 ±0.003	0.1553 ±0.004	0.2182 ±0.006
Inclusive individuals (2015)	Original	Random	0.500 ±0.00	0.500 ±0.00	0.500 ±0.00	0.1095 ±0.00	0.1129 ±0.00	0.1112 ±0.00	0.2086 ±0.00	0.2060 ±0.00	0.2077 ±0.00
		Baseline	0.6240 ±0.006	0.6228 ±0.006	0.6227 ±0.00	0.1552 ±0.003	0.1590 ±0.003	0.1467 ±0.001	0.1502 ±0.004	0.1552 ±0.004	0.2327 ±0.007
		Black-Box	0.6235 ±0.009	0.6226 ±0.009	0.6223 ±0.009	0.1547 ±0.003	0.1585 ±0.003	0.1463 ±0.001	0.1545 ±0.003	0.1599 ±0.003	0.2428 ±0.005
	Synthetic	Random	0.500 ±0.00	0.500 ±0.00	0.500 ±0.00	0.1164 ±0.00	0.1223 ±0.00	0.1213 ±0.00	0.1838 ±0.00	0.1889 ±0.00	0.1983 ±0.00
		Baseline	0.6186 ±0.006	0.6188 ±0.006	0.6186 ±0.006	0.1657 ±0.003	0.1606 ±0.003	0.1465 ±0.001	0.1620 ±0.003	0.1592 ±0.003	0.2169 ±0.005
		Black-Box	0.6236 ±0.006	0.6237 ±0.006	0.6236 ±0.006	0.1646 ±0.003	0.1595 ±0.003	0.1456 ±0.001	0.1626 ±0.003	0.1596 ±0.003	0.2259 ±0.006
Exclusive individuals (2015)	Original	Random	0.500 ±0.00	0.500 ±0.00	0.500 ±0.00	0.1095 ±0.00	0.1129 ±0.00	0.1112 ±0.00	0.2086 ±0.00	0.2060 ±0.00	0.2077 ±0.00
		Baseline	0.5269 ±0.009	0.5198 ±0.009	0.5201 ±0.009	0.0830 ±0.001	0.2116 ±0.005	0.1279 ±0.001	0.0829 ±0.003	0.1779 ±0.008	0.2182 ±0.02
		Black-Box	0.5272 ±0.005	0.5195 ±0.005	0.5199 ±0.005	0.0817 ±0.001	0.2100 ±0.005	0.1280 ±0.001	0.0804 ±0.003	0.1693 ±0.008	0.2283 ±0.02
	Synthetic	Random	0.500 ±0.00	0.500 ±0.00	0.500 ±0.00	0.1164 ±0.00	0.1223 ±0.00	0.1213 ±0.00	0.1838 ±0.00	0.1889 ±0.00	0.1983 ±0.00
		Baseline	0.5268 ±0.009	0.5198 ±0.009	0.5201 ±0.009	0.0825 ±0.001	0.2116 ±0.005	0.1279 ±0.001	0.0829 ±0.003	0.1779 ±0.008	0.2182 ±0.02
		Black-Box	0.5272 ±0.005	0.5195 ±0.005	0.5198 ±0.005	0.0817 ±0.001	0.2100 ±0.005	0.1280 ±0.001	0.0804 ±0.003	0.1693 ±0.008	0.2283 ±0.02

Comparing the row "Original" for the three individuals sets and across all three sets of sensitive attributes (columns), we see that the attack is less successful for the "Exclusive" individuals who were unseen in the training data of the classifier. This fact might lead us to wonder whether training the classifier on synthetic data might lead to less successful attacks, since the individuals in the training data would be in some way "different" with the target individuals. This, however, turns out not to be the case. Comparing the row "Synthetic" for the three individuals sets and across all three sets of sensitive attributes (columns), we see that if the training data is synthesized using the original training data, the model is just as susceptible to attack as when trained on the original data. This point is less surprising when we take into account the high success of the "Random" attack. This attack recovers sensitive attributes of individuals without access to the trained machine learning model. Instead, priors are used. We assume that the information of the priors is also retained in the trained model. These results demonstrate the magnitude of the challenge that we face, if we wish to release a trained machine learning model publically.

6   Conclusion and Future Work

In this paper, we have investigated an attack on a machine learning model trained to predict individual's propensity-to-move i.e., in the next two years. for individuals in the training data as well as for "unseen" individuals. However, we observed that for "unseen" individuals, the attribute inference attack is somewhat less successful. This result is consistent with the training data used to train ML model having a different distribution than the "unseen" individuals.

To explore the ability of synthetic data to protect against attribute inference attack, we created fully synthetic data using CART model. The ML model trained on synthetic data maintained prediction performance, but was found to leak in the same way as the original classifier. This result is not particularly surprising. Synthetic data mimics properties of the original data including overall structure, correlation between features, and the joint distributions [25].

Our results is interesting because until now The SDC community working with synthetic data has mainly focused on measuring the risk of identity disclosure rather than attribute disclosure [26]. In the identity disclosure literature, synthetic data has been shown to provide protection [7,27].

Our work draws attention to the fact a lot of work is still needed to protect against attribute disclosure [2]. A potential solution to protect against attribute inference attack is to apply privacy-preserving techniques during synthesis, e.g., data perturbation or masking sensitive attributes. Also, it would be interesting to explore different combinations of ML and conventional models to synthesize and carry out attribute attacks. From an evaluation perspective, future work should look at other metrics [15] (e.g., from SDC and/or ML perspective) to evaluate and quantify the success of attribute inference attack for a given target individual. Finally, future research should expand the threat model that we have adopted in this research (Sect. 2) and other attack scenarios in which the attacker has access to more limited resources, e.g., assuming that attacker does not have access to all attributes in data.

References

1. Mehnaz, S., Dibbo, S.V., Kabir, E., Li, N., Bertino, E.: Are your sensitive attributes private? Novel model inversion attribute inference attacks on classification models. In: 31st USENIX Security Symposium (USENIX Security), Boston, MA. USENIX Association (2022)
2. Andreou, A., Goga, O., Loiseau, P.: Identity vs. attribute disclosure risks for users with multiple social profiles. In: Proceedings of the IEEE/ACM International Conference on Advances in Social Networks Analysis and Mining, pp. 163–170. ASONAM (2017)
3. Burger, J., Buelens, B., de Jong, T., Gootzen, Y.: Replacing a survey question by predictive modeling using register data. In: ISI World Statistics Congress, pp. 1–6 (2019)
4. Coulter, R., Scott, J.: What motivates residential mobility? Re-examining self-reported reasons for desiring and making residential moves. Popul. Space Place 21(4), 354–371 (2015)
5. Crull, S.R.: Residential satisfaction, propensity to move, and residential mobility: a causal model. In: Digital Repository at Iowa State University (1979). http://lib.dr.iastate.edu/
6. De Cristofaro, E.: A critical overview of privacy in machine learning. IEEE Secur. Priv. 19(4), 19–27 (2021)
7. Domingo-Ferrer, J.: A survey of inference control methods for privacy-preserving data mining. In: Aggarwal, C.C., Yu, P.S. (eds.) Privacy-Preserving Data Mining, pp. 53–80. Springer, Boston (2008). https://doi.org/10.1007/978-0-387-70992-5_3
8. Drechsler, J.: Synthetic Datasets for Statistical Disclosure Control: Theory and Implementation, vol. 201. Springer, New York (2011). https://doi.org/10.1007/978-1-4614-0326-5
9. Drechsler, J., Reiter, J.P.: An empirical evaluation of easily implemented, non-parametric methods for generating synthetic datasets. Comput. Stat. Data Anal. 55(12), 3232–3243 (2011)
10. Fackler, D., Rippe, L.: Losing work, moving away? Regional mobility after job loss. Labour 31(4), 457–479 (2017)
11. Fredrikson, M., Jha, S., Ristenpart, T.: Model inversion attacks that exploit confidence information and basic countermeasures. In: Proceedings of the 22nd ACM Conference on Computer and Communications Security (SIGSAC), CCS 2015, pp. 1322–1333 (2015)
12. Fredrikson, M., Lantz, E., Jha, S., Lin, S., Page, D., Ristenpart, T.: Privacy in pharmacogenetics: an end-to-end case study of personalized warfarin dosing. In: 23rd USENIX Security Symposium (USENIX Security), San Diego, CA, pp. 17–32. USENIX Association (2014)
13. Heyburn, R., et al.: Machine learning using synthetic and real data: similarity of evaluation metrics for different healthcare datasets and for different algorithms. In: Data Science and Knowledge Engineering for Sensing Decision Support: Proceedings of the 13th International FLINS Conference, pp. 1281–1291. World Scientific (2018)

14. Hidano, S., Murakami, T., Katsumata, S., Kiyomoto, S., Hanaoka, G.: Model inversion attacks for prediction systems: without knowledge of non-sensitive attributes. In: 15th Annual Conference on Privacy, Security and Trust (PST), pp. 115–11509. IEEE (2017)
15. Hittmeir, M., Mayer, R., Ekelhart, A.: A baseline for attribute disclosure risk in synthetic data. In: Proceedings of the 10th ACM Conference on Data and Application Security and Privacy, pp. 133–143 (2020)
16. Hundepool, A., et al.: Statistical Disclosure Control. Wiley, Hoboken (2012)
17. de Jong, P.A.: Later-life migration in the Netherlands: propensity to move and residential mobility. J. Aging Environ. **36**, 1–10 (2020)
18. Kleinhans, R.: Does social capital affect residents' propensity to move from restructured neighbourhoods? Hous. Stud. **24**(5), 629–651 (2009)
19. Liu, B., Ding, M., Shaham, S., Rahayu, W., Farokhi, F., Lin, Z.: When machine learning meets privacy: a survey and outlook. ACM Comput. Surv. **54**(2), 1–36 (2021)
20. Reiter, J.P., Mitra, R.: Estimating risks of identification disclosure in partially synthetic data. J. Priv. Confid. **1**(1) (2009)
21. Reiter, J.P., Wang, Q., Zhang, B.: Bayesian estimation of disclosure risks for multiply imputed, synthetic data. J. Priv. Confid. **6**(1) (2014)
22. Rigaki, M., Garcia, S.: A survey of privacy attacks in machine learning. arXiv preprint arXiv:2007.07646 (2020)
23. Salter, C., Saydjari, O.S., Schneier, B., Wallner, J.: Toward a secure system engineering methodology. In: Proceedings of the 1998 Workshop on New Security Paradigms, pp. 2–10. NSPW (1998)
24. Shokri, R., Stronati, M., Song, C., Shmatikov, V.: Membership inference attacks against machine learning models. In: Symposium on Security and Privacy (SP), pp. 3–18. IEEE (2017)
25. Stadler, T., Oprisanu, B., Troncoso, C.: Synthetic data-anonymisation groundhog day. In: 29th USENIX Security Symposium (USENIX Security). USENIX Association (2020)
26. Taub, J., Elliot, M., Pampaka, M., Smith, D.: Differential correct attribution probability for synthetic data: an exploration. In: Domingo-Ferrer, J., Montes, F. (eds.) PSD 2018. LNCS, vol. 11126, pp. 122–137. Springer, Cham (2018). https://doi.org/10.1007/978-3-319-99771-1_9
27. Templ, M.: Statistical Disclosure Control for Microdata: Methods and Applications in R. Springer, Cham (2017). https://doi.org/10.1007/978-3-319-50272-4
28. Tramèr, F., Zhang, F., Juels, A., Reiter, M.K., Ristenpart, T.: Stealing machine learning models via prediction APIs. In: 25th USENIX Security Symposium (USENIX Security 2016), Austin, TX, pp. 601–618. USENIX Association (2016)
29. Yeom, S., Giacomelli, I., Fredrikson, M., Jha, S.: Privacy risk in machine learning: analyzing the connection to overfitting. In: 31st Computer Security Foundations Symposium (CSF), pp. 268–282. IEEE (2018)

Case Studies

A Note on the Misinterpretation
of the US Census Re-identification Attack

Paul Francis[✉]

Max Planck Institute for Software Systems (MPI-SWS), Saarbrücken and
Kaiserslautern, Germany
francis@mpi-sws.org

Abstract. In 2018, the US Census Bureau designed a new data reconstruction and re-identification attack and tested it against their 2010 data
release. The specific attack executed by the Bureau allows an attacker
to infer the race and ethnicity of respondents with average 75% precision for 85% of the respondents, assuming that the attacker knows the
correct age, sex, and address of the respondents. They interpreted the
attack as exceeding the Bureau's privacy standards, and so introduced
stronger privacy protections for the 2020 Census in the form of the Top-
Down Algorithm (TDA).

This paper demonstrates that race and ethnicity can be inferred *from
the TDA-protected census data* with substantially *better* precision and
recall, using *less* prior knowledge: only the respondents' address. Race
and ethnicity can be inferred with average 75% precision for 98% of
the respondents, and can be inferred with 100% precision for 11% of
the respondents. The inference is done by simply assuming that the
race/ethnicity of the respondent is that of the majority race/ethnicity
for the respondent's census block.

We argue that the conclusion to draw from this simple demonstration
is NOT that the Bureau's data releases lack adequate privacy protections. Indeed it is the Bureau's stated purpose of the data releases to
allow this kind of inference. The problem, rather, is that the Bureau's
criteria for measuring privacy is flawed and overly pessimistic. There is
no compelling evidence that TDA was necessary in the first place.

1 Introduction

The US Census Bureau releases privacy-protected statistics from the decennial
census. In past decades, this data was protected using aggregation and swapping:
occasionally exchanging an individual response from one geographic area, or
block, with that in another block.

In 2019, the US Census Bureau reported on a new re-identification
attack, developed by the Bureau, against these traditional *swap-protected* data
releases [3]. The attack was demonstrated on the 2010 release. The Bureau considered the attack serious enough that they developed a new privacy protection

© Springer Nature Switzerland AG 2022
J. Domingo-Ferrer and M. Laurent (Eds.): PSD 2022, LNCS 13463, pp. 299–311, 2022.
https://doi.org/10.1007/978-3-031-13945-1_21

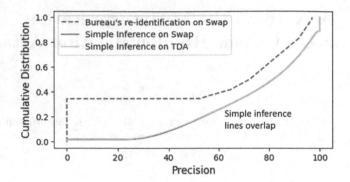

Fig. 1. CDF of precision for our simple inference "non-attack" run on the US Census Bureau's new disclosure protection mechanism (TDA) as well as the Bureau's prior mechanism (Swap). For comparison, the precision for the Bureau's re-identification attack on the prior swap mechanism is also shown. Points to the lower-right mean a more effective "attack". Note that effectiveness of the non-attack on swap-protected data is virtually identical to that of the TDA-protected data (lines overlap).

method. Called the Top-Down Algorithm (TDA), the new method uses aggregation and noise addition: perturbing counts with random noise from a normal distribution [5]. The 2020 census release is *TDA-protected*.

The Bureau also prepared a TDA-protected release of the 2010 census so that stakeholders could evaluate data quality[1].

The specific re-identification attack demonstrated by the Bureau has two parts. First they reconstruct the original data from the swap-protected data. The reconstructed attributes are block, age, sex, race, and ethnicity (Hispanic or not). Next they link externally-derived data (address, age, and sex) with the reconstructed data to infer race and ethnicity of the re-identified respondents. The re-identification attack achieved 75% precision (75% of race/ethnicity inferences were correct assuming correct prior knowledge).

This paper demonstrates that race/ethnicity inferences with *better precision and recall* can be made against the 2010 *TDA-protected release* using *less prior knowledge* (only address instead of address, age, and sex). Our demonstration yields better than 95% precision for 23% of respondents, and virtually 100% precision for 11% of respondents (see Table 1).

Our demonstration operates by merely predicting that the race/ethnicity of any given respondent is that of the majority race for the corresponding block[2]. In Sect. 4.1, we argue that it is in fact the intention of the Bureau that the majority race/ethnicity of any block can be accurately inferred. If this is so, then our demonstration is not an attack at all. Rather, it simply utilizes the

---

[1] https://www.census.gov/programs-surveys/decennial-census/decade/2020/planning-management/process/disclosure-avoidance/2020-das-development.html.
[2] An idea borrowed from Ruggles et al. [13].

Table 1. Summary of the Bureau's attack and our simple inference non-attack. The reported precision and recall for the re-identification attack are for all blocks, blocks with between 10 and 49 respondents, and blocks with between 1 and 9 respondents respectively. The reported precision and recall for the non-attack are limited to blocks where the majority race has at least 5 persons.

	Bureau's re-identification attack	Our simple inference "non-attack"
Released data	2010 swap-protected release	2010 TDA-protected release
Prior knowledge	Address, age, sex	Address
Inferred information	Race and ethnicity	Race and ethnicity
Linking attributes	Address/block, age, sex	Address/block
Mechanism	Constraint-solver reconstruction	Simple table lookup
Ground truth	Census' internal raw data	2010 swap-protected release
Precision/Recall	P = 75%, R = 85% (all)	P = 75%, R = 98%
(block size)	P = 92%, R = 17% (10-49)	P >= 95%, R = 23%
	P = 97%, R = 1.5% (1-9)	P = 100%, R = 11%

statistical inferences that census data is supposed to enable. As such, we refer to our demonstration as a simple inference *non-attack*.

In 2018, Ruggles et al. [12] argued that the Bureau's reconstruction attack is not particularly effective, and that TDA is not necessary. In 2021, Ruggles and Van Riper [13] simulated a simple statistical random reconstruction from national-level statistics, and showed that it can be roughly as effective as the Bureau's reconstruction attack (see Sect. 4.3).

The contribution of this paper is that it is a much more concrete demonstration of the mismeasure of the Bureau's attack, for two reasons. First, our non-attack uses the 2010 census data as high-quality ground truth. By contrast, Ruggles and Van Riper use simulated data. Second, our non-attack runs on the TDA-protected data itself. This demonstrates directly that either the Bureau does not intend to protect against this inference, or that TDA fails to provide the intended protections.

We argue that it is the former. Indeed it is important to note that the Bureau as far as we know has not run its own reconstruction attack against the TDA-protected release. We suspect that doing so would yield results similar to the same attack on the 2010 swap-protected release.

Finally, note that this paper is intentionally narrow in scope. It pertains only to inferring race/ethnicity. Other types of inferences (i.e. age) would not have similarly high precision. Likewise we say nothing about other reconstruction attacks that may exist and may be more effective than that demonstrated by the Bureau. Finally, we make no recommendations as to how the Bureau may better define its privacy measures and criteria.

Section 2 describes the Bureau's re-identification attack in more detail. Section 3 describes our non-attack. Section 4 explores the question of whether our non-attack represents a meaningful privacy loss or (more likely) not.

The code and data for our non-attack may be found at https://gitlab.mpi-sws.org/francis/census-misinterpretation.

2   The Bureau's Re-identification Attack

The Bureau's re-identification attack, as well as our non-attack, combine *prior knowledge* with *released data* to infer the race and ethnicity of a *target* individual (the census respondent). Table 1 summarizes both attacks.

The re-identification attack works as follows (Fig. 2):

1. Derive a set of constraints from the swap-protected release (per block).
2. Use a constraint solver to reconstruct the original data per block. The resulting reconstructed records are pseudonymous (not linked to identified persons).
3. Link the reconstructed records to externally-derived prior knowledge data on attributes shared by the reconstructed data and the prior knowledge data. This serves to identify the persons in the reconstructed data.
4. Infer the unknown attributes from the so-identified reconstructed data.

A good overview of the re-identification attack and its results can be found in Abowd [4]. A general description of the constraint solver approach can be found in Garfinkel et al. [6].

In the attack demonstrated by the Bureau, the reconstructed data consisted of attributes *block, age, sex, race, and ethnicity*. Blocks are geographical areas ranging from zero persons to several thousand persons. There are 6M blocks in the USA. There are two ethnicity values, *Hispanic* and *Not Hispanic*. There are 63 race values. The values are built from six basic categories (*white, black, asian, native, island, and other*), either individually or in combinations (mixed race). The majority of race values, however, are white, black, or asain. In total there are 126 race/ethnicity combinations.

The Bureau ran the attack twice using two different sources of prior knowledge. One source consists of commercially-available data, and can therefore be run by anybody. The accuracy of the commercial data is questionable, leading to some uncertainty as to whether any lack of attack effectiveness is due to errors in reconstruction or errors in the prior knowledge. The second source consists of the Bureau's own internal data, and is therefore a perfect match. This represents the worst-case scenario. For the purpose of comparing our non-attack with the Bureau's re-identification attack, we focus only on these worst-case results. This second internal source is referred to as CEF (Census Edited File) in Abowd [4].

A reconstructed record is correct when its name, address, age (within one year), sex, race, and ethnicity match with a record in the Bureau's internal data.

Abowd [4] provides measures of *precision* and *recall*. Precision is defined as the percentage of correct records to linked records, and we use this definition as well (i.e. in Table 1 and Fig. 3). Abowd's recall is defined as the percentage of correct reconstructed records to all prior knowledge records. This definition doesn't make sense to us because the attacker does not know whether a reconstructed record is correct or not, and therefore has no basis on whether to make a prediction or not. We therefore use a different measure of recall: the fraction of linked records to total prior knowledge records. This definition is used in Table 1.

There are 279,179,329 prior knowledge records (Abowd's Table 2, column *"Records with PIK, Block, Sex, and Age"*, row *"CEF"*). Abowd's Table 6 gives the number of linked records (column *"Putative Re-identifications (Source: CEF)"*) and the number of correct records (column *"Confirmed Re-identifications (Source CEF)"*) for each binned block size as well as for all blocks taken together. Abowd's Table 6 provides the precision measures. The recall measures can be computed by dividing the putative re-identifications by the total prior knowledge records.

Three of the precision and recall values are given in (this paper's) Table 1. One is for all blocks, and the other two are for the block sizes with the highest precision.

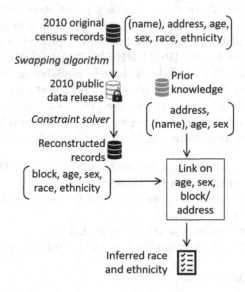

Fig. 2. Bureau's re-identification attack.

3   Our Simple Inference Non-attack

Our inference mechanism (non-attack) requires only the street address of the target individual as prior knowledge. The information needed to link street address to census block is public information[3].

To infer a person's race and ethnicity, we simply look up the *majority* race and ethnicity for the person's census block, and infer that this is the race/ethnicity of the person. The majority race/ethnicity for a given block is that with the highest count of all 126 race/ethnicity values.

---

[3] For instance https://geocoding.geo.census.gov.

We define *precision* as the majority race/ethnicity count divided by the block count. This is simply the statistical probability that any given person in a block indeed has the majority race/ethnicity.

To measure precision, we use the 2010 swap-protected release as the ground truth. This is not a perfect ground truth, but we believe that it is close enough for our purpose, which is merely to show that the Bureau's re-identification attack and our simple inference non-attack are in the same ballpark.

In the 2010 swap-protected release, the block count is an exact count (not noisy, see paragraph 37 of Abowd [4]). Because of swapping, race/ethnicity counts may not be exact, but we assume that the majority race/ethnicity count is almost always exact. This is because swapping generally occurs with households that are unique, i.e. those with rarer race/ethnicities. Our non-attack depends only on the count of the majority race/ethnicity, which is less prone to distortion from swapping. Note that the exact parameters for swapping are not published [9] so we are not completely certain of this assumption.

The block-level 2010 swap-protected and TDA-protected releases were compiled into a single table[4] by IPUM NHGIS (National Historical Geographic Information System). The block-level data is available as per-state tables, so we merged the tables for all states plus DC and Puerto Rico for our measure. We measured only voting-age (over 18 years) counts to better compare with the Bureau's attack, which also uses voting age data (paragraph 5 from Abowd [4]).

Our procedure for measuring per-block precision goes as follows:

1. Find the block's majority race/ethnicity MR_{TDA} from the TDA-protected data (the race/ethnicity with the highest count).
2. Set the majority count MC_{GT} as the count for race/ethnicity MR_{TDA} from the swap-protected data (ground truth).
3. If $MC_{GT} < 5$, set precision as zero (see Sect. 3.1).
4. Otherwise, set the block count BC_{GT} as that of the swap-protected data (ground truth).
5. Set precision as MC_{GT}/BC_{GT}.

In most cases, the TDA-protected data and the swap-protected data have the same majority race/ethnicity ($MR_{TDA} = MR_{SWAP}$). In this case, there is no precision penalty incurred by the noise from TDA. The majority race/ethnicity is different in blocks where the majority race/ethnicity and the second race/ethnicity have similar counts. In these cases, the noise from TDA can be enough to promote the second race/ethnicity to the majority. The precision penalty in these cases, however, is relatively small since the counts are not very different.

For example, suppose that the ground-truth count for the majority race/ethnicity for a block with 100 persons is 51, and for the second race/ethnicity is 49.

---

[4] https://www.nhgis.org/privacy-protected-2010-census-demonstration-data, Vintage 2021-06-08.

However, the noise from TDA causes the second race/ethnicity to be the majority. In this case, the measured precision for TDA will be 49% instead of the correct 51%—the precision penalty paid for the noise is relatively small.

The *recall* for a given precision is measured as the fraction of records with the given precision or better for blocks where there are at least 5 persons with the majority race/ethnicity. Table 1 gives several illustrative precision/recall values.

Figure 1 is a CDF showing the precision of all records for our non-attack executed on both the TDA-protected and swap-protected releases. A precision of zero is conservatively assigned when the associated block has fewer than 5 persons with the majority race. The precision of the non-attack on the swap-protected data is virtually identical to that of the TDA-protected data. In other words, the TDA protection does not affect the ability to (statistically) infer peoples' race and ethnicity.

Also shown in Fig. 1 are the precision measures taken from Abowd [4]. Each point represents a different range of block sizes. Prior-knowledge records without a match among the reconstructed records are assigned a precision of zero. From Fig. 1 we see that simple inference is substantially more effective than the Bureau's re-identification.

Note that in our non-attack, the attacker knows roughly what precision any given block has. Because of noise added to TDA-protected counts, an attacker does not know the exact precision for a given block. The amount of noise, however, is relatively small, so the attacker has a good estimate of precision.

The race/ethnicity of a substantial fraction of the population (11%) can be inferred with virtually 100% precision. This is for the simple reason that many blocks have only a single race/ethnicity.

Note that this result is supported by Kenny et al. [8], which also predicts individual race and ethnicity, but additionally uses analysis of names to help predict race and ethnicity. Kenny et al. found that TDA (using a version of TDA with more noise than the final version we tested) did not degrade the quality of these predictions.

Figure 3 shows the non-attack precision as a whisker plot per block size group (1–9 persons, 10–49 persons etc.). This figure shows that the noise of TDA substantially distorts the data for the smallest blocks (9 or fewer persons), but not for blocks larger than that. We assume that the near-perfect precision measure for the swap-protected data for the smallest blocks is because small blocks tend to be very homogeneous, and because swapping removes most of what little non-homogeneity remains. Note in particular that the relative distortion due to swapping is greater for small blocks. As such, the near-perfect precision measure for small blocks for swap-protected data does not accurately reflect reality. In other words, it would not necessarily be correct to conclude from this figure that swapping fails to protect privacy for small blocks.

Figure 4 shows the absolute error between swap-protected and TDA-protected counts. Most of the comparisons are for the majority race/ethnicity. When the two releases have different majority race/ethnicity, the comparison uses the majority race of the TDA-protected data. The error is relatively small,

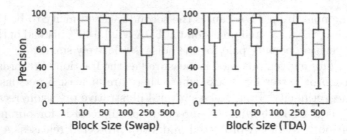

Fig. 3. Non-attack precision by block size for TDA-protected and swap-protected releases.

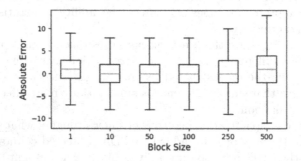

Fig. 4. Absolute error in count introduced by TDA noise between swap-protected and TDA-protected data for the same race/ethnicity in each block. Measured as TDA-count minus swap-count.

less than plus or minus 10 in most cases. The standard deviation in error for the smallest blocks is 2.8, and 4.0 for the largest blocks. This supports the observation that most of the loss of precision occurs for the smallest blocks. We don't know enough about TDA or swapping to understand why the median error is greater than zero, or why the range of error increases with larger blocks.

3.1   Effect of Majority Race/Ethnicity Threshold

In our non-attack, we ignore blocks where the majority race/ethnicity has fewer than 5 persons. The idea here is that a group of 5 persons is a reasonable privacy threshold. In other words, revealing that there are at least 5 persons with a given race and ethnicity does not unduly compromise the privacy of those 5 persons since they are not race/ethnicity uniques. The measures given in Table 1 and Fig. 1 use this threshold.

The choice of 5 is our own, and others may feel that a larger threshold is required, or that a smaller threshold is adequate. We therefore give the cumulative distribution of precision for several different thresholds in Fig. 5. Note that even with a threshold of 20, a substantial fraction of blocks allow correct inference with 100% precision.

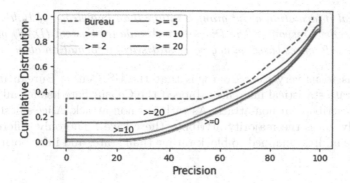

Fig. 5. CDF of precision for our non-attack for different thresholds for the number of persons with the majority race/ethnicity. Blocks with fewer than the threshold number of persons are conservatively assigned a precision of zero. For comparison, the precision for the Bureau's re-identification attack on the prior swap mechanism is also shown.

4   Discussion and Conclusion

In this paper, we show that the TopDown Algorithm (TDA) used by the US Census Bureau for the 2020 census in no way prevents attackers from inferring race and ethnicity with high accuracy for a substantial portion of census respondents.

Here we discuss two questions. First, does the ability to make this inference constitute a privacy violation of some sort? Note that it not up to us to answer this question. Rather, this is a matter for US Census and US Government policy makers. Nevertheless, if the answer is no, then the US Census has incorrectly measured the effectiveness of their attack. If the answer is yes, then TDA is an inadequate defense.

Second, assuming the answer is no (there is no privacy violation), then what did the Bureau do wrong in its privacy measure?

4.1   Has Privacy Been Violated?

The US Census Bureau has quite clearly stated that it is a goal to allow for accurate statistical inferences. The following is from [5]:

> *Some inferences about confidential information can be achieved with purely statistical information (especially for blocks with many identical records). These inferences rely on aggregate statistical information about groups and do not rely on any individuals' confidential census responses. For example, suppose Alice is trying to learn how Bob responded to the race question, and she already knows Bob lived in Montana at the time of the 2010 Census enumeration. Alice could then review the 2010 Census tables, and because she can find that 89.4 percent of respondents reported "White Alone" in Montana, Alice can guess with high confidence that Bob's census response was "White Alone." This is an example of an inference based on aggregate*

statistical information about groups, rather than knowledge of Bob's confidential census response. The Disclosure Avoidance System (DAS) permits accurate inferences based on aggregate statistical information about groups.

It seems to us very clear from this that the US Census Bureau intends to allow accurate statistical inferences. Indeed the Census here has for all practical purposes described our non-attack. As with our non-attack, Alice's best strategy is to simply guess the majority attribute (here race). The only differences are that we use address mapped to block rather than state, and infer both race and ethnicity.

4.2   What is Wrong with the US Census Privacy Measure?

It is important to point out that the Bureau doesn't use inference as its measure of privacy. In Abowd [4], two measures of privacy are given, neither of which is inference.

The first measure is simply the fraction of correctly reconstructed records. Referring to Fig. 2, this would be the fraction of reconstructed records that match the original census records on block, age, sex, race, and ethnicity.

The first seven pages of Abowd [4] pertain to the reconstruction measure. These seven pages conclude with:

> *Consequently, the new technology-enabled possibility of accurately reconstructing HDF microdata from the published tabular summaries and the fact that those reconstructed data do not meet the disclosure avoidance standards established at the time for microdata products derived from the HDF demonstrate that the swapping methodology as implemented for the 2010 Census no longer meets the acceptable disclosure risk standards established when that swapping mechanism was selected for the 2010 Census.*

In other words, the mere ability to reconstruct with some level of success (see Abowd's Table 1), *whether or not the reconstructed records can be matched to named persons*, is the *only* criteria required by the Bureau.

A simple thought experiment, however, shows that reconstruction alone is not a valid measure of privacy. Imagine, for instance, a table with one column, sex, with values 'M' and 'F' over thousands of rows. Given noisy counts of the two values, it would be easy to reconstruct the table with high accuracy. Clearly this in and of itself is not a violation of privacy. At a minimum, the number of unique values in the table is also important. The census data used for the attack has only 57% uniques on the five attributes.

In any event, the Bureau recognizes the importance of being able to link reconstructed records with named persons, and so the subsequent 12 pages of Abowd [4] focus on the second measure, that of re-identification. This measures the fraction of reconstructed records that can be successfully linked to prior knowledge of named persons. Referring again to Fig. 2, this is the box labeled "Link on age, sex, block/address".

Strictly speaking, this re-identification measures the ability to identify name, address, age, sex, race, and ethnicity with some level of success. Abowd [4] reports that the Bureau finds that this also fails its criteria for privacy:

The Data Stewardship Executive Policy Committee (DSEP) determined that the simulated attack success rates in Table 6 were unacceptable for the 2020 Census. Decennial census data protected by the 2010 disclosure avoidance software is no longer safe to release.

Although Abowd [4] never uses the word 'infer', given that re-identification requires prior knowledge of name, age, sex, and address, it seems perfectly reasonable to describe the re-identification as inferring race and ethnicity from name, age, sex, and address. The Bureau is implicitly saying that this is not acceptable. On the other hand, they are saying that it is ok to infer race and ethnicity from address, because they very intentionally release data designed to do this.

There are perhaps three possible responses that the Bureau could make to this apparent contradiction:

1. Re-identification using prior knowledge and inference using the same or less prior knowledge are different and can't be compared.
2. It is not this specific re-identification per se that is a problem, but the fact that re-identification can happen in general.
3. The re-identification doesn't really matter, since in any event the reconstruction alone failed the privacy criteria.

The first seems non-sensical.

Regarding the second, it would be helpful if the Bureau identified cases of re-identification that revealed substantially more information than what the data release is supposed to reveal statistically.

4.3 The Ruggles and van Riper Reconstruction

Regarding the third, Ruggles and Van Riper [13] provide evidence that even the Bureau's reconstruction reveals nothing more than what is meant to be revealed statistically. In their demonstration, Ruggles et al. used the following statistical information as the basis for reconstruction (taken from the 2010 census):

1. The national distributions of ages, sexes, and block sizes.
2. The fact that 78% of individuals on average have the majority race/ethnicity of their block.

Armed with only this knowledge, Ruggles et al. built 10000 synthetic blocks by randomly assigning block size according to the national distribution, and then randomly assigning individuals to the blocks with age and sex following the national distributions. They then mimicked reconstruction of the synthetic blocks by fresh random assignments age and sex following the national distributions. Finally, they assumed that on average the race and ethnicity would be correct 78% of the time.

The result is that on average 41% of records matched on all five attributes (block, age, sex, race, and ethnicity), compared to 45% for the Bureau's attack (using commercially obtained name, age, sex, and address). These two reconstruction measures are certainly in the same ballpark.

The idea that the Bureau's reconstruction is little better than random is also supported by Muralidhar [11]. He shows that the Bureau's reconstruction can produce a large number of different solutions. Any given solution chosen by the Bureau is effectively a random choice among many.

4.4   Conclusion

In conclusion, we believe that the Bureau has not adequately demonstrated a meaningful privacy threat against the 2010 swapping method. The threat may exist, but has not been demonstrated. We also believe that the criteria used by the Bureau to measure privacy is flawed in that it does not take the intended released statistical knowledge into account. This is demonstrated partially by the Ruggles and Van Riper reconstruction, and more definitively by the inference non-attack of this paper.

Note that implementing TDA has been costly both in terms of data quality and timely data release. Numerous studies point to problems for a variety of research tasks and government functions, including redistricting [8], health [7, 14,15], and demographics [10,15,16]. (Note that some of these studies may be based on earlier proposed versions of anonymization with more noise than the final version.)

The state of Alabama filed a (failed) lawsuit in part to force the Census Bureau return to the former low-distortion method of anonymization [1], and a second lawsuit to force the Bureau to release delayed housing data is ongoing as of this writing (Spring 2022) [2]. The Bureau has yet to release all of the tables that it normally releases.

Although this paper does not make any concrete proposals on how better to measure privacy, it seems clear to us that more research is needed, especially regarding the role that expected statistical inference plays in measuring privacy. We hope that this paper serves to motivate that research, and that it leads to a more circumspect approach to measuring privacy loss in statistics organizations.

References

1. Brennan Center for Justice. Court Rejects Alabama Challenge to Census Plans for Redistricting and Privacy (2020). https://www.brennancenter.org/our-work/analysis-opinion/court-rejects-alabama-challenge-census-plans-redistricting-and-privacy

2. Brennan Center for Justice. Fair Lines America Foundation vs. US Dept of Commerce (2022). https://www.brennancenter.org/our-work/court-cases/fair-lines-america-foundation-v-us-dept-commerce

3. Abowd, J.: Staring down the database reconstruction theorem (2019). https://www2.census.gov/programs-surveys/decennial/2020/resources/presentations-publications/2019-02-16-abowd-db-reconstruction.pdf

4. Abowd, J.: Second declaration of john m. abowd, fair lines versus US dept. of commerce, appendix b - 2010 reconstruction-abetted re-identification simulated attack (2021). https://www2.census.gov/about/policies/foia/records/disclosure-avoidance/appendix-b-summary-of-simulated-reconstruction-abetted-re-identification-attack.pdf

5. U.S. Census Bureau: Disclosure avoidance for the 2020 census: an introduction (2021). https://www2.census.gov/library/publications/decennial/2020/2020-census-disclosure-avoidance-handbook.pdf

6. Garfinkel, S., Abowd, J.M., Martindale, C.: Understanding database reconstruction attacks on public data: these attacks on statistical databases are no longer a theoretical danger. Queue **16**(5), 28–53 (2018)

7. Hauer, M.E., Santos-Lozada, A.R.: Differential privacy in the 2020 census will distort COVID-19 rates. Socius **7**, 2378023121994014 (2021)

8. Kenny, C.T., Kuriwaki, S., McCartan, C., Rosenman, E.T., Simko, T., Imai, K.: The use of differential privacy for census data and its impact on redistricting: the case of the 2020 us census. Sci. Adv. **7**(41), eabk3283 (2021)

9. McKenna, L., Haubach, M.: Legacy techniques and current research in disclosure avoidance at the U.S. census bureau (2019). https://www.census.gov/library/working-papers/2019/adrm/legacy-da-techniques.html

10. Mueller, T., Santos-Lozada, A.R.: The 2020 US census differential privacy method introduces disproportionate error for rural and non-white populations. SocArXiv (2021)

11. Muralidhar, K.: A re-examination of the census bureau reconstruction and reidentification attack. Priv. Stat. Databases (2022)

12. Ruggles, S., et al.: Implications of differential privacy for census bureau data and scientific research. Minnesota Population Center, University of Minnesota, Minneapolis (Working Paper 2018-6) (2018)

13. Ruggles, S., Van Riper, D.: The role of chance in the census bureau database reconstruction experiment. Popul. Res. Policy Rev. **41**(3), 781–788 (2021). https://doi.org/10.1007/s11113-021-09674-3

14. Santos-Lozada, A.R.: Changes in census data will affect our understanding of infant health. Socius **7**, 23780231211023640 (2021)

15. Santos-Lozada, A.R., Howard, J.T., Verdery, A.M.: How differential privacy will affect our understanding of health disparities in the united states. Proc. Natl. Acad. Sci. **117**(24), 13405–13412 (2020)

16. Winkler, R.L., Butler, J.L., Curtis, K.J., Egan-Robertson, D.: Differential privacy and the accuracy of county-level net migration estimates. Popul. Res. Policy Rev. **41**, 1–19 (2021)

A Re-examination of the Census Bureau Reconstruction and Reidentification Attack

Krishnamurty Muralidhar$^{(\boxtimes)}$ (iD)

University of Oklahoma, Norman, OK 73019, USA
krishm@ou.edu

Abstract. Recent analysis by researchers at the U.S. Census Bureau claims that by reconstructing the tabular data released from the 2010 Census, it is possible to reconstruct the original data and, using an accurate external data file with identity, reidentify 179 million respondents (approximately 58% of the population). This study shows that there are a practically infinite number of possible reconstructions, and each reconstruction leads to assigning a different identity to the respondents in the reconstructed data. The results reported by the Census Bureau researchers are based on just one of these infinite possible reconstructions and is easily refuted by an alternate reconstruction. Without definitive proof that the reconstruction is unique, or at the very least, that most reconstructions lead to the assignment of the same identity to the same respondent, claims of confirmed reidentification are highly suspect and easily refuted.

Keywords: Disclosure · Reconstruction · Reidentification

1 Introduction

According to the declaration by Dr. John Abowd, Chief Scientist and Associate Director for Research and Methodology at the United States Census Bureau, the disclosure prevention procedures used in the 2010 Census did not prevent the ability of an adversary to identify the Census respondents and results in the confirmed reidentification of as many as 179 million respondents to the 2010 Census. (Abowd 2021a, p. 12). The complete procedure used by the Census Bureau (hereafter, REID) consisted of three steps:

(1) Reconstruct microdata (individual level data) for all respondents in the US by using publicly available Census data.
(2) Link the reconstructed microdata to a commercial database using (Age, Sex) and assign a name and address to the reconstructed microdata records.
(3) Compare the enhanced microdata with the original Census data to confirm the identity of the respondents.

The Census collects both individual and household level data. The original data that is gathered is edited for errors and any other issues and the Census Edited File (CEF) is created. All personally identifiable information is removed and replaced with a unique

J. Domingo-Ferrer and M. Laurent (Eds.): PSD 2022, LNCS 13463, pp. 312–323, 2022.
https://doi.org/10.1007/978-3-031-13945-1_22

identifier, the Protected Identification Key. Statistical disclosure limitation procedures (privacy protection measures) are applied to CEF which results in the creation of the Hundred Percent Detail File (HDF). All Census publications are produced from HDF.

The Census data released to the public are released in two categories, personal and household. No linkage between the two categories is provided. The personal level data that is released to the public consists of (Age, Sex, Race, and Ethnicity). Household level data is similar but has additional information (number of individuals in household, relationship to the householder) and the data release also provides additional information on the Age variable (average and median).

The Census releases data at different geographic levels: nation, state, county, tract, block group, and block. The final three are census-defined constructs and do not necessarily correspond to traditional geographic classification. For personal level data, the data at the smaller geographic level is aggregated to the next higher level, that is, the results at the block level are aggregated to block groups, block groups are aggregated to tracts, etc. The multiple tables that are released (Total Population, Sex by Age, Total Races, and others) are all aggregations of the most detailed data release (Age by Sex, by Race, by Ethnicity). The different tables released form the basis of the reconstruction of the respondent microdata.

Every respondent record consists of both personal information and information about how the respondent is related to the householder (the primary individual in the household in whose name the housing unit is owned or rented). Information regarding the relationship variable is only released in the household tables and not as a part of the individual level data. During the reconstruction, the REID team did not recreate the entire record for the respondent, but only the variables released in individual level tables, namely, Age, Sex, Race, Ethnicity (Abowd 2021a). REID procedure is implemented using two external data files: commercial data sources and CEF as the external data file.

Unfortunately, it is impossible for anyone outside the Census to have access to either of these two files. As far as the commercial data, while the sources of the data have been identified, the accuracy of these specific data files were never identified (Rastogi and O'Hara 2012). Since the reidentification claims are directly affected by the accuracy of the external data, the accuracy of the external data source must be verified. However, it is very difficult (if not impossible) to recreate this external data file to serve as a comparison. According to Abowd (2021a), using CEF as the external data file is a worst-case scenario since it is the most accurate data that an adversary can have. As a result, the choice of CEF as the external data favors the results and claims of REID. Again, unfortunately, the Census Bureau does not provide access to CEF without special authorization. Thus, we have a situation where it is practically impossible to gain access to the data to verify the results of REID. To overcome this impossible situation, I have chosen to generate a hypothetical CEF based on the characteristics of the available data and as the "external data" file. As with the use of the true CEF, it is assumed that there is no inaccuracy between the true and external data.

The analysis in this paper is based on data from Tract 5.01, Laramie County, Wyoming (https://data.census.gov/cedsci/advanced). It should be noted that the choice of this tract was simply a matter of convenience. Similar data are available in practically every county in every state in the nation. The tract consists of a total of 148 blocks, 127 occupied blocks, and a total population of 8164. The largest block in the tract has a population of 450 and the smallest occupied block has a population of 1. Table 1 shows the race and ethnicity breakdown for this tract (AIAN represents American Indian or Alaskan Native, NHPI represents Native Hawaiian or Pacific Islander and Multiple represents two or more races).

Table 1. Race breakdown for Tract 5.01, Laramie County, Wyoming

Ethnicity	White	Black	AIAN	Asian	NHPI	Other	Multiple
Not Hispanic	6420	210	56	115	26	14	155
Hispanic	688	32	20	4	3	304	117

2   Reconstruction of Respondent-Level Data

The basic premise underlying the entire REID experiment can be summarized by the following statement: "While the statistical and computer science communities have been aware of this vulnerability since 2003, only over the last few years have computing power and the sophisticated numerical optimization software necessary to perform these types of reconstructions advanced enough to permit reconstruction attacks at any significant scale." (Abowd 2021a, p. 14)".

This is incorrect. The Census data files from 2010 (and even 2000) could be used reconstruct the microdata for every respondent in the nation very easily. At the tract level, Census releases tables of count by individual year of age, sex, race, and ethnicity (PCT12A-O). To reconstruct the data at the tract level is just a matter of creating a list of individuals based on the counts provided in the tables. The difference between the description above and the reconstruction in the REID experiment is the level of geography. The reconstruction above is at the tract level while the REID reconstruction is at the block level. At the block level, the Age variable is grouped (except for ages 20 and 21). In addition, other than White respondents, Age by Sex is only provided for the Race category as a whole and breakdown by Ethnicity is not provided. The reconstruction procedure at the block level must be applied twice (first for Hispanic respondents followed by non-Hispanic respondents) for all respondents who are not White.

Within each tract, the reconstruction can be performed independently for each Sex and Age Group (23 in total). My analysis in this paper focuses on Males in the Age Group (25–29) in Tract 5.01 in Laramie County, Wyoming. At the tract level, there are a total of 338 respondents in this Age Group in 87 different blocks. From Tables P8 and P9, the tract level race breakdown for this Age Group is shown in Table 2. From Tables P8 and P9 at the block level, the adversary can also create a similar table for each of the 87 blocks (except for Ethnicity for non-White individuals as noted earlier). Table 2 provides information for one such block (Block 4000) in Tract 5.01.

Given the information for Block 4000 in Table 2, reconstructing the individuals in Block 4000 is simply a matter of creating a list as shown in Table 3. The individuals in all other blocks can be similarly reconstructed with the exception that this reconstruction presents age only as a group (25–29), rather than individual year of age.

Table 2. Race breakdown for ages (25–29), Tract 5.01, Laramie County, Wyoming

		White	Black	AIAN	Asian	NHPI	Other	Multiple
Tract 5.01	Not Hispanic	263	13	3	6	1	1	3
	Hispanic	28	0	2	0	1	12	5
Block 4000	Not Hispanic	9	0	0	1	0	0	0
	Hispanic	3	0	0	0	0	0	0

Table 3. Reconstructed records for Block 4000 (with Age Groups only)

Tract	Block	Sex	Race	Ethnicity	Age
5.01	4000	Male	White	Not Hispanic	(25–29)
5.01	4000	Male	White	Not Hispanic	(25–29)
5.01	4000	Male	White	Not Hispanic	(25–29)
5.01	4000	Male	White	Not Hispanic	(25–29)
5.01	4000	Male	White	Not Hispanic	(25–29)
5.01	4000	Male	White	Not Hispanic	(25–29)
5.01	4000	Male	White	Not Hispanic	(25–29)
5.01	4000	Male	White	Not Hispanic	(25–29)
5.01	4000	Male	White	Not Hispanic	(25–29)
5.01	4000	Male	White	Not Hispanic	(25–29)
5.01	4000	Male	White	Hispanic	(25–29)
5.01	4000	Male	White	Hispanic	(25–29)
5.01	4000	Male	White	Hispanic	(25–29)
5.01	4000	Male	Asian	Not Hispanic	(25–29)

For all Males in Age Group (25–29) in Tract 5.01, individual year of age breakdown by (Race and Ethnicity) can be obtained from PCT12A-O and is reconstructed below in Table 4.

This reconstruction can be performed independently for each Sex and Age Group in the tract. This is precisely why the reconstruction problem "is massively parallel in tracts" (Abowd 2018, p. 16). Note that the values for (Sex, Race, Ethnicity) reconstructed for every individual in every block in Tract 5.01 (like Block 4000 in Table 3) will always satisfy all the additivity constraints for these three variables at the tract level. The only missing variable in the reconstructed data is the individual year of age. Hence, the entire reconstruction problem reduces to one of assigning individual year of age values at the tract level for a given (Sex, Race, Ethnicity) to the respondents in the blocks with the same (Sex, Race, Ethnicity).

Table 4. Individual year of age by (Race and Ethnicity) for Tract 5.01

Ethnicity	Age	White	Black	AIAN	Asian	NHPI	Other	Multiple
Not Hispanic	25	39	1	0	0	0	0	1
	26	53	7	1	4	0	1	1
	27	56	2	0	1	0	0	0
	28	57	2	1	1	1	0	1
	29	58	1	1	0	0	0	0
Hispanic	25	3	0	1	0	0	5	1
	26	6	0	0	0	0	1	0
	27	6	0	0	0	0	1	0
	28	9	0	0	0	1	1	2
	29	4	0	1	0	0	4	2

For the purposes of illustration, consider the reconstruction of (Male, Black, non-Hispanic) respondents in the Age Group (25–29) in Tract 5.01. Table 5 shows the distribution of the 13 (Male, Black, non-Hispanic) respondents in the (25–29) Age group in the blocks in Tract 5.01.

Table 5. Male, Black, non-Hispanic respondents in different blocks in Tract 5.01

Block	3014	3017	3019	3021	4002	4003	4012	4021	4026
Respondents in (25–29) Age Group	1	1	1	1	1	1	4	1	2

The REID approach to the reconstruction of individual year of age for these 13 respondents is to express the problem as a system of linear equations to find individual year of age assignment at the block level which satisfies individual year of age frequencies at the tract level. This system of linear equations is solved using Gurobi optimization software. The purpose of optimization software is to find the best possible solution (optimal) from among many solutions that satisfy the mathematical equations (feasible), where the best possible is evaluated based on the objective function. In some cases, there are multiple optimal solutions, but usually only a few. The reconstruction problem does not have an objective function (there is no reason to treat one reconstruction as being superior to any other), and every feasible solution is an acceptable solution. Hence, the number of potential solutions remains very large.

Even for this small group of 13 individuals, the number of possible solutions runs in the hundreds of thousands. The single age 25 value can be assigned to any one of the 13 respondents in the block in 13 different ways. The seven age 26 values can be assigned to the remaining 12 respondents in 792 different ways. The two age 27 values can be assigned to the remaining five respondents in 10 different ways. The two age 28 values can be assigned to the remaining three respondents in three different ways. Finally, there is only way to assign the single age 29 value to the single remaining respondent. In total, there are $(13 \times 792 \times 10 \times 3 \times 1 =) 308{,}880$ different assignments of age values across this small group of individuals. Every one of these assignments is a feasible solution to the system of linear equations representing (Male, Black, non-Hispanic) respondents in Age Group (25–29) in Tract 5.01. This process can then be repeated for each Age Group in each (Race, Ethnicity) combination for each Sex in each Tract.

The reconstruction reduces to the problem of assigning individual year of age at the block level while preserving the frequency of the respective age at the tract level. A simpler way to achieve this is to create a vector of individual years of age with the same frequency as at the tract level, randomly sort this vector, and assign them to individuals in each block. Every random sort of the age vector satisfies the age frequencies at the tract level and represents a feasible solution to the system of linear equations. Table 6 shows five different individual year of age assignments for the (Male, Black, non-Hispanic) individuals in Tract 5.01.

Table 6. Five different reconstructions of age

Block	Respondent	I	II	III	IV	V
3014	1	25	27	26	28	29
3017	1	26	26	27	26	25
3019	1	27	27	28	26	26
3021	1	26	26	26	26	26
4002	1	26	26	26	26	26
4003	1	26	28	29	26	28
4012	1	29	26	26	29	26
	2	27	28	26	27	26
	3	28	29	26	25	26
	4	28	26	27	26	26
4021	1	26	25	26	27	27
4026	1	26	26	25	26	28
	2	26	26	28	28	27

White, non-Hispanic respondents constitute 263 of the 338 respondents in Tract 5.01, of whom 39 respondents are of age 25. The possible assignments of just the 39 values of age 25 across the 263 respondents is greater than 10^{46}. The reconstruction of each race and ethnicity combination is performed independently of all others. As a result, the number of feasible assignments is the product of the possible solutions for each race and ethnicity combination. Suffice it to say that the number of alternative reconstructions is practically infinite. Every one of these infinite reconstructions represents a feasible solution to the system of linear equations representing the data and consistent with all the tables used by the REID team both at the block and tract level.

3   Putative Reidentification

The Census reidentification attack is explained in detail in Abowd (2021a). The process is as follows:

Identify the corresponding census block for every address in the source file. Then, looping through all the records in the reconstructed microdata file produced from the reconstruction, find the first record in the source file that matches exactly on block, sex, and age. Once this step is completed, run through the remaining unmatched records from the reconstructed microdata and find the first unmatched record from the source file that matches exactly on block and sex, and matches on age plus or minus 1 year (Abowd 2021a, Appendix B, p. 7).

When a match is found, it represents a putative identification, and the identification information is harvested from the external data file and appended to the reconstructed data. Abowd (2021b, Table 2) reports a putative identification of 77% (238,175,305 out of 308,745,538).

This procedure was applied to the 10 different reconstructions for Tract 5.01 and the putative identification results are provided in Table 7. These results are consistent with the national level putative identification rate of 77% observed by Abowd (2021b). When these results are viewed independently, these results seem to provide strong support for the REID reidentification results.

Table 7. Aggregate putative identification rates for Tract 5.01

Reconstruction	1	2	3	4	5	6	7	8	9	10
Putative identification (%)	80	77	81	76	76	78	80	79	79	80

These results, however, cannot be viewed independently. Each row of the 338 reconstructed records in this data set represents a unique individual (as does the reconstructed record of each of the 308,745,538 respondents across the nation). The objective of the REID reconstruction and reidentification experiment is to assign identity to each of these unique individuals. The ten different reconstructions presented above for the 338 individuals in Tract 5.01 are the result of applying different values of age (from 25 to 29) to these unique individuals. To conclude that matching with the external data results in putative reidentification, it is necessary that the same record in the reconstructed data is assigned the same identity on every reconstruction. Assigning different identities to the same reconstructed record on different reconstructions implies uncertainty in the validity of the reidentification. The fallacy in the REID approach is to treat a single reconstruction as definitive proof of reidentification.

It is important to note that this analysis can be performed by the adversary with knowledge of the external data (sans race and ethnicity) and the tabular data released by the Census. Any intelligent adversary would realize that making claims of reidentification based on a single reconstruction can be immediately refuted by the Census Bureau by presenting one of the infinite alternate reconstructions. Yet valid reidentification based on a single reconstruction is precisely the claim that the REID team is making.

Generally, the extent of agreement between two alternate reconstructions represents a simple measure of the similarity between two reconstructions. Agreement between two reconstructions was computed as the number of respondents for whom the same identity was assigned in both reconstructions. If most of the reconstructions show strong agreement, that would be evidence that conclusions based on multiple reconstructions will not be very different from one another. Analyzing the 10 reconstructions for Tract 5.01 indicates that this is not the case. The average agreement between any two reconstructions is 16% with a minimum of 10% and a maximum of 21%. There is little confidence that any two reconstructions will lead to the same conclusion.

Table 8 shows the number of times the same identity from the external data file was assigned to the same record in the reconstructed data over all 10 reconstructions for all 338 records. If the same identity is assigned to the same reconstructed record every time, it supports the conclusion that reidentification has occurred, which is not the case. Not a single record was assigned the same identity across all reconstructions (or even nine out of 10 reconstructions). Furthermore, only three reconstructed records were assigned the same identity in eight of the 10 reconstructions. For the adversary to be confident in the putative identifications, it is necessary that the same identity is assigned to the same individual on all (or at least most reconstructions). These results provide little or no confidence in the putative identification.

Table 8. Number of records for which the same identity was assigned

Number of reconstructions where the same identity was assigned	1	2	3	4	5	6	7	8	9	10
Number of records		6	96	96	66	41	14	16	3	0	0

Table 9, which shows the number of different identities assigned to the same record, provides a different perspective but leads to the same conclusion. The most important observation from this table is that, not one of the reconstructed records had the same identity assigned to it in all 10 reconstructions. Almost 75% of the reconstructed records were assigned at least four different identities. From the adversary's perspective, the inability to consistently assign the same identity to the same individual across all 10 reconstructions implies that assigning identity based on the reconstructed data is unreliable.

The number of identities assigned is dictated by the size of the block. When the block size is k, there can be no more than (k + 1) identities (number of individuals in the block + no match) that can be assigned. This implies that of the 35 records for which only two identities were assigned, 26 of them belonged to a block with a single individual.

Table 9. Number of identities assigned to the reconstructed records in Tract 5.01

Number of different identities assigned over 10 reconstructions	1	2	3	4	5	6	7	8	9	10
Number of records	0	35	52	90	45	48	28	27	12	1

For every block with 10 or more individuals, at least four identities were assigned to each record. In the largest block, 27 of the 29 records were assigned at least six different identities.

It is also illustrative to analyze results pertaining to a single block. Consider Block 4012 with a total of 10 individuals: four (White, non-Hispanic), four (Black, non-Hispanic), one (AIAN, Hispanic), and one (White, Hispanic) with unique identifiers (RR226 to RR236 for the reconstructed records and 226 to 336 for individuals in the external data). Table 10 provides the assignment of identity for these individuals on the 10 different reconstructions. This table shows that no reconstructed record is assigned the same identity more than five times. Every reconstructed record is also assigned between four and eight different identities.

Table 10. Identity from external data assigned to each reconstructed record in Block 4012 (The numbers in the table represent the ID of the original record. Blank cells indicate no match was found)

Reconstructed record	1	2	3	4	5	6	7	8	9	10
RR226	228	227	229	227	229	226	234	234	227	227
RR227	227	234	234	234	228	229	229	231	229	232
RR228	226	232	226	232	226	232	226	228	234	226
RR229	234	233	232	229	230	227	227	227	231	228
RR230	232		233	226	231	228	235	232	228	230
RR231	230	228	235	228	227	233	232	230	226	
RR232	231	230	230	230	233	234	230	233	230	233
RR233	236	235	227	231	234	230	233	226	232	234
RR234	229	231	228	235	235	231	228		233	231
RR235	233	226	231	233		235	231			229

Table 11 shows the race and ethnicity assigned to every individual (identified by the ID) in the external data file from the 10 reconstructions. There is no individual from the external data who is assigned the same race and ethnicity across all reconstructions. Furthermore, every record is assigned at least three different race and ethnicity combinations. The table shows that the race and ethnicity of: (a) every individual from the external data is designated as (Black, non-Hispanic) at least once; (b) at least eight different individuals are designated (AIAN, Hispanic) at least once; (c) and seven different individuals are designated (White, Hispanic) at least once.

Interestingly, individuals with IDs 226, 232, and 234, who are most frequently designated as (Black, non-Hispanic), all happen to be (White, non-Hispanic). Individuals with IDs 227, 228, 230, and 233, who happen to be (Black, non-Hispanic) are least frequently designated as (Black, non-Hispanic). If the adversary were to rely on the reconstruction to designate race and ethnicity to the individuals in the external data (which lacks this information), the probability of a correct designation is no better than designating them randomly. If the intent of the team that designed the release of the 2010 Census tabular data was to prevent disclosure, they were extremely successful indeed!

Table 11. Race and ethnicity assigned to individuals in Block 4012 from the 10 reconstructions (W = White, not Hispanic; WH = White, Hispanic; B = Black; A = AIAN Hispanic. Blank cells indicate no match was found)

ID	1	2	3	4	5	6	7	8	9
226	B	W	B	B	B	W	B	B	WH
227	B	W	B	W	WH	A	A	A	W
228	W	WH	W	WH	B	B	W	B	B
229	W		W	A	W	B	B		B
230	WH	W	W	W	A	B	W	WH	W
231	W	W	W	B	B	W	W	B	A
232	B	B	A	B		B	WH	B	B
233	W	A	B	W	W	WH	B	W	W
234	A	B	B	B	B	W	W	W	B
235		B	WH	W	W	W	B		

Given that even putative identification is highly questionable, any analysis regarding confirmation of identification is entirely moot.

4 Conclusions

Abowd (2021a) claims that the objective of REID analysis "a modern database reconstruction-abetted re-identification attack can reliably match a large number of 2010 census responses to the names of those respondents – a vulnerability that exposed information of at least 52 million Americans and potentially up to 179 million Americans" (p. 8). To make this claim based on a single reconstruction, it is necessary to prove that the reconstruction was unique. The first section of this paper definitively shows that the reconstruction is not unique.

In the absence of a unique reconstruction, the only other way for the REID team to make a claim of confirmed reidentification is to show that all (or at least most) reconstructions result in the same assignment of identity to the reconstructed records. Without this minimal level of proof, the reidentification is essentially the result of random matching. Thus far, the REID team has provided no such proof. Using real Census data and the same analysis as the REID team, my analysis clearly shows that multiple reconstructions result in multiple identities being assigned to the same record in the reconstructed data, which is all that is required to refute any claim of meaningful reidentification. This study adds support to Ruggles and van Riper (2021) and Francis (2022) who showed that a random assignment performs as well as the Census reconstruction.

Some may consider results based on a single tract, in a single county, in a single state, or that only 10 reconstructions were performed, no better than a single reconstruction. But this would be missing the point. It is the Census Bureau which made the claim "Internal research has conclusively proven the fundamental vulnerabilities of the 2010 swapping methodology (Abowd 2021c)". I am simply refuting this claim by showing that, for any given track, a practically infinite number of such reconstructions exist, each reconstruction provides different results about the identity of a respondents, which casts serious doubt on this claim. It is up to the Census Bureau researchers to show either that the reconstruction was unique or, at the very least, that most reconstructions lead to the same conclusion regarding the identity of a respondent. Without such proof, claims of confirmed reidentification are highly suspect and easily refuted.

Any claims made by an adversary based on a single reconstruction can be refuted by the data administrator by issuing the following challenges:

(1) The adversary is challenged to provide the identity of the (Male, White, Hispanic, Age Group 25–29) record in Block 4012. Based on the first reconstruction, the adversary identifies this record as belonging to the individual with ID 230 in the external data. Using the same data, the data administrator counters by showing that, based on the remaining nine reconstructions, that this record could also belong to individuals with IDs (228, 235, 227, 233, 232, 230, 226) or not identified at all (reconstruction 10).

(2) The adversary is challenged to identify the (Black, non-Hispanic) individuals in Block 4012. Based on the first reconstruction, the adversary can only identify individuals with ID (226, 227, 232) in the external data as being (Black, non-Hispanic) since an age match was not found for one reconstructed record. Using the same data, the data administrator counters with the remaining nine reconstructions to show that any of the 10 individuals could be identified as (Black, non-Hispanic).

The interesting fact is that, to prevent any disclosure, the administrator never actually confirms or denies the adversary's claim. The administrator simply shows that there exist reconstructions that present alternative solutions that refute the adversary's claim. Faced with these facts, the adversary has no recourse but to acknowledge that the claims of reidentification cannot be substantiated. As the data administrator, it would be the Census Bureau's duty to challenge the claims made by the REID team, which they have not. I am doing so on behalf of the public.

.

Acknowledgement. I would like to thank Margo Anderson and Connie Citro for their helpful comments and suggestions, and Carolyn Jensen, Business Communication Center, Price College of Business, for her editorial assistance.

References

Abowd, J.M.: Staring Down the Database Reconstruction Theorem. Joint Statistical Meeting, Vancouver B.C., Canada (2018)

Abowd, J.: Declaration of John Abowd, State of Alabama v. United States Department of Commerce. Case No. 3:21-CV-211-RAH-ECM-KCN (2021a). http://vhdshf2oms2wcnsvk7 sdv3so.blob.core.windows.net/thearp-media/documents/Declaration_of_John_M._Abowd. pdf. (downloaded 14 Jan 2022)

Abowd, J.: Supplemental Declaration of John M. Abowd, State of Alabama v. United States Department of Commerce. Case No. 3:21-CV-211-RAH-ECM-KCN (2021b). https://www. brennancenter.org/sites/default/files/2021-06/M.D.%20Ala.%2021-cv-00211%20dckt%200 00116_001%20filed%202021-04-26%20Abowd%20declaration.pdf. (downloaded 14 Jan 2022)

Abowd, J.: Second Declaration of John M. Abowd, Fair Lines America Foundation Inc. V. United States Depart of Commerce and United States Bureau of the Census. Case No. Civ. A. No. 1:21-cv-01361 (ABJ) (2021c). https://www2.census.gov/about/policies/foia/records/disclosure-avo idance/abowd-fair-lines-v-commerce-second-declaration.pdf. (downloaded 15 Mar 2022)

Francis, P.: A Note on the Misinterpretation of the US Census Re-identification Attack. Privacy in Statistical Databases (2022). (Forthcoming)

Rastogi, S., O'Hara, A.: 2010 Census Match Study Report, 2010 CENSUS PLANNING MEMO NO. 247, United States Census Bureau, Washington DC (2012). https://www.census.gov/lib rary/publications/2012/dec/2010_cpex_247.html. (downloaded 27 Nov 2021)

Ruggles, S., Van Riper, D.: The Role of Chance in the Census Bureau Database Reconstruction Experiment, Population Research and Policy Review (published online) (2021). https://link.spr inger.com/article/10.1007%2Fs11113-021-09674-3. (downloaded 27 Nov 2021)

Quality Assessment of the 2014 to 2019 National Survey on Drug Use and Health (NSDUH) Public Use Files

Neeraja Sathe[1]([⊠]), Feng Yu[1]([⊠]), Lanting Dai[1], Devon Cribb[1], Kathryn Spagnola[1], Ana Saravia[1], Rong Cai[2], and Jennifer Hoenig[2]

[1] RTI International, Research Triangle Park, NC 27709-2194, USA
{nss,fyu}@rti.org
[2] Substance Abuse and Mental Health Services Administration, 5600 Fishers Lane, Rockville, MD 20857, USA

Abstract. National Survey on Drug Use and Health (NSDUH) public use files (PUFs) have been produced using a statistical disclosure control technique named MASSC, which stands for Micro Agglomeration, Substitution, Subsampling, and Calibration, to protect confidentiality and quality of data. To inform researchers that NSDUH PUFs maintain high data quality and comparability with NSDUH restricted-use files (RUFs), about 300 NSDUH published tables (based on RUF data) of substance use and mental health were selected and reproduced using PUF data. Key estimates and their respective standard errors (SEs) produced from the two sets of 2014 to 2019 data files were compared. Summary statistics of ratios of the estimates and ratios of their SEs from PUF and RUF data were produced. Out of 22,000 estimates compared, average ratios for estimated percentages across years were within the 0.99 to 1.01 range, and the average increase in SEs for the estimates produced from PUFs across years was about 7–11%, for both substance use and mental health measures. Multiyear trend comparisons between PUF and RUF estimates were also conducted graphically to demonstrate PUF estimates provide similar trend patterns to RUF estimates across years. This study will provide confidence to researchers and policymakers for making policies and public health decisions based on NSDUH PUFs.

Keywords: Statistical disclosure limitation · MASSC · Quality assessment · Public use file · NSDUH

1 Introduction

The National Survey on Drug Use and Health (NSDUH) is an annual survey that collects data on substance use, mental health, and other health measures among the U.S. civilian, noninstitutionalized population aged 12 years or older. It is sponsored by the Substance Abuse and Mental Health Services Administration (SAMHSA), U.S. Department of Health and Human Services, and is planned and managed by the SAMHSA Center for Behavioral Health Statistics and Quality (CBHSQ). Data collection and analysis are conducted under contract with RTI International. NSDUH data have been used extensively

© Springer Nature Switzerland AG 2022
J. Domingo-Ferrer and M. Laurent (Eds.): PSD 2022, LNCS 13463, pp. 324–346, 2022.
https://doi.org/10.1007/978-3-031-13945-1_23

by researchers to study substance use (e.g., alcohol use, tobacco use, marijuana use, prescription drug misuse, heroin use) and mental health issues (e.g. major depressive episode and mental illness).

NSDUH collects personal information on substance use, substance use treatment, mental illness, and other health-related measures, which are sensitive and private. Disclosure risk is always a concern with such data. Disclosure occurs when an unauthorized individual (an "intruder") tries to link a record in the microdata file to an identifiable respondent (a "target"). Variables used in the identification of the target's record are called identifying variables (IVs), which are usually known to others or can be found elsewhere. Such variables may include age, gender, race, education, income, and so on.

CBHSQ is a statistical unit approved by the U.S. Office of Management and Budget and has the responsibility to protect confidential data from disclosure identification. As such, NSDUH data are protected under the federal law known as the Confidential Information Protection and Statistical Efficiency Act of 2002 (CIPSEA). CIPSEA establishes confidentiality protections for information collected by U.S. statistical agencies, which ensures that all NSDUH data are used for statistical purposes only. During data collection, CIPSEA language is included in the lead letter and informed consent materials that are sent to respondents, informing respondents that all survey responses will be fully protected under federal law by CIPSEA. To protect the respondents' confidentiality, comply with federal regulations, and honor the confidentiality pledge, statistical disclosure treatment has been imposed on all NSDUH public use files (PUFs) to minimize disclosure risk.

To meet the increasing needs of researchers at large and protect data confidentiality and data quality, NSDUH PUFs have been produced using a statistical disclosure control technique called MASSC, which stands for Micro Agglomeration, Substitution, Subsampling, and Calibration [1–3]. In the micro agglomeration step, using a selected set of key IVs, the data are partitioned into risk strata to control for the level of treatment. Then, on a random basis, a sample of records is drawn from each stratum, and variables are substituted from a similar donor record. This substitution step introduces uncertainty about the identity of a record in the database and makes it difficult for an intruder to be certain that any record corresponds to a specific individual, because some of the variables used to identify the record may have come from other individuals. Next, a portion of the records is randomly removed from the file to reduce the probability of determining that any known respondent was in the PUF. This subsampling step introduces further uncertainty about the presence of a target record in the database. These two steps in combination substantially minimize the risk of an individual being identified or targeted. In MASSC, substitution and subsampling are done while simultaneously constraining the resulting file to a minimal increase in bias and a minimal decrease in precision for several substance use outcomes across several domains. In addition, the weights on the final file are recalibrated to known totals from the full analytic RUF to minimize the decrease in precision. Other perturbation techniques like variable collapsing, dropping, and local suppression (called post-MASSC treatment) are also used for NSDUH PUFs.

Quality assessment for the 2002 to 2013 PUFs have been conducted and published on the SAMHSA website [4]. A study was conducted to continue to gauge the impact of NSDUH disclosure avoidance treatment on the data quality of the 2014 to 2019 PUFs. This paper summarizes findings from that study. A more detailed report will be available on the SAMHSA website.

2 Materials and Methods

A subset of NSDUH tables was replicated using PUF data. Those estimates (i.e., estimates based on the restricted-use data) and their associated standard errors (SEs) for key substance use and mental health measures were compared with estimates produced from NSDUH PUFs. In addition, correlations of PUF and RUF estimates and SEs were examined, and multiyear trends for a small subset of key substance use and mental health measures were compared between the PUFs and RUFs.

2.1 Selection of NSDUH Published Tables to Replicate Using PUF Data

About 300 tables of substance use and mental health estimates from the 2014 to 2019 NSDUH substance use and mental health detailed tables (https://www.samhsa.gov/data/) (i.e., estimates based on the RUFs, and referred to as the RUF estimates) were selected and replicated using the corresponding PUFs (referred to as the PUF estimates). These tables were used to determine the impact of NSDUH disclosure treatment on the bias and precision of the PUF estimates compared with estimates from the untreated RUFs. Substance use outcomes included tobacco use, alcohol use, illicit drug use, perceived risk of substance use, and substance use disorder and treatment. Mental health outcomes included past year serious mental illness or any mental illness, major depressive episode, and suicide. Tables included domains such as age, gender, and race/ethnicity and were based on a variety of outcomes and included variables for which partial records were substituted and other variables that were not directly perturbed. Due to subsampling on the PUF, all variables were indirectly affected.

2.2 NSDUH PUFs and Comparison Table Production

NSDUH PUFs are publicly available on a data archive website (https://www.datafiles.samhsa.gov/), and the PUF-estimated percentages produced are based on downloaded data. Estimated percentages from RUFs (available online in NSDUH detailed tables, and referred to as RUF estimates) are displayed alongside PUF estimates. The tables also include SEs of these two sets of estimated percentages and ratios between RUF and PUF estimates and their corresponding SEs. Usual NSDUH precision-based suppression rules [5] are applied for RUF estimates but not for PUF estimates because PUF estimates can be calculated using data in the public domain. Thus, if an estimate is suppressed from a RUF table, the PUF estimate is still retained. However, the ratio of the estimates or corresponding SEs is suppressed.

2.3 Quality Assessment

The quality assessment was conducted by examining ratios of estimates and SEs from the PUF and RUF data on each substance use and mental health measure in the comparison tables. The ratios were calculated using the following equations:

$$\text{Ratio of Estimates} = \frac{\hat{\theta}_{PUF}(i)}{\hat{\theta}_{RUF}(i)} \tag{1}$$

$$\text{Ratio of SEs} = \frac{SE_{PUF}(i)}{SE_{RUF}(i)},\tag{2}$$

where $\hat{\theta}_{PUF}(i)$ and $\hat{\theta}_{RUF}(i)$ are the estimated percentages (referred to as estimates) from the PUF and RUF, respectively, for measure i, and $SE_{PUF}(i)$ and $SE_{RUF}(i)$ are their corresponding SEs.

Correlations between PUF and RUF estimates as well as PUF and RUF SEs were produced. No statistical tests of significance between PUF and RUF estimates were conducted because it was believed that, for most estimates, such tests would be unlikely to detect any significant differences given that the underlying data were almost the same.

Multiyear trend plots for a select set of measures and domain combinations (e.g., age groups, gender) were plotted to visually compare differences in trends between the estimates from the RUF and PUF data. All estimates from 2014 to 2018 were compared with the 2019 estimates.

3 Results and Discussion

3.1 Point Estimate Comparison

Forty-nine tables (from the set of NSDUH detailed tables) showing estimated percentages and their respective SEs were selected for comparison from each year (30 for substance use outcomes and 19 for mental health outcomes) of data from the 2014 to 2019 NSDUHs, resulting in a total of about 300 tables to be reproduced and compared. All ratios of estimates and their corresponding SEs from the PUF and RUF data in the selected tables for substance use and mental health measures were calculated using Eqs. (1) and (2). Unrounded PUF and RUF estimates were used in the calculation of these ratios, then the ratios were rounded to two decimal places in the tables. Examples of tables investigated are presented for substance use in Appendix Table A1 and for mental health in Appendix Table A2. A complete set of tables that were examined will be available in a comprehensive report on the SAMHSA website.

If the ratios of the estimates and SEs were close to 1, one can say that the PUF provided estimates that were similar to the RUF estimates. For example, the RUF and PUF estimated percentages of past year alcohol use among individuals aged 21 or older for 2017 in Appendix Table A1 were 70.7 and 70.6%, respectively, which yielded a ratio of the estimates of 1.00. The associated SEs of this past year alcohol use measure were 0.37 and 0.39%, as produced from the RUF and PUF, respectively, which gave a ratio of the SEs of 1.07. This comparison shows that the 2017 estimates produced from the PUF and RUF for this particular outcome and domain were fairly similar and that there was a 7% increase in the SEs produced from the PUF compared with the SEs produced from the RUF.

Large ratios indicate that PUF estimates have a larger than expected deviation from RUF estimates. Appendix Table A2 shows that estimated percentages of adults with serious mental illness in the past year among Native Hawaiian or Other Pacific Islanders for 2018 were 4.5 and 4.9% from the RUF and PUF, respectively, and SEs were 1.97 and 2.29% from the RUF and PUF, respectively. This comparison resulted in a ratio of estimates of 1.09 and a ratio of SEs of 1.16, which means there was a 9% increase in the PUF estimate and a 16% increase in the PUF SEs.

3.2 Distribution Analysis of the Ratios of the Estimates

To assess the overall impact of disclosure treatment on NSDUH data quality, the distributions of the ratios of estimates and the ratios of SEs were studied to examine the change in estimates and precision. Summary statistics for these ratios of estimates and SEs were produced. Table 1 shows the distribution of the ratios of all estimates (percentages) by year across all substance use and mental health measures considered for this study. Table 2 shows the distribution of the ratios of all associated SEs. For substance use measures, for each year, about 2,400 estimates were examined, and for mental health measures, for each year, about 1,200 to 1,400 estimates were examined. Results show that the average (mean) ratios for estimated percentages across years were within the 0.99 to 1.01 range (Table 1) for both substance use and mental health measures. Also, the average (mean) increase in SEs across years was about 7 to 9% for substance use measures and about 8 to 11% for mental health measures (mean ratios in Table 2 range from 1.07 to 1.11 overall). Because each PUF had about a 20% reduction in sample size (i.e., the PUF sample size is about 80% of the RUF sample size), some increase in SEs for PUF estimates was to be expected. That is, the ratio of PUF SEs and RUF SEs would be roughly around $1/\sqrt{0.8}$ which is 1.12. Thus, the overall PUF estimates were expected to have a 12% increase in SEs compared with the RUF estimates. Tables 1 and 2 demonstrate that in spite of using MASSC to perturb the PUF data, the quality of the PUF data remained similar to that of the RUF data. Across all 22,144 substance use and mental health estimates reviewed using 2014 to 2019 NSDUH data, the average ratio of estimates from the PUF and RUF was 1.00, and the average ratio of SEs from the PUF and RUF was 1.08. Thus, on average, NSDUH PUF and RUF estimates are similar and the decrease in precision is only about 8%.

Table 1. Distributions of ratios of percentages of substance use and mental health measures produced from the PUF and the RUF, by year: 2014–2019 NSDUHs

Description	N	Mean	0% Min	10%	50% Median	90%	100% Max
Substance Use (SU)							
2014	2,362	1.01	0.47	0.97	1.01	1.07	1.46
2015	2,369	1.00	0.17	0.95	1.00	1.05	4.15
2016	2,362	1.00	0.00	0.94	1.00	1.05	3.01
2017	2,363	1.00	0.00	0.96	1.00	1.06	1.50
2018	2,366	1.01	0.00	0.97	1.00	1.06	1.74
2019	2,366	1.00	0.00	0.96	1.00	1.06	1.52
Overall SU: 2014–2019	14,188	1.00	0.00	0.96	1.00	1.06	4.15
Mental Health (MH)							
2014	1,229	1.00	0.00	0.94	1.01	1.07	1.53
2015	1,245	0.99	0.09	0.93	1.00	1.05	1.49
2016	1,257	1.00	0.39	0.95	1.00	1.06	1.46
2017	1,402	1.01	0.51	0.96	1.01	1.08	1.47
2018	1,403	1.00	0.00	0.95	1.00	1.06	1.50
2019	1,420	1.00	0.37	0.95	1.00	1.07	1.34
Overall MH: 2014–2019	7,956	1.00	0.00	0.95	1.00	1.07	1.53
Overall SU and MH 2014–2019	22,144	1.00	0.00	0.95	1.00	1.06	4.15

PUF = public use file; RUF = restricted-use file. Ratio = PUF % ÷ RUF %. Ratios that were suppressed because the RUF estimate was suppressed have been excluded from this summary table.
Source: SAMHSA, Center for Behavioral Health Statistics and Quality, NSDUH, 2014 to 2019.

The distributions of estimate ratios across years in Table 1 appeared to be stable across the timeline. For ratios of estimates, data were centered (median) around 1.00, and most estimates had ratios between 0.94 (lowest 10th percentile) and 1.07 (highest 90th percentile) for substance use measures and between 0.93 (lowest 10th percentile) and 1.08 (highest 90th percentile) for mental health measures. Ratios of percentages ranged from 0.00 to 4.15 for substance use measures and from 0.00 to 1.53 for mental health measures. For the ratios of SEs (Table 2) for substance use measures, data were centered (median) around 1.07 to 1.09 across years, and most ratios of SEs were between 0.90 (lowest 10th percentile) and 1.26 (highest 90th percentile). For the ratios of SEs for mental health measures, data were centered (median) around 1.08 to 1.11 across years, and most ratios of SEs were between 0.91 (lowest 10th percentile) and 1.28 (highest 90th percentile). Ratios of SEs spread from 0.00 to 3.46 and from 0.00 to 1.89 for substance use and mental health measures, respectively.

Table 2. Distributions of ratios of standard errors of percentages of substance use and mental health measures produced from the PUF and the RUF, by year: 2014–2019 NSDUHs

Description	N	Mean	0% Min	10%	50% Median	90%	100% Max
Substance Use (SU)							
2014	2,362	1.08	0.44	0.93	1.08	1.24	1.56
2015	2,369	1.09	0.20	0.93	1.08	1.25	3.46
2016	2,362	1.07	0.00	0.90	1.07	1.24	2.15
2017	2,363	1.07	0.00	0.91	1.07	1.23	1.58
2018	2,366	1.08	0.00	0.92	1.09	1.25	1.74
2019	2,366	1.09	0.00	0.93	1.09	1.26	1.57
Overall SU: 2014–2019	14,188	1.08	0.00	0.92	1.08	1.25	3.46
Mental Health (MH)							
2014	1,229	1.09	0.00	0.93	1.09	1.28	1.69
2015	1,245	1.08	0.10	0.93	1.08	1.24	1.60
2016	1,257	1.08	0.33	0.92	1.08	1.23	1.61
2017	1,402	1.11	0.31	0.94	1.11	1.26	1.65
2018	1,403	1.11	0.00	0.95	1.11	1.29	1.89
2019	1,420	1.09	0.37	0.91	1.08	1.27	1.55
Overall MH: 2014–2019	7,956	1.09	0.00	0.93	1.09	1.26	1.89
Overall SU and MH 2014–2019	22,144	1.08	0.00	0.92	1.08	1.25	3.46

PUF = public use file; RUF = restricted-use file; SE = standard error. Ratio = PUF SE ÷ RUF SE.

Ratios that were suppressed because the RUF estimate and, consequently, its standard error were suppressed have been excluded from this summary table.

Source: SAMHSA, Center for Behavioral Health Statistics and Quality, NSDUH, 2014 to 2019.

3.3 Extreme Ratios

Extreme ratios of the estimates and the SEs from PUF and RUF were identified, which were defined as ratios <0.5 or >1.5 for either the estimates or SEs. Reasons for these extreme ratios are mainly because of zero or near zero prevalence (low prevalence) and small sample sizes. Examples of tables with extreme ratios are presented for substance use in Appendix Table B1 and for mental health in Appendix Table B2.

In Appendix Table B1, in 2015, for nicotine (cigarette) dependence in the past month among Asian persons aged 12 to 17, RUF and PUF estimates were 0.1 and 0.3%, respectively, and RUF and PUF SEs were 0.06 and 0.21%, respectively. The ratio of the unrounded estimates was 4.15, and the ratio of the unrounded SEs was 3.46, which were the maximum ratios for percentages as well as SEs of all substance use measures studied (Tables 1 and 2). These extreme ratios were due to low prevalence rates of the measure and a small domain with RUF and PUF sample sizes being less than 1,000. A similar case was seen among American Indian or Alaska Native persons aged 12 to 17, where the PUF and RUF estimates ratio was 0.43.

In Appendix Table B2, for the 2019 marijuana use disorder in the past year among persons aged 12 to 13 with MDE in the past year, both PUF and RUF estimates were less than 1% (low prevalence) and the sample size was small (i.e., below 1,000), resulting in ratios of both estimates and SEs being <0.5. As a reminder, ratios of estimates close to 1 are desirable.

Another reason for extreme ratios is that SEs for the PUF were larger than SEs for the RUF, but the RUF relative SEs (RSEs) were relatively small (<30%). This indicates that the RUF estimates were fairly stable, but, after disclosure treatment, the PUF estimates became less precise. For example, in Appendix Table B2, for the 2019 illicit drug use disorder in the past year among males aged 12 to 17 with no MDE in the past year, the PUF and RUF estimates were 3.0 and 2.9%, respectively, and the sample size was not small (>6,000). However, the PUF and RUF SEs were 0.41 and 0.27%, respectively, yielding a ratio of the SEs of 1.55. This might be due to the subsampling in MASSC treatment that can cause some inflation in variance. The ratio of their estimates was not affected much (in this example, the ratio of estimates was 1.04). Among the 22,144 ratios of estimates and 22,144 ratios of SEs that were observed for this study, only 58 (23 cases for substance use and 35 cases for mental health) were identified as being in this category, where the RUF estimate was stable (RUF RSE < 30%) and the PUF estimate was less precise (ratio of PUF SE to RUF SE > 1.5), which means only 0.13% of cases were greatly impacted by PUF treatment.

In summary, among the 14,000 estimates (and their corresponding SEs) that were compared for the substance use measures, 97 cases were identified as having extreme ratios (based on either the ratio of estimates or the ratio of SEs). For the approximately 8,000 estimates (and their corresponding SEs) that were compared for the mental health measures, 65 cases were identified with extreme ratios. Thus, across all substance use and mental health measures observed for this study, only a small amount (i.e., 0.4%) of cases had extreme ratios.

3.4   Correlations Between RUF and PUF Estimates

For researchers, agencies, and NSDUH data users to make sound policy decisions based on PUF data, it is important that RUF and PUF data be highly correlated. To verify this, correlations were calculated between the rounded PUF and RUF estimates and SEs across 2014 to 2019 for the selected set of substance use and mental health measures. Results (not shown) indicated that the estimates were all highly correlated, with correlation coefficients approximately equal to 1.00 for both substance use and mental health measures. The SEs were also highly correlated, with correlation coefficients between 0.98 and 0.99 for both substance use and mental health measures.

3.5   Multiyear Trend Comparison

Trend analysis is important in NSDUH for providing information on changes in rates and occurrences of substance use and mental health conditions over time. Such information can be used in prediction, prevention, intervention, and new policy development. To examine whether the PUF estimates provide similar trend patterns to the RUF estimates across years, several measures from the tables that compare the PUF estimates and the RUF estimates were selected. These estimates include tobacco use, alcohol use, illicit drug use, substance use disorder, mental illness, depression, suicidal thoughts and behavior, receipt of mental health services, and receipt of drug use treatment, by age group and gender. Because the underlying data between the PUF and the RUF were almost the same, no tests of significance between RUF and PUF trends were done. However, pairwise t tests were conducted to compare the estimates in 2019 with corresponding estimates in prior years from both the PUF and the RUF. Trends were assessed by plotting the PUF and RUF estimates by year to obtain graphical evidence of the overall direction of both trends (i.e., trends from PUF estimates and trends from RUF estimates). These plots also present whether PUF estimates had changes across time (by comparing 2014 estimates with 2019 estimates, comparing 2015 estimates with 2019 estimates, and so on), as observed from testing conducted using the RUF data. Significance testing was conducted at the 5% level.

Plots of PUF and RUF estimates over time were generated for substance use and mental health measures. Overall, 15 trend plots were produced for select age groups and 10 trend plots were produced by gender, where trends for males and females were plotted separately. For illustration purposes, one plot for a substance use measure and one for a mental health measure across years are displayed in Figs. 1 and 2. The PUF estimates demonstrated similar trends to the RUF estimates. For example, from 2014 to 2019, prevalence tended to decrease with time for past year alcohol use among youths aged 12 to 20 (Fig. 1). Also, the pattern of significant changes observed in the RUF and PUF estimates is the same. Estimates for 2014 and 2015 alcohol use were significantly different from (and higher than) the estimate for 2019 alcohol use among youths, but estimates for 2016 to 2018 alcohol use were not significantly different from the 2019 alcohol use estimate. Thus, conclusions about trends of alcohol use among youths that one could draw from RUF and PUF data are that the trends are similar (Table 3).

For the most part, trend plots by gender for substance use measures and mental health measures, where both males and females were plotted on the same graph for PUF and RUF, show a similar pattern to plots by age groups. That is, the PUF estimates by gender demonstrated similar trends to the RUF estimates by gender for both males and females. Regarding serious suicidal thoughts in the past year among adults aged 18 or older (Fig. 2), some differences in the estimates comparison against 2019 were observed in PUF versus RUF. For example, there was a significant difference between the 2017 and 2019 estimates of serious thoughts of suicide in the past year among RUF females aged 18 or older, but that difference was not significant among PUF females (Fig. 2, Table 4). One possible reason for the dissimilarity is that the SE of the PUF estimate was relatively larger than that of the RUF estimate, which could lead to the PUF test not being significant at the 0.05 level. However, the directionality of the change was maintained for all such cases.

Out of 162 estimate comparisons (paired *t* tests) between 2014 to 2018 estimates and 2019 estimates for selected measures over the 25 trend plots that were examined, nine changes in significance occurred. That is, the pair test went from being significant using RUF data to being nonsignificant using PUF data, or vice versa, which accounted for roughly 6% of the overall comparisons. This testing was done at the 5% level of significance, which means there was a 5% risk of concluding that a difference existed when there was no actual difference. Thus, some of the discrepancies that were seen in the RUF tests and the PUF tests could also be explained due to this reason. However, because PUF SEs were expected to be larger than RUF SEs due to subsampling on the PUF, some differences in tests were expected.

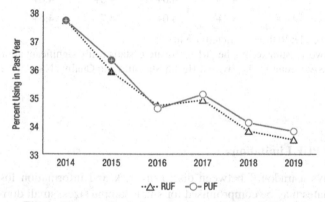

Fig. 1. Past year alcohol use among persons aged 12 to 20

Table 3. Past year alcohol use among persons aged 12 to 20

%	2014	2015	2016	2017	2018	2019
RUF	37.7+	35.9+	34.7	34.9	33.8	33.5
PUF	37.7+	36.3+	34.6	35.1	34.1	33.8

PUF = public use file; RUF = restricted-use file.
+Difference between estimate and the 2019 estimate is statistically significant at the .05 level.
Source: SAMHSA, Center for Behavioral Health Statistics and Quality, NSDUH, 2014–2019.

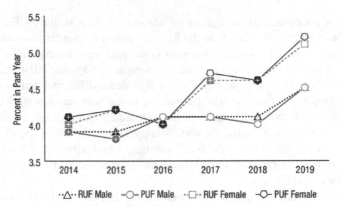

Fig. 2. Had serious thoughts of suicide in past year among persons aged 18 or older

Table 4. Had serious thoughts of suicide in past year among persons aged 18 or older

%	2014	2015	2016	2017	2018	2019
RUF Male	3.9+	3.9+	4.1	4.1	4.1	4.5
PUF Male	3.9+	3.8+	4.1	4.1	4.0	4.5
RUF Female	4.0+	4.2+	4.0+	4.6+	4.6+	5.1
PUF Female	4.1+	4.2+	4.0+	4.7	4.6+	5.2

PUF = public use file; RUF = restricted-use file.
+Difference between estimate and the 2019 estimate is statistically significant at the .05 level.
Source: SAMHSA, Center for Behavioral Health Statistics and Quality, NSDUH, 2014–2019.

3.6   NSDUH PUF Limitations

There is always a trade-off between disclosure risk and information loss. Precision for PUF estimates may be compromised for small sample sizes, small domains, or rare outcomes using a single year of data. However, this can also be true for RUFs. Unreliable RUF estimates are routinely suppressed in NSDUH detailed tables. In such cases, PUF users are encouraged to apply precision-based suppression rules to unreliable estimates from the PUF. Examples of suppression rules for publishing NSDUH estimates can be found on the SAMHSA website [5]. Alternately, multiple years of data may be combined to generate more stable estimates.

To protect confidentiality, geographic identifiers were not included on NSDUH PUFs. Therefore, the PUFs cannot be used to produce estimates at the region, state, or substate levels or to perform related analyses. Similarly, for any other variables that were deleted due to disclosure reasons, estimates cannot be obtained using the PUFs. In such cases, users can request access to the NSDUH RUFs via the SAMHSA research data center website (https://www.samhsa.gov/data/data-we-collect/samhsa-rdc).

4 Conclusions

Based on the comparisons between the 2014 to 2019 PUF and RUF estimates and corresponding SEs, it can be concluded that the NSDUH PUFs continue to provide high-quality data for producing estimates and SEs for substance use and mental health measures. In general, consistent analytic results can be expected between the two types of files even when the actual numbers are slightly different. The MASSC disclosure treatment controls bias and variance so that information loss due to treatment is minimal.

NSDUH data are used by stakeholders at all levels of the U.S. government as well as states, localities, nonprofit organizations, and academic researchers. It is imperative that publicly available NSDUH data produce accurate and high-quality prevalence estimates while maintaining the confidentiality of survey respondents. MASSC treatment allows stakeholders to use NSDUH PUFs to obtain reliable estimates and to track substance use prevalence or rates of mental health conditions over time (i.e., to monitor trends). However, PUF users are cautioned in the analysis and interpretation of near zero or low prevalence rates and of estimates based on small sample sizes or small domains. Thus, PUF users are encouraged to consider employing suppression rules to estimates considered to have low precision.

An extension of this work in the future may involve examining the interrelations among NSDUH variables via bivariate or multivariate analyses from the PUFs and RUFs.

Appendix

Table A1. Alcohol use in lifetime, past year, and past month and binge and heavy alcohol use in past month among persons aged 21 or older, by demographic characteristics: comparison of percentages and standard errors from restricted-use file and public use file estimates, 2017 NSDUH

Demographic characteristic	Lifetime alcohol use			Past year alcohol use			Past month alcohol use			Past month binge alcohol use			Past month heavy alcohol use		
	RUF % (SE)	PUF % (SE)	Ratio of % (Ratio of SE)	RUF % (SE)	PUF % (SE)	Ratio of % (Ratio of SE)	RUF % (SE)	PUF % (SE)	Ratio of % (Ratio of SE)	RUF % (SE)	PUF % (SE)	Ratio of % (Ratio of SE)	RUF % (SE)	PUF % (SE)	Ratio of % (Ratio of SE)
Total	87.3 (0.26)	87.3 (0.25)	1.00 (0.93)	70.7 (0.37)	70.6 (0.39)	1.00 (1.07)	56.8 (0.37)	56.8 (0.42)	1.00 (1.13)	26.5 (0.30)	26.5 (0.32)	1.00 (1.07)	6.7 (0.15)	6.7 (0.20)	0.99 (1.30)
Gender															
Male	90.3 (0.31)	90.4 (0.32)	1.00 (1.04)	73.8 (0.46)	73.7 (0.50)	1.00 (1.10)	61.7 (0.51)	61.7 (0.57)	1.00 (1.13)	31.7 (0.46)	31.7 (0.55)	1.00 (1.20)	9.3 (0.26)	9.2 (0.32)	0.99 (1.22)
Female	84.5 (0.38)	84.4 (0.39)	1.00 (1.01)	67.7 (0.51)	67.8 (0.56)	1.00 (1.10)	52.4 (0.49)	52.4 (0.54)	1.00 (1.09)	21.7 (0.37)	21.7 (0.41)	1.00 (1.10)	4.4 (0.17)	4.4 (0.20)	1.00 (1.23)
Hispanic origin and race															
Not Hispanic or Latino	88.4 (0.28)	88.5 (0.28)	1.00 (0.98)	71.4 (0.39)	71.5 (0.44)	1.00 (1.14)	57.8 (0.41)	57.9 (0.50)	1.00 (1.23)	26.0 (0.33)	26.0 (0.38)	1.00 (1.16)	6.9 (0.17)	6.9 (0.19)	0.99 (1.16)
White	91.6 (0.26)	91.8 (0.27)	1.00 (1.04)	74.2 (0.42)	74.4 (0.46)	1.00 (1.08)	61.2 (0.45)	61.3 (0.51)	1.00 (1.14)	27.1 (0.38)	27.2 (0.44)	1.00 (1.16)	7.6 (0.20)	7.7 (0.23)	1.00 (1.11)
Black or African American	80.0 (0.78)	80.2 (0.80)	1.00 (1.03)	63.6 (0.96)	63.4 (1.04)	1.00 (1.09)	48.8 (0.94)	48.9 (1.06)	1.00 (1.13)	25.6 (0.81)	25.6 (0.87)	1.00 (1.07)	5.0 (0.40)	4.6 (0.45)	0.92 (1.12)
AIAN	88.3 (2.39)	87.7 (2.57)	0.99 (1.08)	57.8 (3.74)	56.4 (3.94)	0.98 (1.05)	45.5 (3.55)	44.3 (3.59)	0.97 (1.01)	30.3 (3.15)	28.7 (3.04)	0.95 (0.97)	7.7 (1.55)	7.9 (1.67)	1.03 (1.08)

(continued)

Table A1. (*continued*)

Demographic characteristic	Lifetime alcohol use			Past year alcohol use			Past month alcohol use			Past month binge alcohol use			Past month heavy alcohol use		
	RUF % (SE)	PUF % (SE)	Ratio of % (Ratio of SE)	RUF % (SE)	PUF % (SE)	Ratio of % (Ratio of SE)	RUF % (SE)	PUF % (SE)	Ratio of % (Ratio of SE)	RUF % (SE)	PUF % (SE)	Ratio of % (Ratio of SE)	RUF % (SE)	PUF % (SE)	Ratio of % (Ratio of SE)
NHOPI	* (*)	78.9 (5.00)	* (*)	58.1 (5.45)	60.1 (6.66)	1.03 (1.22)	48.4 (5.32)	49.6 (6.49)	1.02 (1.22)	28.4 (4.94)	27.1 (4.52)	0.96 (0.92)	6.7 (2.19)	6.5 (2.00)	0.96 (0.91)
Asian	69.0 (1.82)	68.0 (2.20)	0.99 (1.21)	57.7 (1.82)	57.3 (2.11)	0.99 (1.16)	42.1 (1.70)	42.0 (2.05)	1.00 (1.21)	14.1 (1.03)	13.8 (1.05)	0.98 (1.01)	2.2 (0.33)	2.1 (0.40)	0.94 (1.19)
Two or More Races	91.3 (1.40)	90.4 (1.56)	0.99 (1.11)	70.7 (2.60)	70.5 (2.55)	1.00 (0.98)	50.7 (2.63)	50.4 (2.65)	0.99 (1.01)	24.2 (2.01)	24.7 (1.70)	1.02 (0.84)	9.0 (1.46)	9.0 (1.60)	1.00 (1.10)
Hispanic or Latino	81.5 (0.76)	81.2 (0.71)	1.00 (0.93)	66.8 (0.97)	66.1 (0.94)	0.99 (0.96)	51.5 (1.01)	51.1 (0.87)	0.99 (0.86)	29.3 (0.82)	29.1 (0.84)	0.99 (1.03)	5.6 (0.36)	5.7 (0.40)	1.02 (1.10)
Gender/race/ hispanic origin															
Male, White, Not Hispanic	93.1 (0.31)	93.2 (0.39)	1.00 (1.26)	75.6 (0.52)	75.5 (0.56)	1.00 (1.07)	64.8 (0.60)	64.8 (0.70)	1.00 (1.17)	31.7 (0.54)	31.8 (0.65)	1.00 (1.19)	10.4 (0.34)	10.4 (0.37)	1.00 (1.10)
Female, White, Not Hispanic	90.2 (0.39)	90.5 (0.41)	1.00 (1.07)	73.0 (0.58)	73.5 (0.62)	1.01 (1.06)	57.8 (0.59)	58.0 (0.59)	1.00 (1.00)	22.7 (0.48)	22.8 (0.58)	1.00 (1.19)	5.1 (0.22)	5.1 (0.26)	1.00 (1.17)

(*continued*)

Table A1. (*continued*)

Demographic characteristic	Lifetime alcohol use			Past year alcohol use			Past month alcohol use			Past month binge alcohol use			Past month heavy alcohol use		
	RUF % (SE)	PUF % (SE)	Ratio of % (Ratio of SE)	RUF % (SE)	PUF % (SE)	Ratio of % (Ratio of SE)	RUF % (SE)	PUF % (SE)	Ratio of % (Ratio of SE)	RUF % (SE)	PUF % (SE)	Ratio of % (Ratio of SE)	RUF % (SE)	PUF % (SE)	Ratio of % (Ratio of SE)
Male, Black, Not Hispanic	83.3 (1.10)	83.4 (1.33)	1.00 (1.21)	67.2 (1.31)	67.0 (1.50)	1.00 (1.14)	54.7 (1.38)	54.4 (1.57)	0.99 (1.14)	30.3 (1.33)	30.1 (1.19)	0.99 (0.90)	7.6 (0.76)	7.0 (0.75)	0.92 (0.99)
Female, Black, Not Hispanic	77.3 (1.10)	77.5 (1.18)	1.00 (1.08)	60.7 (1.31)	60.5 (1.23)	1.00 (0.94)	44.0 (1.28)	44.3 (1.24)	1.01 (0.97)	21.8 (1.00)	21.8 (1.08)	1.00 (1.08)	2.9 (0.39)	2.6 (0.51)	0.90 (1.29)
Male, Hispanic	88.7 (0.86)	88.4 (0.82)	1.00 (0.95)	73.8 (1.32)	73.5 (1.53)	1.00 (1.16)	58.9 (1.42)	58.6 (1.32)	0.99 (0.93)	37.7 (1.29)	37.4 (1.43)	0.99 (1.11)	8.0 (0.60)	8.1 (0.62)	1.02 (1.04)
Female, Hispanic	74.4 (1.19)	74.1 (1.20)	1.00 (1.01)	59.8 (1.38)	58.8 (1.38)	0.98 (1.00)	44.3 (1.34)	43.7 (1.25)	0.99 (0.93)	21.1 (0.94)	20.9 (1.02)	0.99 (1.08)	3.3 (0.42)	3.4 (0.41)	1.03 (0.97)

* = low precision; AIAN = American Indian or Alaska Native; NHOPI = Native Hawaiian or Other Pacific Islander; PUF = public use file, which refers to estimates produced from a NSDUH PUF; RUF = restricted-use file, which refers to estimates produced from a NSDUH RUF and published in the NSDUH detailed tables; SE = standard error that is associated with the estimate.

NOTE: Ratio of % = PUF % ÷ RUF %; Ratio of SE = PUF SE ÷ RUF SE.

NOTE: Estimates of binge alcohol use include use by those who were heavy alcohol users.

Definitions: Measures and terms are defined in Appendix A of the 2019 NSDUH detailed tables at https://www.samhsa.gov/data/.

Source: SAMHSA, Center for Behavioral Health Statistics and Quality, National Survey on Drug Use and Health, 2017.

Table A2. Serious mental illness in past year among persons aged 18 or older, by age group and demographic characteristics: comparison of percentages and standard errors from restricted-use file and public use file estimates, 2018 NSDUH

Demographic characteristic	Aged 18+			Aged 18–25			Aged 26+			Aged 26–49			Aged 50+		
	RUF % (SE)	PUF % (SE)	Ratio of % (Ratio of SE)	RUF % (SE)	PUF % (SE)	Ratio of % (Ratio of SE)	RUF % (SE)	PUF % (SE)	Ratio of % (Ratio of SE)	RUF % (SE)	PUF % (SE)	Ratio of % (Ratio of SE)	RUF % (SE)	PUF % (SE)	Ratio of % (Ratio of SE)
Total	4.6 (0.12)	4.6 (0.16)	1.00 (1.30)	7.7 (0.25)	7.6 (0.29)	0.99 (1.15)	4.1 (0.13)	4.1 (0.17)	1.01 (1.31)	5.9 (0.18)	5.8 (0.22)	0.99 (1.17)	2.5 (0.19)	2.6 (0.23)	1.04 (1.24)
Gender															
Male	3.4 (0.16)	3.3 (0.20)	0.99 (1.28)	5.4 (0.30)	5.3 (0.35)	0.99 (1.18)	3.0 (0.17)	3.0 (0.23)	0.98 (1.34)	4.4 (0.25)	4.3 (0.33)	0.97 (1.33)	1.7 (0.24)	1.7 (0.26)	1.02 (1.11)
Female	5.7 (0.18)	5.8 (0.23)	1.01 (1.26)	10.0 (0.41)	10.0 (0.45)	0.99 (1.09)	5.1 (0.19)	5.2 (0.25)	1.02 (1.28)	7.2 (0.28)	7.3 (0.32)	1.01 (1.14)	3.2 (0.27)	3.4 (0.33)	1.04 (1.24)
Hispanic origin and race															
Not Hispanic or Latino	4.8 (0.14)	4.8 (0.17)	1.01 (1.26)	8.1 (0.29)	8.0 (0.33)	0.99 (1.12)	4.3 (0.15)	4.3 (0.18)	1.01 (1.25)	6.3 (0.21)	6.3 (0.24)	1.00 (1.14)	2.6 (0.20)	2.7 (0.25)	1.04 (1.22)
White	5.1 (0.16)	5.2 (0.21)	1.02 (1.29)	9.1 (0.35)	9.1 (0.42)	1.00 (1.19)	4.6 (0.18)	4.7 (0.23)	1.02 (1.31)	7.2 (0.27)	7.3 (0.36)	1.01 (1.33)	2.8 (0.24)	2.9 (0.29)	1.04 (1.24)
Black or African American	3.6 (0.31)	3.5 (0.32)	0.97 (1.04)	4.5 (0.55)	4.6 (0.65)	1.02 (1.18)	3.5 (0.36)	3.3 (0.35)	0.96 (0.97)	4.4 (0.43)	4.2 (0.46)	0.94 (1.07)	2.4 (0.59)	2.4 (0.48)	0.99 (0.81)
AIAN	6.4 (1.60)	6.4 (1.51)	0.99 (0.94)	6.3 (2.50)	7.0 (2.76)	1.11 (1.11)	6.5 (1.80)	6.2 (1.73)	0.97 (0.96)	8.5 (2.43)	7.6 (1.66)	0.90 (0.68)	* (*)	5.0 (2.65)	* (*)

(continued)

Table A2. (*continued*)

Demographic characteristic	Aged 18+			Aged 18–25			Aged 26+			Aged 26–49			Aged 50+		
	RUF % (SE)	PUF % (SE)	Ratio of % (Ratio of SE)	RUF % (SE)	PUF % (SE)	Ratio of % (Ratio of SE)	RUF % (SE)	PUF % (SE)	Ratio of % (Ratio of SE)	RUF % (SE)	PUF % (SE)	Ratio of % (Ratio of SE)	RUF % (SE)	PUF % (SE)	Ratio of % (Ratio of SE)
NHOPI	4.5 (1.97)	4.9 (2.29)	1.09 (1.16)	* (*)	3.2 (2.81)	* (*)	4.9 (2.33)	5.3 (2.64)	1.08 (1.13)	* (*)	5.8 (3.34)	* (*)	* (*)	4.4 (3.97)	* (*)
Asian	2.1 (0.29)	2.0 (0.26)	0.93 (0.90)	5.8 (1.10)	4.9 (1.01)	0.84 (0.92)	1.4 (0.29)	1.4 (0.29)	0.99 (1.00)	2.2 (0.47)	2.2 (0.47)	0.98 (0.99)	0.1 (0.08)	0.1 (0.08)	1.03 (1.02)
Two or More Races	7.1 (0.96)	6.9 (1.05)	0.97 (1.10)	14.5 (2.02)	13.5 (2.24)	0.93 (1.11)	5.5 (1.05)	5.5 (1.06)	1.00 (1.01)	8.9 (1.39)	9.2 (1.61)	1.04 (1.15)	3.2 (1.46)	3.1 (1.38)	0.97 (0.95)
Hispanic or Latino	3.6 (0.26)	3.5 (0.28)	0.98 (1.07)	6.4 (0.51)	6.4 (0.61)	1.00 (1.21)	3.0 (0.30)	2.9 (0.31)	0.97 (1.05)	3.9 (0.39)	3.7 (0.43)	0.95 (1.12)	1.4 (0.47)	1.5 (0.50)	1.04 (1.07)
Current employment															
Full-Time	3.9 (0.15)	3.9 (0.20)	1.00 (1.39)	7.6 (0.38)	7.3 (0.44)	0.96 (1.15)	3.4 (0.16)	3.5 (0.22)	1.02 (1.41)	4.2 (0.20)	4.2 (0.26)	0.99 (1.31)	2.0 (0.27)	2.2 (0.34)	1.09 (1.26)
Part-Time	5.8 (0.32)	5.7 (0.32)	0.98 (1.02)	8.5 (0.53)	8.3 (0.59)	0.99 (1.13)	4.8 (0.39)	4.7 (0.45)	0.98 (1.15)	8.1 (0.65)	7.9 (0.56)	0.97 (0.86)	1.9 (0.46)	2.0 (0.48)	1.02 (1.05)

(*continued*)

Table A2. (continued)

Demographic characteristic	Aged 18+			Aged 18–25			Aged 26+			Aged 26–49			Aged 50+		
	RUF % (SE)	PUF % (SE)	Ratio of % (Ratio of SE)	RUF % (SE)	PUF % (SE)	Ratio of % (Ratio of SE)	RUF % (SE)	PUF % (SE)	Ratio of % (Ratio of SE)	RUF % (SE)	PUF % (SE)	Ratio of % (Ratio of SE)	RUF % (SE)	PUF % (SE)	Ratio of % (Ratio of SE)
Unemployed	7.9 (0.69)	7.8 (0.75)	0.99 (1.08)	9.3 (0.95)	9.1 (1.11)	0.99 (1.17)	7.3 (0.93)	7.2 (0.95)	0.99 (1.03)	8.7 (1.05)	8.5 (1.10)	0.98 (1.04)	4.3 (1.84)	4.4 (2.17)	1.02 (1.17)
Other[1]	4.7 (0.24)	4.8 (0.25)	1.02 (1.04)	6.4 (0.49)	6.8 (0.46)	1.06 (0.94)	4.5 (0.26)	4.6 (0.27)	1.01 (1.05)	10.0 (0.55)	10.1 (0.57)	1.00 (1.03)	2.9 (0.29)	2.9 (0.33)	1.01 (1.13)

* = low precision; AIAN = American Indian or Alaska Native; NHOPI = Native Hawaiian or Other Pacific Islander; PUF = public use file, which refers to estimates produced from a NSDUH PUF; RUF = restricted-use file, which refers to estimates produced from a NSDUH RUF and published in the NSDUH detailed tables; SE = standard error that is associated with the estimate.

NOTE: Ratio of % = PUF % ÷ RUF %; Ratio of SE = PUF SE ÷ RUF SE.

NOTE: Serious Mental Illness (SMI) aligns with DSM-IV criteria and is defined as having a diagnosable mental, behavioral, or emotional disorder, other than a developmental or substance use disorder. Estimates of SMI are a subset of estimates of any mental illness (AMI) because SMI is limited to persons with AMI that resulted in serious functional impairment. These mental illness estimates are based on a predictive model and are not direct measures of diagnostic status.

[1] The other employment category includes students, persons keeping house or caring for children full time, retired or disabled persons, or other persons not in the labor force.

Definitions: Measures and terms are defined in Appendix A of the 2019 NSDUH detailed tables at https://www.samhsa.gov/data/.

Source: SAMHSA, Center for Behavioral Health Statistics and Quality, National Survey on Drug Use and Health, 2018.

Table B1. Nicotine (cigarette) dependence in past month among persons aged 12 or older, by age group and demographic characteristics: comparison of percentages and standard errors from restricted-use file and public use file estimates, 2015 NSDUH

Demographic characteristic	Aged 12+			Aged 12–17			Aged 18+			Aged 18–25			Aged 26+		
	RUF % (SE)	PUF % (SE)	Ratio of % (Ratio of SE)	RUF % (SE)	PUF % (SE)	Ratio of % (Ratio of SE)	RUF % (SE)	PUF % (SE)	Ratio of % (Ratio of SE)	RUF % (SE)	PUF % (SE)	Ratio of % (Ratio of SE)	RUF % (SE)	PUF % (SE)	Ratio of % (Ratio of SE)
Total	10.8 (0.20)	10.9 (0.22)	1.01 (1.10)	1.3 (0.11)	1.4 (0.14)	1.08 (1.29)	11.8 (0.22)	11.9 (0.24)	1.01 (1.11)	11.7 (0.34)	11.6 (0.36)	0.99 (1.08)	11.8 (0.24)	12.0 (0.25)	1.01 (1.04)
Gender															
Male	12.0 (0.28)	12.2 (0.30)	1.02 (1.08)	1.3 (0.14)	1.4 (0.15)	1.07 (1.09)	13.2 (0.30)	13.4 (0.34)	1.01 (1.11)	13.1 (0.48)	12.9 (0.49)	0.99 (1.03)	13.2 (0.35)	13.4 (0.37)	1.02 (1.05)
Female	9.7 (0.25)	9.7 (0.28)	1.01 (1.11)	1.3 (0.16)	1.4 (0.22)	1.10 (1.37)	10.5 (0.28)	10.6 (0.30)	1.01 (1.10)	10.3 (0.42)	10.3 (0.46)	1.00 (1.09)	10.5 (0.30)	10.6 (0.32)	1.01 (1.05)
Hispanic origin and race															
Not Hispanic or Latino	11.9 (0.23)	12.1 (0.24)	1.01 (1.05)	1.5 (0.13)	1.6 (0.17)	1.08 (1.32)	12.9 (0.25)	13.1 (0.26)	1.01 (1.06)	13.4 (0.40)	13.3 (0.44)	0.99 (1.09)	12.8 (0.27)	13.0 (0.27)	1.02 (1.00)
White	12.5 (0.27)	12.7 (0.29)	1.02 (1.09)	1.8 (0.17)	1.9 (0.23)	1.10 (1.31)	13.4 (0.29)	13.6 (0.32)	1.02 (1.10)	14.6 (0.49)	14.5 (0.50)	0.99 (1.03)	13.2 (0.31)	13.5 (0.34)	1.02 (1.08)
Black or African American	11.9 (0.57)	11.7 (0.59)	0.99 (1.03)	0.7 (0.21)	0.7 (0.25)	0.94 (1.20)	13.3 (0.64)	13.1 (0.64)	0.99 (1.01)	10.8 (0.76)	10.7 (0.93)	0.99 (1.24)	13.8 (0.75)	13.6 (0.78)	0.99 (1.04)
AIAN	16.6 (2.40)	16.7 (2.66)	1.00 (1.11)	0.8 (0.45)	0.4 (0.28)	0.43 (0.63)	18.6 (2.72)	18.7 (2.95)	1.00 (1.08)	* (*)	19.4 (5.96)	* (*)	18.0 (3.07)	18.5 (3.48)	1.03 (1.14)
NHOPI	8.4 (1.92)	8.8 (1.56)	1.04 (0.81)	* (*)	0.0 (0.00)	* (*)	10.1 (2.33)	10.4 (1.86)	1.03 (0.80)	* (*)	18.1 (3.93)	* (*)	7.8 (2.61)	7.8 (2.06)	1.00 (0.79)

(continued)

Table B1. (*continued*)

Demographic characteristic	Aged 12+			Aged 12–17			Aged 18+			Aged 18–25			Aged 26+		
	RUF % (SE)	PUF % (SE)	Ratio of % (Ratio of SE)	RUF % (SE)	PUF % (SE)	Ratio of % (Ratio of SE)	RUF % (SE)	PUF % (SE)	Ratio of % (Ratio of SE)	RUF % (SE)	PUF % (SE)	Ratio of % (Ratio of SE)	RUF % (SE)	PUF % (SE)	Ratio of % (Ratio of SE)
Asian	4.3 (0.57)	4.3 (0.62)	1.01 (1.08)	0.1 (0.06)	0.3 (0.21)	4.15 (3.46)	4.7 (0.62)	4.7 (0.68)	1.01 (1.09)	5.8 (1.08)	5.6 (1.13)	0.96 (1.05)	4.5 (0.70)	4.5 (0.76)	1.02 (1.09)
Two or More Races	16.2 (1.53)	16.5 (1.53)	1.02 (1.00)	2.3 (0.78)	2.2 (0.59)	0.99 (0.76)	19.0 (1.84)	19.3 (1.83)	1.01 (0.99)	16.4 (2.03)	16.4 (2.42)	1.00 (1.19)	19.8 (2.29)	20.1 (2.24)	1.02 (0.98)
Hispanic or Latino	5.0 (0.32)	5.0 (0.33)	1.00 (1.03)	0.7 (0.19)	0.8 (0.26)	1.10 (1.38)	5.6 (0.37)	5.6 (0.37)	1.00 (0.99)	5.6 (0.51)	5.6 (0.53)	1.00 (1.03)	5.6 (0.44)	5.6 (0.45)	1.00 (1.01)
Family income															
Less Than $20,000	18.8 (0.56)	19.3 (0.58)	1.02 (1.03)	2.2 (0.35)	2.4 (0.45)	1.07 (1.29)	20.4 (0.61)	20.9 (0.64)	1.02 (1.05)	14.4 (0.70)	14.6 (0.63)	1.01 (0.90)	22.3 (0.76)	22.9 (0.85)	1.03 (1.12)
$20,000–$49,999	12.8 (0.36)	12.5 (0.47)	0.98 (1.30)	1.6 (0.24)	1.6 (0.27)	1.03 (1.17)	13.9 (0.40)	13.6 (0.51)	0.98 (1.28)	12.3 (0.56)	11.9 (0.62)	0.97 (1.10)	14.2 (0.45)	13.9 (0.57)	0.98 (1.26)
$50,000–$74,999	9.2 (0.43)	9.3 (0.42)	1.01 (0.98)	1.0 (0.22)	1.2 (0.27)	1.26 (1.19)	10.0 (0.47)	10.1 (0.46)	1.01 (0.97)	10.4 (0.87)	10.7 (1.05)	1.04 (1.20)	9.9 (0.52)	10.0 (0.48)	1.01 (0.93)
$75,000 or More	5.9 (0.23)	6.2 (0.30)	1.05 (1.29)	0.8 (0.13)	0.9 (0.15)	1.10 (1.12)	6.5 (0.26)	6.8 (0.33)	1.05 (1.29)	8.2 (0.55)	7.9 (0.62)	0.96 (1.14)	6.3 (0.28)	6.7 (0.36)	1.07 (1.27)

* = low precision; AIAN = American Indian or Alaska Native; NHOPI = Native Hawaiian or Other Pacific Islander; PUF = public use file, which refers to estimates produced from a NSDUH PUF; RUF = restricted-use file, which refers to estimates produced from a NSDUH RUF and published in the NSDUH detailed tables; SE = standard error that is associated with the estimate.

NOTE: Ratio of % = PUF % ÷ RUF %; Ratio of SE = PUF SE ÷ RUF SE.

Definitions: Measures and terms are defined in Appendix A of the 2019 NSDUH detailed tables at https://www.samhsa.gov/data/.

Source: SAMHSA, Center for Behavioral Health Statistics and Quality, National Survey on Drug Use and Health, 2015.

Table B2. Substance use disorder status in past year among persons aged 12 to 17, by past year major depressive episode (MDE) status and demographic characteristics: comparison of percentages and standard errors from restricted-use file and public use file estimates, 2019 NSDUH

| | No substance use disorder | | | Substance use disorder | | | | | | | | | | | | | | | | | | | | |
| | | | | Illicit drugs | | | Marijuana | | | Opioids | | | Alcohol | | | Both illicit drugs and alcohol | | | Illicit drugs or alcohol | | |
	RUF % (SE)	PUF % (SE)	Ratio of % (Ratio of SE)	RUF % (SE)	PUF % (SE)	Ratio of % (Ratio of SE)	RUF % (SE)	PUF % (SE)	Ratio of % (Ratio of SE)	RUF % (SE)	PUF % (SE)	Ratio of % (Ratio of SE)	RUF % (SE)	PUF % (SE)	Ratio of % (Ratio of SE)	RUF % (SE)	PUF % (SE)	Ratio of % (Ratio of SE)	RUF % (SE)	PUF % (SE)	Ratio of % (Ratio of SE)
Total[1]	95.5 (0.22)	95.3 (0.29)	1.00 (1.32)	3.6 (0.20)	3.8 (0.27)	1.06 (1.37)	2.8 (0.17)	2.9 (0.23)	1.05 (1.33)	0.3 (0.06)	0.4 (0.07)	1.03 (1.21)	1.7 (0.12)	1.7 (0.13)	1.03 (1.09)	0.8 (0.09)	0.8 (0.10)	1.07 (1.08)	4.5 (0.22)	4.7 (0.29)	1.04 (1.32)
MDE	89.5 (0.79)	88.7 (1.01)	0.99 (1.28)	8.3 (0.71)	8.8 (0.80)	1.07 (1.13)	6.1 (0.59)	6.4 (0.65)	1.05 (1.09)	1.0 (0.27)	1.0 (0.37)	1.08 (1.39)	4.6 (0.57)	4.8 (0.70)	1.04 (1.23)	2.4 (0.42)	2.3 (0.41)	0.97 (0.97)	10.5 (0.79)	11.3 (1.01)	1.08 (1.28)
Age group																					
12–13	97.5 (0.90)	97.8 (0.93)	1.00 (1.04)	1.8 (0.79)	1.9 (0.90)	1.04 (1.15)	0.4 (0.26)	0.1 (0.10)	0.37 (0.37)	0.2 (0.16)	0.3 (0.22)	1.34 (1.39)	0.7 (0.43)	0.3 (0.29)	0.46 (0.68)	* (*)	0.0 (0.00)	* (*)	2.5 (0.90)	2.2 (0.93)	0.88 (1.04)
14–15	90.1 (1.23)	89.1 (1.72)	0.99 (1.40)	7.8 (1.12)	8.7 (1.44)	1.11 (1.29)	5.8 (0.92)	6.6 (1.18)	1.14 (1.28)	0.9 (0.40)	1.0 (0.50)	1.12 (1.26)	3.8 (0.80)	4.0 (0.99)	1.05 (1.24)	1.7 (0.57)	1.7 (0.73)	1.02 (1.27)	9.9 (1.23)	10.9 (1.72)	1.10 (1.40)
16–17	85.0 (1.43)	83.7 (1.49)	0.98 (1.04)	11.8 (1.27)	12.5 (1.26)	1.06 (0.99)	9.2 (1.10)	9.5 (1.23)	1.03 (1.11)	1.4 (0.52)	1.5 (0.73)	1.05 (1.40)	7.4 (1.11)	7.9 (1.25)	1.07 (1.12)	4.2 (0.85)	4.1 (0.68)	0.97 (0.80)	15.0 (1.43)	16.3 (1.49)	1.09 (1.04)
Gender																					
Male	89.7 (1.47)	89.0 (1.92)	0.99 (1.30)	8.5 (1.33)	9.4 (1.58)	1.10 (1.19)	6.6 (1.15)	7.1 (1.45)	1.10 (1.25)	1.2 (0.63)	1.5 (0.80)	1.22 (1.26)	4.9 (1.11)	5.0 (1.52)	1.02 (1.37)	3.2 (0.90)	3.4 (1.19)	1.07 (1.31)	10.3 (1.47)	11.0 (1.92)	1.07 (1.30)
Female	89.4 (0.93)	88.6 (1.22)	0.99 (1.31)	8.1 (0.82)	8.6 (0.97)	1.06 (1.18)	5.9 (0.67)	6.2 (0.74)	1.04 (1.11)	0.9 (0.28)	0.9 (0.35)	1.01 (1.25)	4.5 (0.63)	4.8 (0.75)	1.05 (1.19)	2.1 (0.44)	1.9 (0.40)	0.92 (0.91)	10.6 (0.93)	11.4 (1.22)	1.08 (1.31)

(continued)

Table B2. (continued)

| | No substance use disorder | | | Substance use disorder | | | | | | | | | | | | | | | | | | | |
| | | | | Illicit drugs | | | Marijuana | | | Opioids | | | Alcohol | | | Both illicit drugs and alcohol | | | Illicit drugs or alcohol | | |
	RUF % (SE)	PUF % (SE)	Ratio of % (Ratio of SE)	RUF % (SE)	PUF % (SE)	Ratio of % (Ratio of SE)	RUF % (SE)	PUF % (SE)	Ratio of % (Ratio of SE)	RUF % (SE)	PUF % (SE)	Ratio of % (Ratio of SE)	RUF % (SE)	PUF % (SE)	Ratio of % (Ratio of SE)	RUF % (SE)	PUF % (SE)	Ratio of % (Ratio of SE)	RUF % (SE)	PUF % (SE)	Ratio of % (Ratio of SE)
NO MDE	96.6 (0.21)	96.5 (0.32)	1.00 (1.49)	2.7 (0.20)	2.8 (0.31)	1.05 (1.52)	2.2 (0.18)	2.3 (0.25)	1.04 (1.40)	0.2 (0.04)	0.2 (0.05)	1.12 (1.19)	1.1 (0.11)	1.1 (0.13)	1.04 (1.17)	0.4 (0.07)	0.5 (0.09)	1.17 (1.28)	3.4 (0.21)	3.5 (0.32)	1.03 (1.49)
Age group																					
12–13	99.4 (0.13)	99.4 (0.14)	1.00 (1.08)	0.6 (0.12)	0.6 (0.13)	0.96 (1.05)	0.3 (0.10)	0.3 (0.10)	0.81 (0.92)	0.1 (0.04)	0.1 (0.05)	1.11 (1.10)	0.1 (0.04)	0.1 (0.04)	0.83 (0.88)	0.1 (0.04)	0.0 (0.02)	0.53 (0.68)	0.6 (0.13)	0.6 (0.14)	0.98 (1.08)
14–15	96.4 (0.41)	96.4 (0.58)	1.00 (1.41)	3.0 (0.38)	3.1 (0.53)	1.02 (1.39)	2.5 (0.31)	2.5 (0.41)	1.02 (1.31)	0.3 (0.10)	0.3 (0.12)	0.99 (1.19)	0.9 (0.18)	0.9 (0.21)	0.96 (1.17)	0.3 (0.09)	0.3 (0.11)	1.11 (1.17)	3.6 (0.41)	3.6 (0.58)	1.00 (1.41)
16–17	94.0 (0.47)	93.6 (0.57)	1.00 (1.22)	4.7 (0.43)	5.0 (0.57)	1.08 (1.32)	4.0 (0.39)	4.3 (0.52)	1.08 (1.32)	0.2 (0.07)	0.3 (0.10)	1.33 (1.37)	2.4 (0.27)	2.6 (0.31)	1.07 (1.16)	1.0 (0.19)	1.2 (0.25)	1.23 (1.30)	6.0 (0.47)	6.4 (0.57)	1.05 (1.22)
Gender																					
Male	96.6 (0.28)	96.5 (0.43)	1.00 (1.54)	2.9 (0.27)	3.0 (0.41)	1.04 (1.55)	2.5 (0.25)	2.5 (0.38)	1.02 (1.52)	0.2 (0.06)	0.2 (0.07)	1.00 (1.18)	1.0 (0.14)	1.0 (0.17)	1.04 (1.25)	0.4 (0.09)	0.5 (0.12)	1.08 (1.29)	3.4 (0.28)	3.5 (0.43)	1.03 (1.54)
Female	96.7 (0.33)	96.6 (0.44)	1.00 (1.34)	2.5 (0.30)	2.6 (0.43)	1.06 (1.43)	1.9 (0.24)	2.1 (0.30)	1.07 (1.26)	0.2 (0.06)	0.3 (0.07)	1.26 (1.20)	1.3 (0.18)	1.3 (0.21)	1.03 (1.15)	0.5 (0.11)	0.6 (0.14)	1.27 (1.34)	3.3 (0.33)	3.4 (0.44)	1.02 (1.34)

* = low precision; PUF = public use file, which refers to estimates produced from a NSDUH PUF; RUF = restricted-use file, which refers to estimates produced from a NSDUH RUF and published in the NSDUH detailed tables; SE = standard error that is associated with the estimate.

NOTE: Ratio of % = PUF % ÷ RUF %; Ratio of SE = PUF SE ÷ RUF SE.

NOTE: Respondents with unknown past year MDE data were excluded.

[1] The Total row includes respondents with unknown past year MDE information.

Definitions: Measures and terms are defined in Appendix A of the 2019 NSDUH detailed tables at https://www.samhsa.gov/data/.

Source: SAMHSA, Center for Behavioral Health Statistics and Quality, National Survey on Drug Use and Health, 2019.

References

1. Singh, A.C.: Method for statistical disclosure limitation (U.S. Patent Application Pub. No. US 2004/0049517A1). RTI International. The patent was granted in June 2006 (Patent No. US7058638B2) (2002)
2. Singh, A.C., Yu, F., Dunteman, G.H.: MASSC: a new data mask for limiting statistical information loss and disclosure. In: Proceedings of the Joint UNECE/EUROSTAT Work Session on Statistical Data Confidentiality, Luxembourg (Working Paper No. 23, pp. 373–394). United Nations Statistical Commission and Economic Commission for Europe Conference of European Statisticians, European Commission Statistical Office of the European Communities (EUROSTAT), Geneva, Switzerland (2003)
3. Singh, A., Yu, F., Wilson, D.H.: Measures of information loss and disclosure risk under MASSC treatment of micro-data for statistical disclosure limitation. In: Proceedings of the 2004 Joint Statistical Meetings, pp. 4374–4381. American Statistical Association, Section on Survey Research Methods, Toronto, Canada (2004)
4. Center for Behavioral Health Statistics and Quality: National Survey on Drug Use and Health: Quality assessment of the 2002 to 2013 NSDUH public use files (2016). https://www.samhsa.gov/data/report/quality-assessment-2002-2013-nsduh-public-use-files
5. Center for Behavioral Health Statistics and Quality: 2019 National Survey on Drug Use and Health (NSDUH): Methodological summary and definitions. Table 3.2 (2020). https://www.samhsa.gov/data/report/2019-methodological-summary-and-definitions

Privacy in Practice: Latest Achievements of the Eustat SDC Group

Ana Miranda(✉), Marta Mas, and Marina Ayestaran

Basque Statistics Office, Vitoria-Gasteiz, Spain
{ana_miranda,m-mas,m-ayestaranarregi}@eustat.eus

Abstract. Maintaining the privacy of the data providers, preserving the confidentiality of the information they provide and its use only for statistical purposes must be fully guaranteed within the statistical activity. This principle largely underpins the credibility of a statistical organization and must be present in all phases of statistical production [2]. Since 1998, research work has been carried out in this field at Eustat: the provision of a scholarship for the study of the most common techniques [3] for protecting microdata files and statistical tables, the application of specific protection measures to real data, the establishment of standard criteria for the protection of statistical information and the widespread dissemination of microdata. In 2018, an expert group was created at Eustat that coordinates and promotes all these tasks within the Organisation. This paper collects the main works and results obtained by the group over these latest five years. Firstly, the preparation and updating of the criteria document on confidentiality and data protection in statistical dissemination is described. The aim is to provide the staff with a guide to basic confidentiality criteria when preparing statistical products for dissemination. Next, the type of analysis that is carried out to prepare secure microdata for dissemination (public use files) is shown. As an example, the analysis carried out for the microdata of the Population Survey in Relation to Activity (PRA) is included. Finally, a solution for the automatic protection of statistical tables by τ-Argus [5] using a SAS macro is presented. Specifically, the application to tables of the Directory of Economic Activities of Eustat is shown.

Keywords: Statistical disclosure control · Official statistics · Microdata protection · Statistical tables protection

1 Eustat Confidentiality Document for Data Release

1.1 Context

The European Statistics Code of Practice [1], adopted by the European Statistical System Committee on November 16, 2017, establishes in its Principle 5 on Statistical Confidentiality and Data Protection, the need to provide staff with guidelines and instructions on the protection of statistical confidentiality throughout all statistical processes, and that the confidentiality policy is publicly available.

© Springer Nature Switzerland AG 2022
J. Domingo-Ferrer and M. Laurent (Eds.): PSD 2022, LNCS 13463, pp. 347–360, 2022.
https://doi.org/10.1007/978-3-031-13945-1_24

This Principle is what inspires the development of this document, which basically refers to the establishment of a general written regulation, which serves as a guide in the preparation of microdata files and tables in the statistical dissemination of Eustat. Since 1998, Eustat has been working on establishing criteria for the protection of statistical data in the data dissemination stage and preparing and updating the document that sets these criteria and serves as a guide for all the staff.

The elaboration of this regulation is the result of the analysis of the statistical data protection treatments that are carried out in Eustat. It also addresses the existing regulations in Europe, Spain, and in the Basque Country, in terms of data protection, and looks to the treatment of confidentiality regarding microdata and statistical tables available in other statistical offices and in Eurostat to stablish protection criteria.

During the last year, the SDC group in Eustat has carried out a profound update of this document that has addressed the following aspects:

- Updating of the regulations on the protection of personal data
- Review and update of the protection criteria applied to the statistical tables
- Review and update of the protection criteria applied to microdata for public use
- Data protection in other dissemination products (specific requests, GIS, etc.)

1.2 Regulations on Personal Data Protection

Statistical operations collect and process data of different types, to which different regulations are applied, even more than one, depending on the scope of application (Fig. 1):

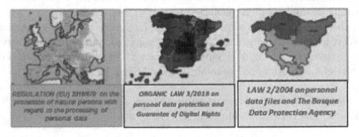

Fig. 1. Regulations by geographical scope.

However, it must be borne in mind that in the pyramid of sources of law, EU Regulations are always at the top. Therefore, Regulation (EU) 2016/679 [6] of the European Parliament and of the Council of 27 April 2016 on the protection of natural persons with regard to the processing of personal data and on the free movement of such data, and repealing Directive 95/46/EC, will always be applied directly and in preference to any other regulation (Fig. 2):

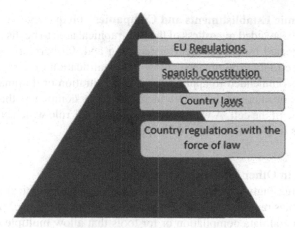

Fig. 2. Pyramid of legal sources.

In addition, the document also includes the main functions of the Basque Data Protection Agency, which is the supervisory body for compliance with data protection regulations at the level of the Autonomous Community of the Basque Country. The document also highlights the figure of the Delegate of Data Protection of the Administration of the Autonomous Community of Euskadi that acts as a contact point between the Agency and the Administration itself.

1.3 Protection Criteria Applied to Statistical Tables

The main product of statistical dissemination is the statistical data tables. Given the different nature of the data and statistical units represented, a distinction is made between population and housing tables and tables of economic establishments and companies in order to define the protection criteria within the confidentiality document:

Population and Household Tables. Frequencies of less than 3 statistical units will be avoided in tables of small geographical areas (municipal and sub-municipal). These cells will be protected by concealment or recoding when it is considered that there is a reasonable risk of indirect identification and, therefore, a possible disclosure of data or individual characteristics that should be protected by statistical secrecy. To assess this risk, it will be necessary to analyse, in each specific case, various factors such as:

- Proportion of cells with low frequency in the table.
- Population size of the geographical areas to be disseminated.
- Identifying power of the crossing variables.
- Sensitivity of the information provided.

Excluded from the application of protection measures are tables in which population totals are given involving only sex, age or both at the same time, provided that age is represented by intervals of sufficient width.

Tables of Economic Establishments and Companies. Frequencies lower than 3 statistical units will be avoided regardless of the geographical area to be disseminated. Said cells will be protected by concealment or recoding and, furthermore, when variables intervene that, due to their dominance, allow the identification of companies or establishments, it is recommended to apply rules of concentration or dominance, that is, to locate those cells or intersections where there are a few companies that contribute to excess to the value of the cell. A well-known rule is the p% rule which consider the two main contributors to the cell and is widely applied.

Data Protection in Other Statistical Products
The need for increasingly detailed information, both at a geographical and conceptual level of the variables represented, makes it necessary to apply 'ad-hoc' protection measures for non-general data compilation or for tools that allow multiple data selections (i.e.: GIS). Eustat confidentiality document also includes these cases and recommends the application of data modification treatments (for example: replacement of a range of values by the midpoint of that range) or the hiding of geographic layers with information about few statistical units. In the case of applying perturbation techniques, the effect on the totals and variable distribution is checked.

2 Micro-data Protection

2.1 Generation of Ready-to-Access Microdata

The microdata are the individual data of the informants that are used to prepare tables of results. They are normally presented as tables in which each row ("record") stores the information of a unit and each column ("field") is a variable.

The microdata files that are available for public access will be protected, that is, they will not include direct identification data and will have been processed in such a way that the possible disclosure of data based on indirect identifiers is greatly hindered.

To protect a microdata file, the first phase consists of evaluating which records can be easily identified and the second phase consists of applying some protection measure. The assessment of the statistical disclosure risk (or statistical risk) of microdata sets is based on measuring in some way the occurrence of rare records.

A key combination is a selection of certain variable values that are considered identifiers for records because they are rare in some way. In short, a key combination allows to detect the rare records in the data set. Those records that have the values established in the key combination will be the records that can reveal confidential information because they are easily identifiable and therefore measures will have to be taken to protect the information.

There is no systematic procedure to establish the combinations of key variables, and these must be established by those responsible for each operation or data set. However, for statistics belonging to the same area, the combinations of key variables analyzed will be very similar, since these combinations often include characteristics common to all of them (for example: sex, age, geographical areas, in sociodemographic statistics or sector of activity or employment strata in economic statistics).

Once the combinations to be used have been determined, they are applied to the microdata and the frequencies of the combinations are observed in the file. If the frequencies are lower than a set limit, some protection measure must be applied.

Protection with information restriction methods is based on reducing the amount of information offered, either because it is directly suppressed, or because it is given at a less detailed level.

The most common method is global recoding, this method consists of giving the information with a lower level of disaggregation: for example, at the province level instead of the municipal level, ages in five-year groups instead of year by year, economic activity at a digit instead of two etc. Global recoding applies to the entire file, not just the records to be protected, and can be applied to both qualitative and quantitative variables. Disaggregation thresholds can be established for the variables common to different statistical operations within the same scope (eg: maximum geographical disaggregation, age intervals, economic sectorization, etc.).

We can also resort to protection with disturbance methods, this is based on altering the information offered, trying to maintain the global characteristics of the whole. There are several techniques that can be applied, the most used is the exchange of records (data swapping). The exchange of data between records consists of changing certain characteristics of the records to make them non-identifiable. Closeness criteria are normally established between the records to be exchanged so as not to alter the global characteristics of the microdata set, for example, exchanging records that are in the same municipality and have the same age, or have the same number of employees, or that they are in the same branch of activity etc.

If it is decided to apply disturbance methods, the user must be warned of the application of such methods for reasons of statistical secrecy, but no details will be given about the records affected neither about the parameters of the protection method.

In general, the microdata files that are disseminated will not present geographic identifiers that refer to areas with less than 10,000 inhabitants. This threshold is considered a suitable limit for the basque geographical context. Therefore, geographic variables will be added to meet this criterion, this includes those referring to place of birth, place of residence, etc.

2.2 Example of Application to PRA Microdata

The growing demands from researchers, policy makers and others for more and more detailed statistical information leads to a conflict. The respondents are only willing to provide a statistical office with the required information if they can be certain that their data will be used with the greatest care, and in particular will not jeopardise their privacy. So statistical use odata by outside users should not lead to a compromise of confidentiality. However, making sure that microdata cannot be misused for disclosure purposes requires, generally speaking, that they should be less detailed, or modified in another way that hampers the disclosure risk.

This is in direct conflict with the wish of researchers to have as detailed data as possible. Much detail allows not only more detailed statistical questions to be answered, but also more flexibility: the user can lump together categories in a way that suits his purposes best. The field of statistical disclosure control in fact feeds on this trade-off: How should a microdata set be modified in such a way that another one is obtained with acceptable disclosure risk, and with minimum information loss? How exactly can one define disclosure risk? How should one quantify information loss? Once these problems have been solved - no matter how provisionary - the question is how all this wisdom can actually be applied in case of real microdata. If a certain degree of sophistication is reached the conclusion is inescapable: specialised software is needed to cope with this problem and μ−Argus [4] is such software.

Producing safe micro data is not a trivial issue. It should first be explained when microdata are considered safe or unsafe. It should also be explained how unsafe data can be modified to become safe.

All the microdata that we publish in the Eustat have been and continue to be subject to review in order to provide the maximum information with the minimum risk. The first survey we analyzed was: PRA Population In Relation To Activity. (Labour Force Survey).

The Population Survey in Relation to Activity operation is a continuous source of information on the characteristics and dynamics of the workforce of the A.C. from Euskadi. It includes the relationship with the productive activity of the population residing in family households, as well as the changes produced in their employment situation; prepares indicators of quarterly variations on the evolution of the active population; it also estimates the degree of participation of the population in activities that are not economically productive. It offers information at the level of historical territories and capitals.

We start from the dataset with the PRA microdata of the last available quarter, to determine the risk of the same we have used the μ-Argus program. The objective of Argus is to hinder the re-identification of individuals represented in the data to be published, that is, to prevent the disclosure of confidential data (disclosure). When a file is considered insecure, anonymization techniques (SDC) will be applied, which will produce modifications in the data, so that an adequate level of security is reached, that is, adequate depending on the use that is going to be made of them: public or scientific.

What concepts must we handle to understand μ-Argus?

- Key variable: variables that allow the informants to be identified. Important: they must be defined as qualitative variables (Categorical).
- Combination: crossing of variables that forms a table.
- Dimension (dimension): number of variables that cross in a table.
- Threshold: the limit at which a frequency or risk, for a combination, is considered safe or unsafe: values below the threshold will be unsafe, above safe.

We are going to carry out a first individual registration risk analysis. The variables that we are going to consider as their potentially identifying crosses are:

- TERH - Province of residence
- SEXO - Sex
- LNAC - Place of birth
- EDAD - Age
- NACI - Nacionality

The weight variable that we use to calculate the risk is ELEV2.

Taking those variables into account directly without any recoding what we get in μ-Argus is (Fig. 3):

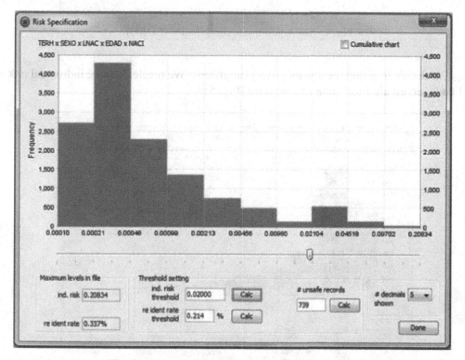

Fig. 3. μ-Argus risk chart without global recoding.

The dataset has 12,749 records, the re-identification ratio is low (0.337%) and so is the number of unsafe records (739), considering a risk greater than 0.02% unsafe.

We review the frequencies and see that age is the variable with the most insecure records in all dimensions (Fig. 4):

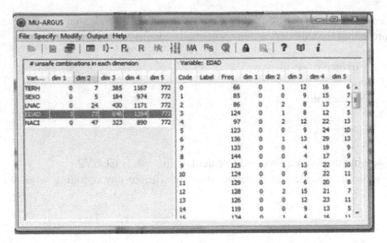

Fig. 4. Mu-Argus unsafe combinatios (EDAD).

We decide to group the age into five-year groups. We recalculate the individual risk of each record after recoding and we have (Fig. 5):

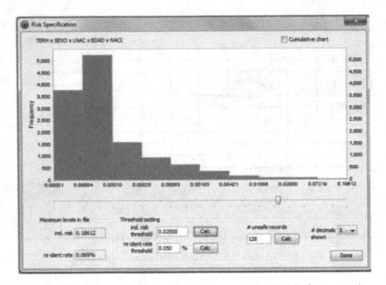

Fig. 5. Mu-Argus risk chart with global recoding (EDAD-five groups).

The risk of re-identification for all registries has dropped considerably, as has the re-identification ratio (0.069%) and the number of insecure registries (128). Right now we would assume that 128 of the records could be identifiable with reasonable effort.

The next variable with the most insecure records is NACI (nationality), we are going to group this into two categories: Spanish/Foreign (Fig. 6).

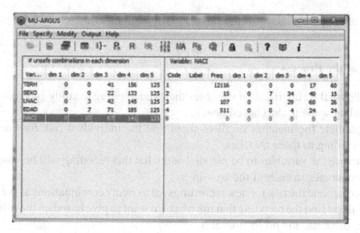

Fig. 6. μ-Argus unsafe combinatios (NACI).

Once recoded, we calculate the individual risk of re-identification and we have (Fig. 7):

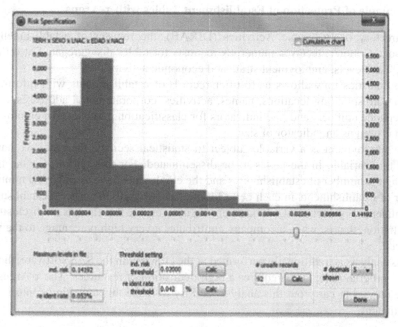

Fig. 7. μ-Argus risk chart with global recoding (EDAD and NACI)

We now have half as many insecure records as in the previous step. The risk of re-identification has also decreased (0.053%), right now we would assume that 92 of the records could be identified with reasonable effort. However, μ-Argus offers the possibility to suppress values in the unsafe combinations at the end of the protection process.

Summary of the Process

Step 1: Choose the classification variables that we are going to study (you can think of adding some to the chosen ones).

Step 2: Calculate the matches of all of them and the individual risk for each record according to those matches.

Step 3: Choose the variables to be recoded and what this recoding will be based on the frequencies in each of the crossings.

Step 4: Recalculate the risk for new recordings for as many combinations as we consider until we find the recoding that fits what we want to give based on the information we offer and the protection of it.

Step 5: Generate the protected file, deleting information, or not, and review which records are problematic.

3 Protection of Statistical Tables

3.1 Example of Protection of Establishment Tables with τ-Argus

The Directory of Economic Activities (DIRAE), the register of establishments in the Basque Country, receives numerous requests for tables disaggregated at different geographical levels, employment strata and economic activity.

The Statistics Law allows us to offer records of establishments with information on their corresponding locations, names, activities, corporate email addresses, corporate telephone numbers and size indicators for classification purposes. An employment stratum is used as an indicator of size.

The employment is a variable subject to statistical secrecy, therefore we have to protect this variable in the tables to be disseminated. For this reason, for the tables including the number of establishments and the employment, we must have a minimum number of establishments in each cell (3 units, according to the criteria established in our confidentiality document). In addition, we must ensure that there is no disclosure by concentration, that is, when a company contributes a very high percentage to the value of the cell.

This analysis is traditionally performed at the Office with the SAS software, through 'ad hoc' programming. This requires a high dedication of resources. The τ-Argus tool has been tested to carry out this analysis and try to make this unveiling control phase more efficient.

The necessary files have been prepared for the analysis with τ-Argus, microdata file and metadata file, and the employment variable has been analyzed in the following crosses:

- th*A10
- th*A64
- comarca*A10

where:

- A10, A64 are different classifications of the economic activity variable (NACE)
- th (province) and comarca (region) are different levels of geographical disaggregation

We evaluate all three tables with the same restrictions (Fig. 8):

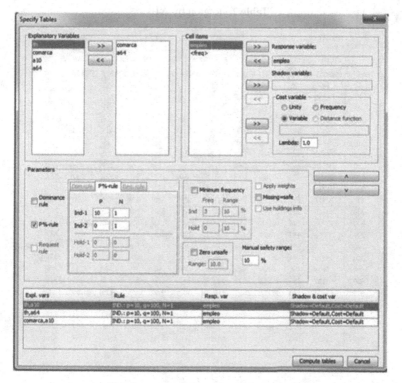

Fig. 8. τ-Argus Specify Tables option

The results are the following:

Province by A10
No unsafe cell appears (Table 1).

Table 1. Province by A10

th x a10										
	-Total	01	02	03	04	05	06	07	08	09	10
-Total	889853	12474	180041	50505	231678	20887	16781	4948	129303	203833	39403
01	148666	3690	39417	6345	33720	2416	2047	527	17988	36904	5612
20	296327	4057	70628	15875	78022	5055	5069	1532	35559	66845	13685
48	444860	4727	69996	28285	119936	13416	9665	2889	75756	100084	20106

Region by A10

The unsafe cells according to the applied criteria appear in red, the blue ones correspond to secondary suppression (Table 2).

Table 2. Region by A10

comarca x a10										
	-Total	01	02	03	04	05	06	07	08	09	10
-Total	889853	12474	180041	50505	231678	20887	16781	4948	129303	203833	39403
01	1496	25	15	267	533	-	-	-	656	-	-
20	2235	-	7	563	110	-	7	-	1537	2	9
48	1664	6	11	549	803	-	-	-	292	-	3
0101	3721	370	2217	107	577	2	5	4	126	275	38
0102	116096	992	25121	4976	27493	2325	1767	466	15403	32523	5030
0103	1440	223	318	58	160	-	13	-	29	620	19
0104	7893	1412	3544	287	1189	19	88	22	315	940	77
0105	4847	242	2444	148	1146	10	19	4	227	484	123
0106	13173	426	5758	502	2622	60	155	31	1232	2062	325
2001	26782	467	3988	1513	11417	194	371	176	1926	5496	1234
2002	20412	345	7185	747	4737	184	339	63	1804	4020	988
2003	30910	325	13875	967	5665	328	335	65	2932	5715	703
2004	145735	794	18065	7951	40376	3927	3123	1005	22843	39321	8330
2005	26006	442	11892	1371	5121	110	325	59	1460	4503	723
2006	17918	540	6902	1090	3964	69	218	60	972	3454	649
2007	26329	1144	8714	1673	6632	243	351	104	2085	4334	1049
4801	7909	356	3426	428	1767	38	90	39	324	1231	210
4802	344157	960	38513	22633	94004	12742	8370	2483	65669	82298	16485
4803	45251	435	16517	2029	11878	339	472	144	5699	6269	1469
4804	8012	680	1162	560	2071	18	124	42	926	2056	373
4805	13501	1159	3120	854	3360	81	285	44	948	3131	519
4806	8933	771	3546	366	1789	38	102	12	390	1652	267
4807	15433	360	3701	866	4264	160	222	125	1508	3447	780

To see more information about any of the cells, by positioning yourself in any of them, its description appears: (the following image refers to one of the insecure cells, region 48 A10 01) (Fig. 9):

Fig. 9. τ-Argus cell information

Province by A64
Similar to the previous table, unsafe cells according to the applied criteria appear in red and blue cells correspond to secondary suppression (Table 3):

Table 3. Province by A64

a64 x th	- Total	01	20	48
- Total	889853	148666	296327	444860
01	8824	3572	2886	2366
02	1266	111	312	843
03	2384	7	859	1518
04	347	51	98	198
05	13111	3676	5216	4219
06	1501	208	636	657
07	2765	631	989	1145
08	3413	399	1767	1247
09	2946	350	1058	1538
10	1025	-	-	1025
11	3694	972	1012	1710
12	660	162	90	408
13	13519	5016	3016	5487
14	4094	1467	1347	1280
15	16505	4444	4431	7630
16	42733	7591	17291	17851
17	6299	426	4089	1784
18	8295	903	2962	4430
19	20156	2118	14282	3756
20	11818	6163	3116	2539
21	6862	1648	2862	2352
22	4808	837	2209	1677

3.2 Automatization of the Table Protection Process

A batch file has also been generated that automatically reproduces the entire protection process.

Subsequently, the application of a sas macro - SAS2Argus - developed by Statistics Sweden, whose purpose is to facilitate the integration of τ-Argus with the SAS software, has been analyzed.

The macro prepares the necessary input files for τ-Argus, and is also capable of starting a τ-Argus batch job. The macro can also import the results of the τ-Argus execution into SAS, thus integrating the risk analysis phase and/or the deletion of cells or another protection method that is decided into the production environment.

4   Conclusions and Future

The SDC group together with the support of the different production and methodological areas of our Statistical Office has made a great effort during the last years to implement security and confidentiality measures in the Office. As a result, guidelines and protection criteria exist in all phases of statistical production and are available to all staff.

In addition, a wide range of microdata, duly anonymized and processed, has been made available to the general public. It is intended to expand this offer over the next few years with the inclusion of new periods for existing microdata and new files for other statistics. It is also one of our main objectives to improve and expand the metadata accompanying the files and to make available to the user in several formats and support for reading them.

Finally, the use of the SAS2Argus macro is already being extended to other statistics produced by the Office, such as the R&D Statistics and the Tourism Survey, which require massive table protection processes, both for general dissemination and for ad hoc requests.

References

1. European Statistics Code of Practice for the National Statistical Authorities and Eurostat (EU statistical authority) Adopted by the European Statistical System Committee. 16th November 2017
2. Hundepool, A., et al.: Handbook on Statistical Disclosure Control. Essnet SDC (2010). https://research.cbs.nl/casc/SDC_Handbook.pdf
3. Hundepool, A., et al.: Statistical Disclosure Control (Wiley Series in Survey Methodology). Wiley (2012)
4. Hundepool, A., et al.: μ−Argus User Manual of Version 5.1.3 (2014)
5. Hundepool, A., Peter- De Wolf, P.P., Giessing, S., Salazar, J.J., Castro, J.: τ-Argus User Manual of Version 4.2 (2014)
6. Regulation (Eu) 2016/679 of the European Parliament and of the Council Of 27 April 2016 on the protection of natural persons with regard to the processing of personal data and on the free movement of such data (2016)

How Adversarial Assumptions Influence Re-identification Risk Measures: A COVID-19 Case Study

Xinmeng Zhang[1]([✉]) [iD], Zhiyu Wan[2] [iD], Chao Yan[2] [iD], J. Thomas Brown[2] [iD],
Weiyi Xia[2] [iD], Aris Gkoulalas-Divanis[3] [iD], Murat Kantarcioglu[4] [iD],
and Bradley Malin[1,2,5] [iD]

[1] Department of Computer Science, Vanderbilt University, Nashville, TN, USA
Xinmeng.zhang@vanderbilt.edu
[2] Department of Biomedical Informatics, Vanderbilt University Medical Center, Nashville, TN,
USA
[3] IBM Watson Health, Cambridge, MA, USA
[4] Department of Computer Science, University of Texas at Dallas, Dallas, TX, USA
[5] Department of Biostatistics, Vanderbilt University Medical Center, Nashville, TN, USA

Abstract. The COVID-19 pandemic highlights the need for broad dissemination of case surveillance data. Local and global public health agencies have initiated efforts to do so, but there remains limited data available, due in part to concerns over privacy. As a result, current COVID-19 case surveillance data sharing policies are based on strong adversarial assumptions, such as the expectation that an attacker can readily re-identify individuals based on their distinguishability in a dataset. There are various re-identification risk measures to account for adversarial capabilities; however, the current array insufficiently accounts for real world data challenges - particularly issues of missing records in resources of identifiable records that adversaries may rely upon to execute attacks (e.g., 10 50-year-old male in the de-identified dataset vs. 5 50-year-old male in the identified dataset). In this paper, we introduce several approaches to amend such risk measures and assess re-identification risk in light of how an attacker's capabilities relate to missing records. We demonstrate the potential for these measures through a record linkage attack using COVID-19 case surveillance data and voter registration records in the state of Florida. Our findings demonstrate that adversarial assumptions, as realized in a risk measure, can dramatically affect re-identification risk estimation. Notably, we show that the re-identification risk is likely to be substantially smaller than the typical risk thresholds, which suggests that more detailed data could be shared publicly than is currently the case.

Keywords: Data sharing · Re-identification risk · COVID-19 · Health data · Data privacy

1 Introduction

The Coronavirus Disease 2019 (COVID-19) outbreak caused a global pandemic that has resulted in devastating and sustained health and economic crisis [1]. As of May 2022,

© Springer Nature Switzerland AG 2022
J. Domingo-Ferrer and M. Laurent (Eds.): PSD 2022, LNCS 13463, pp. 361–374, 2022.
https://doi.org/10.1007/978-3-031-13945-1_25

there have been over 80 million confirmed cases (i.e., a person with laboratory confirmation of COVID-19 infection) in the United States and over 500 million worldwide. Though expected to become endemic at some point, COVID-19 continues to be a public health problem with waves of infection that are likely to reoccur for some time [2]. In this respect, it provides a clear justification for the creation of more timely case reporting strategies and surveillance efforts.

Public health departments typically rely on a case surveillance process to routinely collect information that is critical for disease control and prevention [3]. Case surveillance reports contain data on various infected individuals, including demographics, symptoms, epidemiologic characteristics (e.g., case confirmed date and location), health conditions, characteristics of hospitalizations, clinical outcomes, and exposure history. When surveillance data is made accessible at the population scale, it can enable faster responses to health emergencies and support data-driven public health research [4, 5]. Over the past several years, several resources of COVID-19 case surveillance data have been made available for public use. For instance, the World Health Organization (WHO) requests all member states to report data at a fidelity no less than national-level aggregated counts of confirmed cases, deaths, and hospitalizations within 48 h of detection [6]. In the United States, the Centers for Disease Control and Prevention (CDC) reports aggregate case and death counts, as well as person-level data that includes age, race, ethnicity, state, and county of residence of those infected [7, 8].

Despite the need to share COVID-19 case surveillance data, concerns about privacy have been raised due to the sensitive nature of the information [9–11]. There are particular concerns that the identities of the corresponding individuals could be inadvertently exposed. In public datasets, typically referred to as anonymised or de-identified data, it is obvious that direct identifiers, such as personal names, national ID numbers, and detailed residential addresses must be removed. However, it is possible that indirect or, what is often referred to as, quasi-identifiers (QIDs) [12], such as the demographic data shared in the CDC's COVID-19 datasets, can indicate small groups of patients in a de-identified dataset, which creates an opportunity for re-identification [13].

It is anticipated that attackers will rely upon QIDs to attempt to match de-identified records to accessible identified datasets through record linkage mechanisms [14, 15]. Prior studies have measured re-identification risks for QIDs by considering the degree of distinguishability within the de-identified dataset [16, 17]. The notion of k-anonymity [13] leads to a typical risk threshold applied in this case, whereby a de-identified dataset is considered protected if, and only if, each combination of QIDs appears at least k times in the dataset. Currently, the CDC relies on this notion of privacy and releases two datasets for COVID-19 case surveillance—one for public use and the other for scientific use—at a level of 11- and 5-anonymity, respectively [7, 18].

The CDC's data publication policies are based on strong adversarial assumptions. Measures of privacy that focus solely on the degree of distinguishability within the dataset to be shared (as k-anonymity does) assume that the recipient of the data is aware that a named individual of interest is in the sample. However, this is a worst-case scenario and weaker adversarial scenarios can be, and in many cases are, considered [19]. Specifically, distinguishability in the de-identified only creates a potential for intrusion. For a re-identification attack to be successful, the recipient of the data either needs to know the identity of the corresponding individuals according to some prior experiences (i.e., background knowledge) or they need to demonstrate re-identification by linking the records to an external, identified dataset through QIDs [19, 20]. This is important to recognize because the estimation of risk in these situations could be quite lower than in the worst-case scenario. In recognition of this fact, alternative approaches estimate re-identification risks based on population uniqueness [21–23]. This perspective, realized in the k-map model [24] for instance, assumes the attacker only knows that the targeted individual in the sample was drawn from a broader population of individuals, such that uniqueness in the dataset is insufficient to claim re-identification success. This model is used when the data sharer has a reasonable expectation of the identified resources that will be leveraged for an attack.

The aforementioned risk measures assume that all individuals below a threshold are equally at risk; by contrast, the marketer risk measure assumes a record's risk is inversely proportional to the number of records it relates to [25]. This risk measure typically assumes that the de-identified dataset is a subset (or a sample) of the identified dataset. However, in reality, both the de-identified and the identified datasets are samples from a broader population, and they do not necessarily demonstrate a sub-/super-set relationship. As a consequence, and as we show in this paper, there can be combinations of QIDs in the de-identified dataset that do not exist in the identified dataset. Similarly, the number of people with a certain combination of QIDs in the de-identified data could be larger than that observed in the identified dataset. For example, imagine that there are 10 patients in the de-identified dataset who are male and 50 years old, but that there are only 5 individuals present in the identified dataset who exhibit the same combination of QIDs. This raises a question about how missing records should be handled in the risk calculation. To the best of our knowledge, current re-identification risk measures do not explicitly address such real world challenges.

Our study introduces novel re-identification risk measures to fill in the gap between the previously proposed risk estimation methods and challenges caused by missing records. Our study extends traditional risk measures to address missing record challenges and allow data sharers to evaluate re-identification risk under various assumptions of an attacker's capability. To demonstrate how different assumptions could affect the estimation of risks, we perform a re-identification risk analysis for case surveillance data of COVID-19 and voter registration records in the United States. Our findings indicate that the re-identification risks vary according to adversarial assumptions. Using an actual record linkage test, we show that the external re-identification risk is likely to be substantially smaller than 0.09, which corresponds to the CDC's intended threshold of 11-anonymity. Our findings suggest that more detailed data could be shared publicly than the current generalization level.

2 Methods

In this paper, the *internal* dataset refers to the de-identified patient-level data to be shared. Formally, this is represented as a set D of n individuals $d_1, d_2, \ldots, d_n$ defined over a set of quasi-identifying features $Y_1, Y_2, \ldots, Y_m$. The records for these individuals can be partitioned into a set of equivalence classes (i.e., the set of unique combinations of quasi-identifying values) $q_1, q_2, \ldots, q_J$. Let f_i be the number of records in D for the equivalence class q_j associated with record d_i.

In addition, we assume there exists one or more *external* datasets that potentially contains the identities of the individuals whose records in the internal dataset are at risk for re-identification. A typical example of such a dataset in the US that has been leveraged for re-identification purposes is a voter registration list [15]. Set E of N individuals $e_1, e_2, \ldots, e_N$ is defined over the same set of quasi-identifying features $Y_1, Y_2, \ldots, Y_m$. Let F_i be the number of records in E for the equivalence class q_j associated with record d_i in D. Records from the internal dataset and the external dataset are linked if they share the same set of quasi-identifying features $Y_1, Y_2, \ldots, Y_m$.

We represent q_j with f_i larger than F_i as invalid classes, denoted as $q_{j\_invalid}$. There are n_r individuals from D who are in $q_{j\_invalid}$.

2.1 Internal Marketer Risk Measure

Based on the formulation introduced by Dankar *et al.* [24, 25], we define an ***Internal Marketer (IM) Risk*** measure:

$$IM\ Risk(D) = \left(\frac{\sum_{i=1}^{n} \frac{1}{f_i}}{n} \right) = \frac{J}{n} \tag{1}$$

which corresponds to the probability that a record in a de-identified dataset can be correctly linked to a targeted individual through QIDs. This measure assumes the adversary knows that a specific individual is in the de-identified dataset. As a result, it represents a worst-case scenario for the data sharer.

2.2 Record Linkage and External Risk Measures

As alluded to earlier, the external dataset is typically a sample from a larger population and, for some equivalence classes, patients in the internal dataset could be linked to a fewer number of identified persons. This incorrectly implies that the probability of correct re-identification is larger than 1. We introduce three new measures to correct this sampling issue under specific adversarial assumptions.

Conservative External Marketer (CEM) Risk: In this scenario, we assume that, if a person exists in the internal dataset, he should also be included in the external dataset. However, for some $q_{j\_invalid}$, the f_i may be larger than the corresponding F_i in the external dataset. Thus, we add dummy records to the external dataset so that the equivalence class is of the same size as that observed in the internal dataset. We leave the external dataset unchanged for all q_j, where f_i is no larger than the corresponding F_i.

Figure 1 depicts a situation in which a de-identified patient dataset is linked to an identified voter registration list. In this figure, there are three male patients who were born in 1959, but there are only two voters in the same equivalence class. Thus, to account for the "missing" patient, we add one voter record ("Imputed for Male 1959" in the upper section of Fig. 1) to the identified dataset. We assume that the attacker has the same prior knowledge about individuals in $q_{j\_invalid}$ and in q_j. This yields an upper bound for re-identification risk, which is calculated as follows:

$$CEM\ Risk(D, E) = \frac{\sum_{i=1}^{n} \frac{1}{\max(F_i, f_i)}}{n} \tag{2}$$

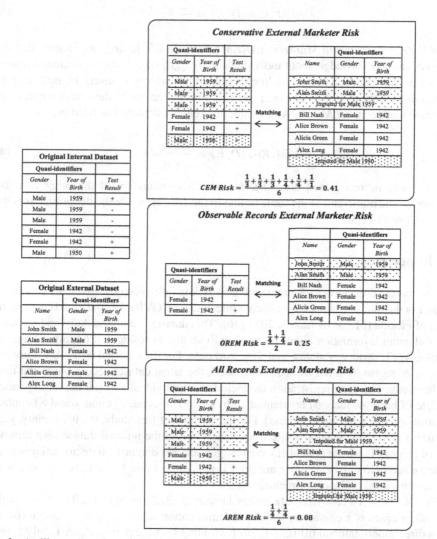

Fig. 1. An illustration of record linkage and risk computation for *CEM* (upper), *OREM* (middle), and *AREM* (lower).

Observable Records External Marketer (OREM) Risk: In this setting, we assume that patients in the internal dataset with no corresponding records in the external dataset (as defined by their QID) are protected by their lack of presence. As a result, in this measure, we assume they are not at risk of re-identification. Thus, these patients are removed from the computation. As shown in the middle section of Fig. 1, three male patients who were born in 1959 and one male patient who was born in 1950 are removed from the internal dataset in the linkage process. This measure yields a risk that is no greater than the upper bound and is calculated as follows:

$$OREM\ Risk(D, E) = \frac{\sum_{i=1}^{n-n_r} \frac{1}{F_i}}{n - n_r} \tag{3}$$

All Records External Marketer (AREM) Risk: In this setting, we assume that the attacker has no knowledge about individuals in $q_{j\_invalid}$ (i.e., in the equivalence classes that do not have enough corresponding records in the external dataset). In the examples shown in the bottom section of Fig. 1, we add dummy records to the external dataset in the same manner as *CEM* risk. As a result, this risk is calculated as follows:

$$AREM\ Risk(D, E) = \frac{\sum_{i=1}^{n-n_r} \frac{1}{F_i}}{n} \tag{4}$$

It should be recognized that *AREM* risk is a combination of the other two risk measures. The numerator is the same as that in the *OREM*, while the denominator is the same as that in *CEM*.

3 Experiments

We use two real datasets to demonstrate how risk is influenced by adversarial assumptions. For the internal dataset, we use case line data for COVID-19 confirmed cases in the state of Florida (FL) as of June 3, 2021 [26]. This dataset is updated daily and includes the following information about infected individuals: 1) residential county, 2) age, 3) gender, 4) FL residency status, and 5) record date. For the external dataset, we use the FL's voter registration list as of June 8, 2020, the latest dataset accessible at the time of this study. The voter registration list includes an individual's 1) full name, 2) gender, 3) date of birth, 4) race, 5) residential address, 6) ZIP code, 7) county and 8) contact information (such as email address). For the purpose of this study, we use county, year of birth (YOB), and gender as quasi-identifiers. From the internal dataset, we remove 5% of records 1) whose patient ID, county, gender, and diagnosis date are unknown, 2) have an age below 0, or 3) those are not FL residents. Table 1 provides a summary of the datasets used in the experiments.

The COVID-19 case-line data covers January 5, 2020, to June 1, 2021. Since rapid growth in cases is a characteristic of pandemic patient-level data, we evaluate risk at each three-month interval till June 1, 2021, yielding six time points: April 1, July 1, and October 1, 2020, and January 1, April 1, and June 1, 2021.

To investigate how different policies affect the risk across demographic groups, we designed 12 alternative case-reporting policies, as shown in Table 2. These are defined by the QID generalization levels, such that policies P1 through P11 vary in their generalization of age and sex. The policies include six potential generalizations of age and two potential generalizations of sex. Here, *suppressed* indicates that the corresponding

Table 1. Summary of the dataset used in this study. *a b c* represents the first quartile, median, and third quartile. *d* ±*e* represents the mean and one standard deviation. *f g%* indicates that the percentage of *f* patients (in a given category) is *g%* among all patients.

Characteristic	Distribution			
	COVID-19 Case line		Voter Registration	
Age	23 38 55	39.5±20.6	NA	
Date of Birth	NA		1912-12-12 to 2020-05-31	
Race/Ethnicity	NA			
Non-Hispanic White	NA		9,009,488	65.7%
Hispanic	NA		2,420,628	17.6%
Non-Hispanic Black	NA		1,950,476	14.2%
Other Races/Ethnicities	NA		328,628	2.3%
Gender				
Male	441,413	46.4%	6,346,193	46.2%
Female	508,316	53.5%	7,363,027	53.7%
Number of counties	67		67	
Event date	2020-01-05 to 2021-06-01		NA	

QID is reported as a null value for all individuals and, thus, the corresponding QID is not used in the linkage experiments. The *current* policy corresponds to the generalization level for the actual COVID-19 case line data from the FL Department of Health.

The re-identification experiments are composed of four steps: 1) apply the policy to the COVID-19 data, 2) harmonize the patients' demographic characteristics in the COVID-19 database with the FL voter database, 3) match the de-identified patient database with the identified voter database, and 4) compute the re-identification risk measures.

In these experiments, we link patients by their county, YOB, and gender. YOB is not directly available in the COVID-19 database and could be inferred from the age of the patient at the COVID-19 positive test event date, but there is an ambiguity in the transformation. For example, imagine that we observe a patient who is 30 years old, for whom the event date is March 1st, 2021. This patient's date of birth could be as early as March 1, 1990, but as late as March 1, 1991. To address this issue, we create a YOB range for each patient with (*event date* − *age* − 1, *event date* − *age*). In the situation where age is generalized to a range in different case-reporting policies, such as a 30-year range, the aforementioned example's age is generalized to 30–59. The YOB lower bound is 1961 and the upper bound is 1991. In general, the *YOB lower bound* is (*event date* − *age range upper bound* − 1 and the *YOB upper bound* is (*event date* − *age lower bound*). We compare the YOB of voter records to *YOB lower bound* and *YOB upper bound* of patient records for the linkage.

Table 2. Case-reporting generalization policy rules.

Policy	Age generalization level	Sex generalization level
P1	Suppressed	Suppressed
P2	60 year range: 0–59, 60 +	
P3	30 year range: 0–29, 30–59, 60–89, 90 +	
P4	15 year range: 0–14, 15–29, 30–44, …	
P5	5 year range: 0–4, 509, 10–14, 15–19, …	
P6	1 year range	
P7	Suppressed	Male/Female
P8	60 year range: 0–59, 60+	
P9	30 year range: 0–29, 30–59, 60–89, 90+	
P10	15 year range: 0–14, 15–29, 30–44, …	
P11	5 year range: 0–4, 509, 10–14, 15–19, …	
Current policy	1 year range	

4 Results

4.1 Internal Risk Evaluation

We evaluate the *IM Risk* at the end of each time period starting on the date of the first confirmed case. As time proceeds, both the number of patients and the number of unique QIDs groups grow, but at different rates. Figure 2 shows the risks for each policy. Following the U.S. Institute of Medicine report [27] and European Medicines Agency guidelines [28], we set a risk threshold of 0.09 (which corresponds to 11-anonymity for a public dataset) as an acceptable level of risk for our following analysis. It can be seen that in April 2020, all of the policies exhibited risks higher than the threshold. In July 2020, October 2020, and January 2021, policies P1-P3, P1-P4, and P1-P5 satisfy the requirement, respectively. After April 2021, all of the policies were under the threshold.

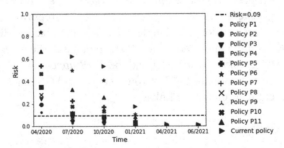

Fig. 2. *IM Risk* evaluated as a function of time.

4.2 External Risk Evaluation

We compare the *CEM*, *OREM*, and *AREM* risks by linking the FL COVID-19 case-line data to FL's voter registration list. Risk is evaluated at each of the six time points, the results of which are summarized in Fig. 3. Recall that the *CEM Risk* is an upper bound of the external marketer risk. In this case, some patients may not be matched to the corresponding voters in the voter registration list, but we added dummy records in the external dataset to acknowledge the existence of the missing records. It can be seen in Fig. 3A that only policies P6 and P11, as well as the current policy, achieve risks that are higher than the 0.09 threshold in April 2020. Still, these risks are lower than the *IM Risk*. By July 2020, policy P6 and the current policy's risks are higher than 0.09. By October 2020, only the current policy has a risk higher than 0.09. After January 2021

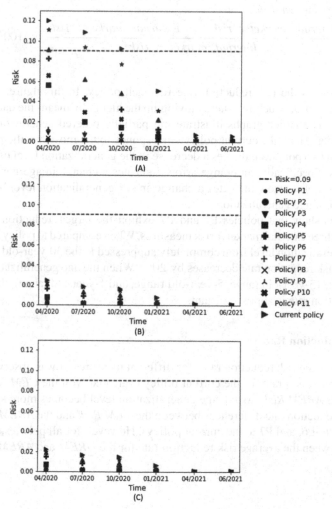

Fig. 3. External risk evaluation with the A) *CEM*, B) *OREM*, and C) *AREM* risks over time.

(one year after the first case-line data was released), all policies exhibit *CEM* risk that is smaller than 0.09.

Next, we analyze the *OREM* (i.e., evaluated with all valid records) and *AREM* (i.e., evaluated with all the records) risks. As shown in Figs. 3B and 3C, none of the policies have risks higher than 0.09 after April 2020.

4.3 Internal vs. External Risks

As anticipated, a comparison of *IM Risk* and external risks shows that risks decrease for all policies once a real identified dataset is factored into the risk assessment. However, the rate of change in risk is not constant across all policies. To illustrate this fact, we defined a risk reduction rate from the *IM Risk* to the external risk as follows:

$$Risk\ reduction\ rate = \frac{Internal\ marketer\ risk\ -\ External\ marketer\ risk}{Internal\ marketer\ risk} \times 100\% \tag{5}$$

Figure 4 shows the risk reduction rate for each policy. In this figure, the arrows indicate a hierarchical structure, where moving up the hierarchy means the data becomes more specific. The lattice graphs illustrate the partially ordered generalization levels between policies. Here, the current policy corresponds to the most specific policy. The arrow between two policies indicates a decrease in the generalization level of one of the QID variables. Specifically, an orange arrow indicates a change in age generalization level, whereas a blue arrow indicates a change in sex generalization level but remains the same level of age generalization.

The results show that policies P1 and P7 exhibit the largest reduction rates with respect to the three external marketer risk measures. When compared to policy P1, change in the age generalization level from completely suppressed to the 30-year-old range (i.e., P3) result in risk reduction rate decreases by 20%. When the age generalization level is less strict (e.g., 15-year-old range, 5-year-old range, and 1-year-old range), the effect on the risk reduction rate is almost constant.

4.4 Risk Reduction Rate

Figure 5 depicts the risk reduction rates for different measures. It was observed that the average risk reduction rate for the current policy evaluated with the *CEM Risk* is 24% larger than the *AREM Risk*. As the age generalization level becomes more specific, the average risk reduction rate differences between the *CEM Risk* and the *AREM Risk* grow (i.e., from P1 to P6, and P7 to the current policy). However, for all policies, there is no difference between the average risk reduction rate for the *OREM* and *AREM* risks.

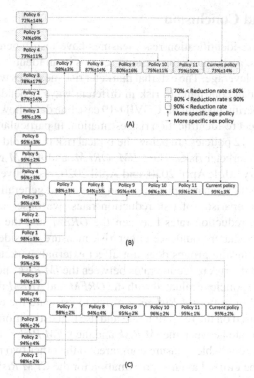

Fig. 4. Risk reduction rates (average of the six time points ± 1 standard derivation) from the *IM* risk to the external risks for policies organized in generalization hierarchy: A) *CEM*, B) *OREM*, and C) *AREM* risks.

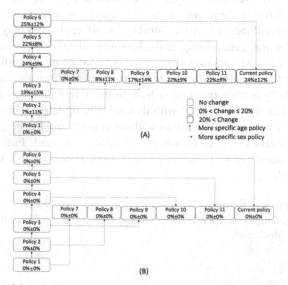

Fig. 5. Change in risk reduction rates (average value across six time points ± 1 standard derivation) between risks evaluated with: A) *CEM* and *AREM* risks and B) *OREM* and *AREM* risks.

5 Discussion and Conclusion

As this work shows, re-identification risk measures have insufficiently addressed real world data challenges, particularly the missing records in an identified resource that an attacker is expected to leverage. The external marketer risk measures we introduced show that missing records can contribute to risk in different ways depending on adversarial assumptions. Our experiments with FL COVID-19 case line data show that such assumptions non-trivially affect re-identification risk estimation. In particular, our results reveal that the risks under all 12 policies are below the typical risk threshold of 0.09 as of April 2021 for the internal marketer risks. The *CEM*, *OREM*, and *AREM* risks are below the threshold as of January 2021, April 2020, and April 2020, respectively. It suggests that more detailed data could be shared publicly than the current generalization policy.

Further, in our comparison of risk reduction rates, we observed that there is no difference in the risk reduction rates between the *OREM* and the *AREM Risk*. This suggests that the data sharer could use either risk measure considering an attacker's decision to target the invalid groups does not affect external marketer risk estimation. We also observed that the risk reduction rates between the *IM Risk* and external risks are relatively stable for all policies evaluated with the *OREM* and *AREM* risks. This suggests that data sharers could use the *IM Risk* as a proxy to estimate the external risks for a data sharing policy based on the reduction rate and use that estimation for other policies. Finally, the reduction rate between the *IM Risk* and the *CEM Risk* is the smallest. Thus, the *CEM Risk* is a more reliable measure compared to the other two measures. The risk reduction rate could be utilized as an approximation for the *CEM Risk* from the *IM Risk* when the external dataset is not accessible.

There are also several limitations we wish to acknowledge as opportunities for future investigations and improvements. First, residency does not always imply the place where an individual currently lives. Our study assumes residency per the voter registration list is equal to the residency in the internal dataset. Second, we only evaluate the risks with COVID-19 case surveillance data from Florida. Yet different states may adopt different approaches to collecting, generalizing, and releasing medical data. Third, the FL's voter registration list is updated on a monthly basis, such that an analysis of the recency of the voter data for the COVID data should be assessed to determine the influence on risk. One particularly notable question is how to maximize the re-identification risk-utility trade-off when data is released with dynamically updated policies [29]. Fourth, our study evaluates re-identification risks without considering the benefits inherent in sharing data and the attacker's gain from the re-identification attack. Future studies could use economic arguments (e.g., based on game theory) to analyze the balance between privacy and utility [30].

Funding. This study was supported by the funding sources: grants CNS-2029651 and CNS-2029661 from the National Science Foundation and training grant T15LM007450 from the National Library of Medicine.

References

1. Sohrabi, C., et al.: World Health Organization declares global emergency: a review of the 2019 novel coronavirus (COVID-19). Int. J. Surg. **76**, 71–76 (2020)

2. Rodriguez-Morales, A.J., et al.: Clinical, laboratory and imaging features of COVID-19: a systematic review and meta-analysis. Travel Med. Infect. Dis. **34**, 101623 (2020)
3. CDC national surveillance. https://www.cdc.gov/coronavirus/2019-ncov/covid-data/faq-surveillance. html#:~:text=CDC%20uses%20national%20case%20surveillance,identify %20groups%20most%20at%20ris. Accessed 20 May 2022
4. Kostkova, P.: Disease surveillance data sharing for public health: the next ethical frontiers. Life Sci. Soc. Policy **14**(1), 1–5 (2018). https://doi.org/10.1186/s40504-018-0078-x
5. Ienca, M., Vayena, E.: On the responsible use of digital data to tackle the COVID-19 pandemic. Nat. Med. **26**, 463–464 (2020)
6. World Health Organization: Global Surveillance for COVID-19 Caused by Human Infection with COVID-19 Virus: Interim Guidance. World Health Organization, Geneva (2020)
7. Lee, B., et al.: Protecting privacy and transforming COVID-19 case surveillance datasets for public use. Public Health Methodol. **136**(5), 554–561 (2021)
8. COVID-19 Case Surveillance Public Use Data. https://data.cdc.gov/Case-Surveillance/COVID-19-Case-Surveillance-Public-Use-Data/vbim-akqf. Accessed 20 May 2022
9. French, M., Monahan, T.: Disease surveillance: how might surveillance studies address Covid-19? Surveill. Soc. **18**(1), 1–11 (2020)
10. Ioannou, A., Tussyadiah, I.: Privacy and surveillance attitudes during health crises: acceptance of surveillance and privacy protection behaviours. Technol. Soc. **67**(101774) (2021)
11. Allam, Z., Jones, D.S.: On the Coronavirus (COVID-19) Outbreak and the smart city network: universal data sharing standards coupled with artificial intelligence (AI) to benefit urban health monitoring and management. Healthcare **8**(1), 46 (2020)
12. Dalenius, T.: Finding a needle in a haystack – or identifying anonymous census record. J. Official Stat. **2**(3), 329–336 (1986)
13. Sweeney, L.: k-anonymity: a model for protecting privacy. Int. J. Uncert. Fuzz. Knowl. Based Syst. **10**(5), 557–570 (2002)
14. Durham, E., Xue, Y., Kantarcioglu, M., Malin, B: Private medical record linkage with approximate matching. In: AMIA Annual Symposium Proceedings 2010, pp. 182–186 (2010)
15. Benitez, K., Malin, B.: Evaluating re-identification risks with respect to the HIPAA privacy rule. J. Am. Med. Inf. Assoc. JAMIA **17**(2), 169–177 (2010)
16. Skinner, C., Holmes, D.: Estimating the re-identification risk per record in microdata. J. Official Stat. **14**(4), 361–372 (1998)
17. Sweeney, L.: Simple demographics often identify people uniquely. Technical Report LIDAP-WP3, Carnegie Mellon University (2000). https://dataprivacylab.org/projects/identifiability/paper1.pdf. Accessed 21 May 2022
18. Centers for Disease Control and Prevention, Agency for Toxic Substances and Disease Registry. Policy on public health research and non-research data management and Access. https://www.cdc.gov/maso/policy/policy385.pdf. Accessed 20 May 2022
19. Xia, W., et al.: Enabling realistic health data re-identification risk assessment through adversarial modeling. J. Am. Med. Inform. Assoc. **28**(4), 744–752 (2021)
20. Xia, W., Kantarcioglu, M., Wan, Z., Heatherly, R., Vorobeychik, Y., Malin, BA.: Process-driven data privacy. In: 24th ACM International on Conference on Information and Knowledge Management (CIKM 2015) Proceedings, pp. 1021–1030. Association for Computing Machinery, New York, NY, USA (2015)
21. Koot, M.R., Noordende, G. van 't, de Laat C.: A study on the re-identifiability of Dutch citizens. In: 3rd Hot Topics in Privacy Enhancing Technologies (HotPETs 2010) Proceedings, pp. 35–49. Berlin, Germany (2010)
22. Golle, P.: Revisiting the uniqueness of simple demographics in the US population. In: 5th ACM Workshop on Privacy in Electronic Society Proceedings, pp. 77–80. New York, NY, USA (2006)

23. Emam, K.E., Buckeridge, D., Tamblyn, R., Neisa, A., Jonker, E., Verma, A.: The re-identification risk of Canadians from longitudinal demographics. BMC Med. Inf. Dec. Mak. **11**(1), 46 (2011)

24. Emam, K.E., Dankar, F.K.: Protecting privacy using k-anonymity. J. Am. Med. Inform. Assoc. **15**(5), 627–637 (2008)

25. Dankar, FK., Emam, KE.: A method for evaluating marketer re-identification risk. In: 2010 EDBT/ ICDT Workshops Proceeding Article 28, pp. 1–10. Association for Computing Machinery, New York, NY, USA (2010)

26. Florida COVID-19 Case Line Data. https://open-fdoh.hub.arcgis.com/datasets/florida-covid19-case-line-data/about. Accessed 20 May 2022

27. Institute of Medicine (IOM): Sharing clinical trial data: Maximizing benefits, minimizing risk. The National Academies Press, Washington, DC (2015)

28. European Medicines Agency: External guidance on the implementation of the European Medicines Agency policy on the publication of clinical data for medicinal products for human use, Revision 4. http://www.ema.europa.eu/ema/index.jsp?curl=pages/regulation/general/general_content_001799.jsp&mid=WC0b01ac0580b2f6ba. Accessed 20 May 2022

29. Brown, J.T., et al.: Dynamically adjusting case reporting policy to maximize privacy and public health utility in the face of a pandemic. J. Am. Med. Inform. Assoc. **29**(5), 853–863 (2022)

30. Wan, Z., et al.: A game theoretic framework for analyzing re-identification risk. PLoS ONE **10**(3), e0120592 (2015)

Author Index

Printed in the United States
by Baker & Taylor Publisher Services